Percy Thrower's
EVERY DAY
GARDENING
in colour

HAMLYN
LONDON·NEW YORK·SYDNEY
TORONTO

Percy Thrower's
EVERY DAY
GARDENING
in colour

on
go
the
re is
has
hole
lting
ay is
ceed
sant

ying
hen-
good
ns.

First published in 1969 by
THE HAMLYN
PUBLISHING GROUP
LIMITED
London · New York
Sydney · Toronto

Astronaut House, Feltham,
Middlesex, England
Eighth impression, 1977

Printed in Spain by
Mateu Cromo, Madrid

ISBN 0 600 44242 X

c
ca
the
ma
V
thi
gar
der
(br
or
On
rul
or
are
pos
Th

Contents

List of Colour Illustrations

Putting First Things First

Ask any experienced garden maker how to begin and he is almost certain to say 'Get your ideas down on paper'—and I would heartily agree. Where I would differ from some is that I would not place too much emphasis on making such paper plans strictly to scale. Attempting such accuracy is off-putting to many, and personal experience has taught me that a roughly scaled plan serves just as well.

On to this plan must go the fixed positions—the house, the entrance gate(s), the drive, and any trees at present on the site. In respect of trees, it is true to say that almost always these are an asset to a new garden, providing focal points and pockets of maturity in a garden that for several years at least can have little else to relieve the flatness and soften the boundaries. One should remember, too, that trees in private gardens may have a preservation order on them, which means

attractive garden there is one common feature: the element of surprise. One's eyes (and footsteps) are led to corners around which there may be goodness knows what treat in store—or no treat at all. For example, a path leading to a little wooden gate alongside a handsome Blue Cedar and mixed shrub plantings may lead to nothing more exciting than the compost heap and rubbish dump but the mind's eye holds out very different prospects. In garden designing this kind of artifice is to be admired and not condemned. My $1\frac{1}{2}$-acre garden has been made from a grass field sloping mostly to the south, but also to the north and east. Naturally, we built the house on the highest ground and facing south, and to leave an uninterrupted view of the garden which would not frequently be spoilt by parked cars I led the drive from the road round to the back (or north) side of the house. It is details like this which should be thought out carefully at this early plan-

The modern style of informal layout makes for ease of maintenance and usually holds the interest for longer than more formal and obtrusive designs

Permanent paths leading to parts of the garden which must be frequently visited in all weathers are an essential element of good design

that permission must be obtained from the local authority before they can be felled or pruned. Of course, some forest trees can completely dominate a small garden and where this is the case then there is really no alternative but to have them removed, if a garden of any value is to be made.

With the fixed positions marked, the next is to decide on the paths. Any part of the which you must visit frequently in winter a permanent path, preferably of stone o expensive for most of us these days), t, concrete or ashes laid on hardcore. want to be dogmatic but as a general to follow the line of the boundary rden features, and where curves make these as soft and open as ny suggestion of fussiness.

Surprise. In almost every

ning stage, for to have them impressed on one later when it is too late to do much about it is extremely frustrating.

Soil Improvement. Many new garden sites consist of little but rough ground and the builder's debris. It is always to be hoped that the good top soil removed during excavations has been placed on one side for subsequent redistribution. In any case, levelling the ground and carrying out any necessary soil improvement is very much the order of the day. Many gardens have more than one kind of soil; I, for example, have gravel quite near the surface at the top of my garden while lower down the slope I have heavy clay. In the case of light sandy or gravelly soils with poor moisture retention properties it will always pay to dig in such humus-forming material as garden compost, peat, farmyard manure (less and less easy to obtain nowadays), leafmould, spent hops,

Much thought should be given to the vistas obtained from the house. The path curving out of sight gives the impression that the garden is larger than it is

straw or shoddy, to improve their structure. Slow-acting fertilisers such as bonemeal and hoof and horn will also do a lot of good. A cheap and easily used soil-testing outfit (costing a few shillings from a garden sundries store) will tell you whether your soil is acid or alkaline, over-acidity being corrected by application of lime.

The most common soil type is clay, which, to say the least, can be difficult to work. But there are few better growing mediums than a clay soil which has been worked on and improved over a matter of years. With clay soils drainage is of paramount importance for in the nature of things they are extremely retentive of moisture, the individual soil particles being very small indeed. The humus-forming materials which I have just mentioned in connection with light sandy soils are just as useful for clay soils for the opposite reason—namely that they break up the close-packed soil particles and so allow excess moisture to drain freely to lower levels. One of the key factors in the successful mastery of clay soils is timing one's cultivations correctly. Working them in wet weather for instance is folly of the worst kind, for the compaction which results can do enormous harm. On the other hand, digging during good weather in autumn so that the surface can be left rough for the alternating frosts and rains of winter to break down is just about the ideal. Where ericaceous or other lime-hating plants (such as ericas, rhododendrons, azaleas, camellias and magnolias) are not going to be grown, dressing with lime or gypsum will also help to improve the soil—and once that happy state of affairs is more than a dream it will grow almost anything. But it is a long-term business.

Perhaps this is a good time to emphasise that it is far better to grow those plants suited to your type of soil than to do things the other way round and try to make the soil suit the plants' require-

ments. I am referring now, of course, to fundamental things like having the wrong kind of soil or the wrong aspect and not to soil improvement or the removal of branches from trees to let in more light.

One can nowadays grow lime-hating plants in calcareous soil by treating the latter with iron sequestrene, to release the iron previously unavailable to the plants, but the necessity for further applications at intervals and the resulting expense does not make this a practical proposition for most of us if large numbers of plants are involved. It is, however, a very useful technique if one wishes to grow a few lime-hating plants of special interest in calcareous soil.

At this early stage, too, it is essential to determine if the natural drainage is sufficient to cope with all likely weather conditions. If there is any reason to doubt this it is a good idea to dig 2- to 3-ft.-deep holes in several parts of the garden and then see if these hold water after heavy rain. Land which is low-lying or very heavy is liable to suffer from bad drainage and the waterlogging associated with this will result in the roots of plants dying and often the complete loss of the plants. Three-inch-diameter drainpipes, put 2ft. underground and leading to a suitable outfall, provide the best remedy but a trench partly filled with rubble, covered first with a layer of turf and then with soil to surface level is usually quite effective ever possible because it is cheaper, but a good lawn can be made quite quickly by this means.

Meadow Grass to Lawn or Starting from Scratch? Another kind of situation which can face a new home owner (as in my own case) is to have most, or some, of his plot of land covered with undisturbed meadow grass. Provided the ground is reasonably even it is possible, by regular mowing, feeding and treating with selective weedkiller, to turn the grass into a very satisfactory lawn. All my lawns are from such field grass with the exception of the areas adjoining the house where I was keen to have especially good turf and where I sowed a mixture of selected fine grasses.

Once the main features have been decided and their positions marked out it is possible to ahead with grass seed sowing or turf laying at appropriate time (see pp. 22 to 25), but th another way of setting about this task wh its merits. This is to sow grass seed over th area and then cut beds and so on in the sward. The advantage of doing things that the development of the garden at a slower pace, for the grass will as a feature in its own right.

The other way of making a l turves. I would recommend s ever possible because it is c lawn can be made quite qui

Foundation Planting

I believe we have to thank the Americans for that excellent term 'foundation planting'. It means what it says: getting in the permanent plants, trees, shrubs, hedges, etc., which will form the framework of the garden. Especial care is necessary here for once trees or specimen shrubs are planted one does not want to have to move them because they are badly sited. I would always give priority to the vistas of the garden obtained from the house, for it is surprising what a large percentage of one's pleasure in a garden comes from seeing plants from that viewpoint. A warning which should not be needed but unfortunately often is, is to avoid planting trees too near the house. The slender stem and small head of a young tree must be related to the mature specimen it will surely become. A tree with a spread of 4ft. when planted may have one of 30ft. in fifteen years time.

Forsythia suspensa, its yellow flowers having a delightful foil in mellow stone. Walls offer the gardener excellent opportunities to widen his repertoire

Concerning Hedges. Many shrubs and trees can be used for hedging purposes and I have described a good selection of these on pp. 70 to 72. I have already referred to the fast-growing *Cupressocyparis leylandii,* a superb conifer for a windbreak, and I would be loath to do without beech in my garden; it is such a lovely green in summer and the glorious golden-brown of the leaves in winter is a constant source of pleasure. Those who have a field with cattle in it abutting their garden, as I have, would not find a better boundary hedge than hawthorn—nor a better one for keeping out children, dogs, etc. as well. *Lonicera nitida,* the Chinese Honeysuckle, makes a fine dense hedge, likewise *Chamaecyparis lawsoniana* (Lawson's Cypress) and, of course, yew, although that is slow-growing. The advantage of all those that I have mentioned is that they need clipping only once a year, whereas the ubiquitous privet, green or golden, must be given such attentions at least four times annually. For a quick-growing hedge, though, the last-mentioned is in a class of its own.

For hedges within the garden I see much merit in flowering shrubs like forsythias, berberis, flowering currant, fuchsias, pyracantha and taller-growing roses like Queen Elizabeth for the colour is always welcome and if the shrubs chosen for such purposes are deciduous, and therefore less effective as screen in winter, this does not usually matter at that time of year. Indeed, a flowering hedge can add much gaiety to a garden.

Special Features

A fairly recent development is the growing of herbaceous perennial plants, annuals, biennials, bulbs and selected shrubs in what are called 'island beds', which in addition to looking attractive in almost any setting have the considerable advantage of allowing the plants growing in them to be viewed from all sides. The beauty of these is that they can be cut out of the grass to any shape or size suited to their surroundings and they give the gardener much more scope than the traditional border with its straight sides and backdrop of hedge, wall or fence.

Rock Gardening. It is only in recent years, too, that rock plants have been liberated from the traditional rock garden to grace many gardens in flat beds, scree beds, peat beds and the like. But rock gardens still have their place, of course, and the way I grow my rock plants, in conjunction with a water feature (two pools, one above the other, with a waterfall in between and an electric pump to circulate the water), provides me with a rock-water feature which is colourful for many months of the year, is easy to look after and has the kind of informality which accords well with this modern age. My pools are made of concrete, but I have no objections to the modern fibreglass and plastic, moulded pools as long as the edges are masked with plants and stone.

Using House Walls to Advantage. Provided there are beds up against the house walls, any gardener worthy of the name is going to make good use of these. The warm south and west sides of the house will naturally make the perfect sites for rather tender plants, and the north and east will prove excellent for camellias (whose late-winter flowers are so easily damaged by early morning sun on frost-covered petals), the Morello cherry, honeysuckles and other decorative plants.

Covering-up Operations. They may not be very large but manhole covers are just about the biggest eyesores at ground level to be found in any garden. I always feel like wincing when their ugliness intrudes on some carefully contrived and otherwise attractive garden scene. An easy way round this particular problem is to plant one of the prostrate conifers like *Juniperus sabina tamariscifolia* just to one side so that the flat folds of growth blanket the ironwork. The well-known Herringbone Cotoneaster, *C. horizontalis*, is another shrub I find excellent for this particular job. The small-leaved *Cotoneaster conspicua* is splendid, I find, for masking the bottom of drain pipes, and at the same time provides the necessary shade for the roots of the large-leaved clematis which so handsomely clothe the same pipes in their upper reaches.

The Charm of a Pergola. To me a pergola always has an old-world charm, and a small one need not take up very much room. Just a few of the plants I grow on mine are the lovely white rose Madame Alfred Carrière, the carmine-pink thornless rose Zéphirine Drouhin, honeysuckle, wisteria and clematis, which provide us with continuity of colour and much joy. I also grow the Passion Flower (*Passiflora caerulea*) in a sheltered

and flower seeds are still remarkably cheap in relation to most other commodities nowadays.

Siting a Greenhouse

I think too much stress has been put by some authorities on the way a greenhouse is orientated. North to south, so that you get the maximum benefit from the sun in the morning and afternoon, is the ideal, but there are often reasons why such orientation is not possible, especially in a small garden. In an east-west-facing house the plants which need sun can be placed on the south side, those that appreciate shade on the north.

Much more important is to have one's greenhouse and frames near to the house so that they are easy to get to in bad weather and there will be no problems about laying on electricity supplies, should these be needed. In Chapter Ten I discuss the different types of greenhouse available and

The author tying in growths on the rose Zéphirine Drouhin, one of many fine plants he grows on his pergola. A small pergola is a practical possibility in many gardens

Spectacular colour effects can be achieved from a mixed planting of annuals, sown here with bedding geraniums. They include alyssum, antirrhinum, lobelias and marigolds

corner, I have described this lovely, slightly tender climber in the chapter on greenhouse plants (p. 243), for it is, of course, most suited for such conditions.

Quick, Temporary Colour

I describe very fully later the many annuals which can be grown with profit by home gardeners (see Chapter Five) but what I want to explain now is their great value to the new garden owner. Provided the ground can be prepared for spring sowing or planting (in the case of half-hardy annuals when all danger of frost has passed) then a splendid display can be had throughout the summer of the first year while other more permanent features get under way. There is, too, a wonderful lot of pleasure to be gained from starting with packets of seeds and getting results so quickly—

the kind of plants which can be grown most rewardingly. Even a small greenhouse, with a little thought, can be made to provide a supply of colour throughout the year.

A conservatory, or sun lounge as it is now called by many people, can be a source of enormous pleasure. It needs to be south- or west-facing, of course, and with french windows into the living room can bring the garden indoors in winter. And what about those windy or wet days at other times of the year when it is pleasant to sit in comfort among one's flowers!

Summary for Owners of New Gardens

To recapitulate, then, I would say that the following points are some of the most important to watch if you are making an entirely new garden:

1. Sort out in your own mind the kind of garden features you want to have and consider carefully how these are going to relate to the size and physical surroundings of your plot.

2. When you are ready to start work put first things first and make sure that jobs which cannot be easily attended to later (like laying land drains, where these are necessary) are given precedence.

3. Do not rush your fences but plan the full development of the garden over four or five years if you have a large area to cope with, less of course for smaller areas.

4. Make sure that the garden you design and develop is the kind you can enjoy over many years, and remember that it is the least pretentious layouts and planting schemes which usually hold one's interest longest.

5. Always buy first-class plants from good nurseries. There is no profit at all in running a garden for plant invalids.

would be to hasten slowly. Formulate one's thoughts carefully, see where modifications can be made to make unsatisfactory features fully acceptable, and only when no solutions seem possible eliminate existing features entirely. If the latter is necessary, however, act boldly and make a really first-class job of it.

I think that the main danger of taking over some other person's creations is that one will act hastily and unwisely. If the garden is unpleasing, the object of displeasure may be immediately obvious, but it is more likely to be more difficult to pin down – like a badly shaped bed in the wrong position, unsympathetic curves or the wrong plant associations. It will often be found that even quite modest changes in emphasis will give a garden quite a different character.

There can be no doubt that making an existing garden accord more nearly to one's own tastes can be just as stimulating as making a new garden.

A greenhouse should be sited as near the house as possible for convenience, where electricity supplies can be laid on with minimum expense

A lightweight, plastic pneumatic sprayer makes pest and disease control an easy chore. The many sprayers available include sizes and kinds to suit all tastes

Taking over an Existing Garden

Sometimes this can present as many problems as starting a new garden but quite often one is lucky enough to follow a keen gardener with good ideas suited to the surroundings. I would always be very cautious indeed about making major alterations unless I was quite sure that what I was doing was right and necessary to develop the full potential of the site. If trees and large shrubs, in particular, are removed these can leave gaping holes which will take a very long time to fill, and they should certainly not be discarded lightheartedly. On occasions, it may be possible to find new sites for good plants which will fundamentally alter the appearance of the garden. Perennials can, of course, be moved without trouble, and shrubs can be re-sited successfully if dealt with carefully.

Once again, as so often in gardening, my advice

Garden Equipment

Everybody when starting to garden wants to know the basic tools necessary to work efficiently. I would name the spade, fork, rake and hoe as the four essentials, for with these it is possible to do most of the cultivating; and next come shears (for grass and hedge cutting) and secateurs for pruning. Chrome armour and stainless steel tools cost more but in the long run it is money well spent for they last longer, are easier to use and easier to keep clean. Another important acquisition, of course, is a sprayer to control greenfly, black spot, mildew and other troubles which are found in every garden at some time or other. There are so many different kinds of sprayer available today, often made of lightweight plastic materials, that finding the right one for your particular needs is no problem whatsoever. A good pneumatic sprayer with a strong plastic container and a spray rod of adequate length will be found ideal and the longest lasting.

A small hand trowel and fork can also be considered almost essential for there are countless times when they are needed for one little job or other. A good, well-balanced watering-can is a prime need.

Lawn Tools. Almost every garden has a lawn and the smallest can be cut very efficiently and easily with a hand mower of the side wheel or cylinder type, but larger lawns need a power mower and I much prefer the battery-type electric machine to those powered from the mains or by internal combustion engines. The battery type mower is so easy to use, absolutely safe and quite obviously the mower of the future; mains powered machines with trailing cables which can easily be cut have, of course, nothing like the freedom of action of the other type. Petrol-

Cultivating Tools. I find that new gardeners often buy spades which are too large to use comfortably. The best to get is a Number Two, the average size, which one can use for long periods without fatigue. It is always best, too, to get a fork of modest size and weight. Of hoes, the Dutch hoe is the most useful for using among flowering plants and vegetables and the draw hoe, of course, is also excellent for working heavy ground. Another advantage of the Dutch hoe is that one walks backwards as one hoes so the cultivated ground is not walked on. With vegetables, the drag hoe is useful for making drills for seed sowing, earthing up potatoes and so on, but the first job can also be done with the corner of a rake, and the second with a spade or fork. A garden line can be made from strong twine and two pieces of stick if you do not want the expense of buying the excellent manufactured article;

If the garden includes numerous trees, shrubs and fruit trees a pair of long-handled secateurs is especially useful

A wheelbarrow or a two-wheeled truck is an almost indispensable aid in all but the smallest gardens

engined mowers are now very reliable indeed, but they are rather more complex than the other kind, of necessity, and therefore not quite so easy for the less technically minded and agile to operate. Rotary lawn mowers are much used nowadays and do a first-class job in gardens where long grass and short must be coped with, but for the best 'finish' cylinder type machines are the ones to use. Air-cushion type mowers are very useful on steep grass slopes, as well as on flat surfaces, where they can be easily moved over the grass.

A garden roller is usually completely unnecessary for the average lawn will get enough firming by people treading on it, and by the weight of the roller if a cylinder type mower is used. A rake with flexible prongs is invaluable, however, for teasing out debris from grass in spring and autumn and I would not be without one, but a special spiking tool for lawns can be done without as an ordinary garden fork does the job adequately.

likewise, another home-made piece of equipment is the dibber, an invaluable tool, which can be made from a broken wooden spade or fork handle.

Secateurs I have already mentioned, but if a large number of shrubs and trees are grown, long-handled pruners are invaluable.

Wheelbarrows and Trucks. Last but by no means least the wheelbarrow—and how much effort this can save during the course of a gardening year! Often nowadays these are made of plastic or fibreglass and the lightness of these is a boon beyond measure for lady gardeners. Two-wheeled trucks are also available and have a special appeal, of course, for those gardeners who find weight lifting (for one must to some extent 'shoulder' the weight with a conventional wheelbarrow) rather a problem. In my early days we used a square hessian sheet for moving lawn mowings, leaves and other light materials and this was an asset on many occasions.

A garden, to be completely satisfactory, must be in harmony with its surroundings. Border plants, a flower-margined pool, a lawn and climbing roses match the charm of the period-style house in the picture on the right, while below, the rather severe lines of a very modern house have their perfect-foil in the cool sweep of paving, the small formal pool and fountain, the carefully positioned pot and wall plants

Left. A gently curving line of stepping stones, a broad expanse of grass and two matching willows in association with the other plantings give this garden character and charm. Elegance and the juxtaposition of sympathetic features are the essence of good design in gardening, as in other aspects of living

Right. A small front garden presents the garden designer with perhaps his most difficult problem. In this case, the problem has been cleverly resolved by the use of contrasting paving stones and pebbles, climbers and bedding plants, like pelagoniums and agave. The conifer balances the upward thrust of the clematis without spoiling the scale of the overall planting scheme

Right, below. A sitting-out area is a boon in any garden, and especially when the pleasures of privacy can be enjoyed in the company of tinkling water. A small fountain is a luxury many garden owners feel able to indulge in now that reasonably priced submersible electric pumps are available for this purpose

A carefully tended lawn provides the
finest of all foils for garden plants.
Correct mowing and attention to
routine seasonal chores like feeding,
aeration and weed control are key
factors in obtaining a good sward

Lawn Making and Aftercare

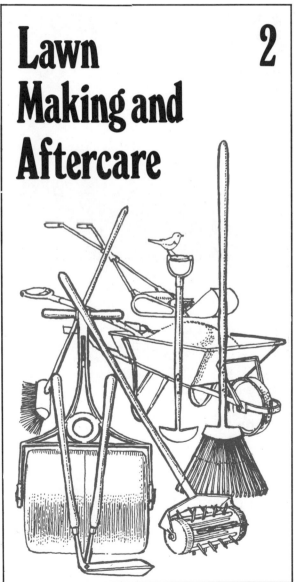

2

For myself, and I am sure for most other people, the lawn is the most important feature in the garden. If the lawn looks good then the chances are that the rest of the garden will look good.

A decision we all have to make when contemplating making a new lawn is whether to use seed or turf. I would recommend seeding whenever possible for it costs about a quarter the money that buying turves does, and the preparatory work for both is just the same. Of course, if the area to be grassed down is not very large and there are children in the family whom it would be difficult to keep off seeded ground then turfing has obvious advantages.

The aids to lawn care available nowadays—tools, machines and chemicals—make maintenance an easy matter, and there is no reason at all why any gardener should not have a lawn of which he can be really proud.

Making a New Lawn

As I have already said in Chapter One, where permanent garden features are concerned it is very important indeed that deficiencies in drainage should be corrected early on. This is certainly true of lawns, for a grass area which holds an excessive amount of water is nothing but a worry in our climate. The grass will be weak and moss, lichen and other weeds will be encouraged, mowing is often delayed because of the state of the ground and in really wet spells it becomes a mud patch if walked on. Heavy clay soils, with a subsoil of pure clay, are the ones most likely to need such attention.

How one treats the ground depends largely on the area involved. If the area is small it is a good idea to remove all the topsoil and put down a layer of clinker, broken bricks, stones or similar material on top of the subsoil to a depth of several inches. This is then rammed down firmly, and covered with another layer of weathered coarse ashes, inverted turves or other bulky material which will effectively stop the topsoil filtering through the drainage material when it is returned.

Another way is to dig a soakaway about 2ft. deep and 2ft. across in the lower part of the area and to fill this with rubble, broken bricks, stones, clinker and so on. Top them up with smaller stones and finish off with at least 4in. of topsoil.

Finally, the most expensive and complicated method of drainage, suitable for large areas, is to lay 3-in. earthenware drainpipes in 2-ft.-deep trenches made with a fall of 1 in 100 and leading to a deep rubble-filled soakaway. The pipes should rest on a bed of small stones or clinker and be covered with brushwood, inverted turves or straw and then clinker to prevent the soil, when returned to the trenches, seeping into the pipes. A more complex system of drainage is to lay a central drain with subsidiary drains leading to it at intervals from each side, herringbone fashion, at intervals of 10 to 40ft., the central pipe having a diameter of 4in. and the subsidiary pipes one of 3in. But it should not normally be necessary to go to such lengths.

Early Preparations for Seed Sowing and Turfing. Whether the lawn is to be made from seed or turf thorough ground preparations are necessary. Having ensured that the drainage is satisfactory the next job is to dig over the soil. For spring sowing (see p. 23 for recommended dates) I like to get this done in the autumn so that the soil is cultivated before the really bad weather arrives, and the latter can be put to work breaking down the rough clods, for it is best at this stage to leave the soil rough to provide maximum exposure to the rain, frost and snow (if this occurs). Remove all roots of perennial weeds as you go along, particularly couch grass, that most pernicious of weeds. Get every bit of couch out or there will be trouble later. Another guard against waterlogging on heavy soil is to work coarse sand into the top few inches of soil—anything up to 14lb. per sq. yd. Digging in fibrous peat is useful on any type of soil, this being worked into the top few inches; and when determining the rate of application work on the principle that the more sandy the soil the heavier should be the dressing. Such additions should be made shortly before seed sowing or turfing.

If you have autumn sowing in mind, or you are planning to lay turf (best done between November and March) then, if you can, let the ground lie fallow during the summer, hoeing it occasionally to break down the lumps and keep newly germinated weeds under control.

Levelling is important for obvious reasons (you cannot do very much about this once the grass is there) and it is worth taking the trouble to peg

out the site and systematically cover the area with board and spirit level. The process consists of starting from a datum point and knocking in the first peg to the required final lawn level. The other pegs are then related to this one for depth so that when the board is rested on any two of them the bubble in the spirit level is dead centre. Of course, on this question of levelling one keeps a sense of proportion and one should not forget that minor irregularities in a lawn surface can add to its attractiveness in some cases. The main thing is that nothing should be done which makes mowing difficult.

All is then ready for the final preparations before seeding or turfing. On a fine day rake the soil down to a firm tilth, and see that any sizeable stones are removed. If any grubs (leatherjackets, wire-worms and so on) are noticed give the soil a dressing of BHC or trichlorphon at this stage.

Firm the soil by treading, with the weight on

Accurate placement of grass seed is greatly aided by marking out areas of 1 sq. yd. with string, or string and bamboo canes, as shown

the heels, so that it is as even as possible and then rake again—but in a direction at right angles to the previous raking. Apply a dressing of a compound fertiliser with a fairly evenly balanced quantity of ingredients and leave the soil alone for about seven to ten days.

Lawns from Seed and Turf

There are many different lawn seed mixtures available, for sunny, open situations, for semi-shady sites, for areas which will get very hard wear, for providing a superb sward and so on. Naturally, the finer the grasses the more expensive the seed will be.

Where appearance is not of first importance and it will be subjected to much wear from children and animals or from general traffic (as on grass paths), a good hard-wearing mixture is the right choice, this including Crested Dogstail, *Cynosurus cristatus*. This will be less close growing than finer mixtures and will require less mowing, but it will need watching to prevent weeds establishing themselves and spreading in the rather open sward.

A fine mixture will contain Fescues and Browntop only and will produce a closely knit turf, consisting mainly of grass with round, wiry blades (the creeping and coarser grasses have flat, wide blades). It will need cutting at least twice a week but produces a first-class lawn.

Finer grasses do best on a sandy, well-drained soil, and stand up to drought slightly better; coarse grasses will do best on slightly heavy soils, although, as indicated, neither will do really well on heavy soil which has inadequate drainage. Heavy shade is a problem and it must be accepted that in such conditions grass is not going to do very well and will probably need fairly frequent

An alternative way of providing a guide is to use a simple wooden frame with 3ft. sides. This is useful when sowing small areas

resowing and patching every year. *Poa nemoralis* grows better than most grasses in shade, though even this tends to die out in the end.

Seed Sowing. The best times for spring sowing are late March and April in the South and April and May in the Midlands and North, but the best time of all in the South is August or September. I have even sown grass seed as late as the end of October with success. A great advantage of August and September sowing is that there are unlikely to be periods of prolonged drought just when the young grass is getting established. In the Midlands and North spring sowing is best for the autumn is more likely to be wet and cold, and trouble from damping-off disease may follow.

Choose a fine, still day, when the soil is neither very wet nor very dry for sowing, and mark out the area into square yards beforehand, if sowing by hand. This is easily done with lengths of string,

or one can use a wooden frame of 1 sq. yd. in area where a small area is involved. This makes accurate, even sowing very easy. Make sure that the soil is broken down to a nice tilth, and use a board to stand on when sowing the seed, so that the soil is not compacted, and one's weight is distributed more evenly. Sow the seed at the rate of $1\frac{1}{2}$ to 2 oz. per sq. yd., sowing half the amount in one direction and half in the other to make its distribution as even as possible. One can also now use a wheeled distributor for seed sowing–the same machine that is used for fertiliser distribution– and this is calibrated to put down the mixture at the required rate. Be careful when using a distributor for sowing that no gaps are left between the seeded rows or, conversely, that the sections do not overlap.

After sowing, rake the soil lightly over the seed, exercising as much care as possible so that the seed is not disturbed. Then string black cotton over the plot if birds are likely to be a nuisance. It is possible to obtain seed treated with bird repellent, or one can use a bird repellent oneself, either to spray the seed with before sowing or to spray the whole area afterwards. Bird scarers such as aluminium strips of foil, bundles of paper and so on can be used also but in many districts these are ineffective. The bird threat is well looked after these days.

Early Treatment. The time taken for the seed to germinate varies, but the first seedlings should start to appear after about 10 days, provided the soil has remained moist and has not dried out during this period. It is likely that, having provided ideal conditions for the grass to germinate, weed seeds in the soil will also be growing, but as the grass seedlings continue to appear, they should eventually overcome the weeds. If a large number show up, it is possible to kill them with a special weedkiller (*not* the ordinary selective weedkillers for lawns) for use within 14 days of germination of the grass seed. This will kill the seedling weeds but not the grass. It is not for use once the grass is older than this, as the weeds will be too strong by then to be affected by this particular type of weedkiller. When the grass is about 2in. high it can be cut for the first time (see p. 26).

Turfing. The best time for turf laying is from November to March, on days when the soil is in workable condition. If the job is done in summer the turves will dry out and shrink, leaving gaps. These gaps will then have to be filled with fine soil to prevent the roots drying out and dying and brown patches appearing. It is advisable–but not always possible, of course–when buying turf, to go and inspect it on the site so that you know exactly what you are getting. What one wants to avoid is poor quality grasses, and weeds. The cost can be lessened considerably by transporting it oneself but turf is very heavy stuff indeed, par-

Left. If the base of a turf is uneven this is easily remedied by placing the bottom side upwards in a special three-sided box and paring the soil level with an old scythe or a cutting edge
Right. When laying turves start every other row with a half turf, so that the joins between the turves are staggered like bricks in a wall

cut oneself, however, may not be so even and these must be trimmed before use. A three-sided box of 1½in. depth and just big enough to take a turf should be made, then the turves can be placed in this, bottom side upwards, and the base pared level with an old scythe or billhook blade, or other suitable cutting edge.

Where one is going to cut one's own turf the procedure then is to mark out the area in foot-wide strips and to cut down the line with an edging iron to a depth of about 1½in. Then make cuts across the grass at 3-ft. intervals and lift the turf with a turfing-iron, a special tool with a roughly triangular-shaped flat blade, like a plate, at the end of a long handle. As the turf is lifted roll it up, then it will not sag and tear under its own weight.

Turf Laying. Having prepared the turf, and made sure that it is free from weeds, it can then be laid. The soil should have been prepared carefully exactly as for seed sowing (see p. 22), and, again as

ticularly when wet, and a saloon car or even an estate car is not suitable for this purpose.

Sea-washed turf produces the finest grass; originally this came from salt marshes in the north of England (Cumberland), but now it may be from any sea-washed marsh. It requires very careful management because it has a different type of soil base to the normal run of turves for lawns, and from this point of view alone it would be better to use grass seed to obtain a really fine lawn.

Turf Cutting and Preparation. Turves are cut in rectangular pieces of 3ft. by 1ft., and when bought are delivered rolled up. They should be unrolled if they are not going to be used immediately, otherwise the grass will turn yellow and will be weak. Rolling turf up also stretches the grass and strains and tears the roots so, for this reason also, it is wise to unroll it immediately. Bought turves are nowadays cut by machine and are, therefore, of uniform thickness. Turves one has

for seed sowing, it is best to choose a dry day for this job when the soil is moist. Turf is laid so that the rows are bonded like bricks, one turf being laid centrally against the division between the two adjacent turves in the next row. The first row should be started by laying half a turf, then the bonding I have just referred to will be achieved when the next, adjacent row is started with a full turf (and, in this case of course, finished with a half turf). Always lay the turf straight up and down the site, or across it.

Always lay turves standing on a board laid on the preceding row of turves. This spreads one's weight, and avoids getting the turves out of alignment. And lay each turf so that it is looped a little. This allows for the slight stretching the turves may have suffered in transit, and, after tapping them gently down when they are all in place, it will ensure that they all fit really tightly. As each row is laid it should be butted up against the previous one with the foot so that each fits really closely. If any bumps or hollows appear in the soil at this stage, this fault can easily be remedied. The importance of making the turves form a level surface does not need to be emphasised.

Always start at the edge of the site with full-sized turves and try to finish in the same way. If a narrow piece of turf is used to finish off an edge, it is much less likely to knit-in satisfactorily, and is easily knocked away or trodden apart, so try to arrange for an exact number of turves to complete the area. To give an example, 100 turves will cover 33 sq. yd. After laying, dress the cracks with a topdressing of either sand or a mixture of loam and peat or leafmould, and brush it well in to encourage the grass to extend fresh roots into the turf next to it, and so knit quickly and strongly. Another way to prevent the turves drying out and withering at the outside edges is to push soil up against them. After about a fortnight, depending

Left. After laying turves, fill in the cracks between them with a topdressing of sand or a mixture of loam, peat and leafmould
Right. A lawn needs mowing twice a week during the main growing season and even three times if the grasses are very fine. The grass should be left about ½in. long

on the prevailing weather conditions, the grass will start to grow again, and can then be cut in the same way as a lawn made from seed.

Mowing

The most important part of caring for a lawn is mowing; any other stretch of grass which is not mown regularly at frequent intervals becomes a field, a paddock, or pasture land. In order to mow correctly it is important to understand the needs of grass. Like any other plant, it requires food and water, and air in the soil, and it requires these things even more than most other plants, because its growth above ground is being constantly and frequently removed. Over the years, by experience and research, it has been gradually ascertained which grasses stand up best to what, after all, is rather drastic treatment. There are not many plants which would survive frequent removal of

their top growth, and fortunately it is the fine grasses which will grow best under these conditions. The coarse grasses gradually die out.

Mowing must be correctly done. The height of cut is important; it should not be so low that the grass is shaved to ground level, brown patches appear and moss and other unwanted plants move in; and it should not be so high that the grass turns brown at the base, while remaining green at the tips. It should be cut at such a height that it is encouraged to send out side-shoots from the base of the main stem—this is known as 'tillering'—so that it becomes thick and mat-like. In practice this means cutting so as to leave the grass about ½in. high. The average lawn certainly does not want cutting closer than this.

Frequency of Cutting. During the main growing season, grass should be cut at least twice a week, and really fine lawns should be mown three times a week—which may seem the counsel of perfec-

Small areas of lawn can be aerated with a garden fork but for larger areas mechanical aids like the spiked roller fitted to the mower shown above are great time savers

A hollow-tined, spring-loaded aerating tool makes light work of lawn aeration. This tool neatly removes cores of soil completely

tion but in fact is necessary. If cutting is not carried out at this frequency and, as all too often happens, is only done once a week, 'bents' will start to appear on the lawn (these are the seed-heads of coarse grasses such as rye grass). More frequent mowing eliminates these coarse grasses or prevents their appearance since it encourages the fine grasses. Another disadvantage of only mowing once a week is that the grass is cut very close, too close, to avoid having to do it more than once, and severe defoliation of this kind weakens the fine grasses over a period of time, so that all sorts of other troubles follow—weeds, coarse grass, moss and so on. Oddly enough, mowing frequently results in the removal of less top growth than if the lawn is mowed once a week.

When mowing is started in the spring—which may be as early as March, if the weather is mild—do not mow too closely the first time, slightly under 1in. will be sufficient, and thereafter reduce it to ½in. In the autumn again make it slightly higher. Cutting can be carried out in the winter, to tidy the grass, and prevent it from getting shaggy; it grows a little at that time of the year, though only very slowly, and on mild days, when the weather is dry, and the ground is not frosted or waterlogged, the grass can be topped to keep it trimmed. The middle of the day is quite often the best time, when the dew has dried off the grass, and any sun there is will be at its warmest. If frost follows this, the grass will not be damaged, but remember, do not mow when there is frost actually on the ground.

Mowing New Lawns. The grass of a lawn newly made from seed should not be cut until the stems are about 2in. high. For the first few mowings the mower must be set at maximum height and the blades must be very sharp, otherwise there is a risk that the young plants will be pulled right out of the ground. The setting of the machine can be lowered progressively as the grass settles down and becomes fully established, but caution is best exercised if there is any doubt about the strength of the grass.

The day before mowing a newly seeded lawn for the first time take the mower (assuming it is of the roller type) over the grass with the cutting blades disengaged, to consolidate the soil round the grass roots and press in any stones on the surface.

The frequency of cutting will depend on the time of year when the grass is sown (spring or late summer) and, by definition, the rate at which it grows.

A lawn made from turves normally has several months to settle down in before the mowing season arrives, but even those lawns laid at the end of the recommended turfing season (March) should be sufficiently well established to allow mowing to be carried out within two or three weeks.

Lawn Maintenance

We cannot all have lawns like bowling greens but we can go a long way towards that goal by careful preparations beforehand (as already detailed) and regular maintenance thereafter. It is all very much worth while.

Aeration. After mowing, probably the next most important part of caring for a lawn—but one which gets very neglected indeed—is maintaining the aeration of the soil. As I have already remarked, grass is like any other plant in that it must have food, water and air, and in a lawn it is inevitable that the soil gradually becomes more and more closely integrated, both because of constant mowing, and of general traffic, particularly if there are children and animals using it. The pore spaces in the soil disappear, the particles of soil get closer

Raking the lawn with a wire-tined rake is also a valuable aid to grass health, for it removes debris which could harbour diseases, and improves surface aeration

Brushing is another useful form of top-growth aeration, especially in the spring. In addition to removing debris and stones it 'lifts' the grass and so assists mowing

and closer together and gradually the air—and oxygen—is driven completely out of the soil. If the soil is heavy there is the added trouble that water is unable to drain away and becomes more and more sour and the grass roots start to rot. As a result the top growth of the grass plant ceases to grow, starts to die off, and eventually unevenly-shaped and irregularly sized patches of brown grass begin to appear all over the lawn.

This type of brown patch is frequently seen on old lawns, where the only care the lawn has is mowing. Brown patches can also occur on new lawns. This is because it takes the soil some years to settle down after being cultivated, and to produce a good crumb structure in which there are plenty of pore spaces and hence room for air and for water to drain away freely. In hot, dry weather, on lawns where aeration has been neglected, grass will turn brown more quickly than on a well-structured soil because, not only are the roots short of air, but the top growth is transpiring water quicker than the roots can obtain it; aeration will help to turn it green as quickly as watering—probably more quickly.

Pricking and Spiking. Aeration can take several forms. The standard method of doing this is by pricking or spiking, *i.e.*, by making holes in the turf to a certain depth and at regular intervals all over the lawn. Compaction as a result of mowing and other traffic may extend to about $2\frac{1}{2}$ to 3in. deep, so that for best results spiking should be to a depth of at least $3\frac{1}{2}$in. to 4in. for preference. An ordinary garden fork can be used, pushing this straight into the lawn at about 6-in. intervals. It can also be done with spiked rollers fitted to mowers, or with separate spiking cylinders which are rolled over the ground; these make dealing with dry soil much less difficult. Aeration can also be carried out with a hollow-tined fork; there is a special one available with springs fitted to it so

that when the tines are pushed into the soil, the springs enable them to be withdrawn much more easily. This also has the advantage that it removes cores of soil completely, and where the soil is very heavy these can then be filled with sand, or a sandy loam mixture. Other sorts of tines can be obtained, from suppliers of turf equipment, depending on the degree of compaction and type of lawn.

Spiking gives its best results in the autumn, and if a hollow-tined fork can be used, so much the better. It may also be carried out in the spring, particularly if the winter has been very wet or very hard, to enable the grass to start into growth more quickly, and it may also be done in emergency during the summer, if the weather is very wet or dry. Some heavy clay soils may require routine spiking several times a year, and if this treatment is followed by filtering coarse sand into the holes, it will be possible to fundamentally alter and improve the structure of the soil over the years.

Raking. Another mild form of aeration is raking, so that light and air can get to the top growth of the grass. This is a spring and autumn job and, if thoroughly carried out, can do a great deal to improve the grass. It gets rid of dead grass, leaves and any decayed vegetation which may be lying on the surface of the soil. Such conditions, too, if allowed to continue, provide a very comfortable home for pests and their offspring, and encourage the spread of fungus diseases. Raking is best carried out in the second half of the season; before this the grass is naturally rather thin, as it is concentrating its energy on producing flowerheads and seeds, and raking can weaken it. After this vegetative growth increases. Autumn raking in particular will not harm it or make it too thin. If a thorough autumn raking is given, then there should not be any need to rake in the spring; the routine brushing before mowing is all that will be required.

A wire-tined rake gives the best results; this is the kind with very springy tines shaped rather like a fan. Rake thoroughly in one direction, and then mow at right angles to the line of raking; then repeat both raking and mowing, again at right angles to each other and in different directions to the first time. If moss is present in the lawn, no raking should be undertaken at all unless the moss has first been killed. Raking tends to spread the moss, particularly the spores, over a much wider area of lawn and will do more harm than good.

For larger lawns it is possible to obtain a special scarifier on a frame which covers a much larger area at one time.

Brushing. Brushing with a stiff broom immediately before mowing is another form of aerating, again of the top growth only. It is a very light method, nevertheless it does help to prevent dead and dying vegetation from settling on the surface, gets rid of any stones which may damage the mower, and brings up the grass if it is flattened. Brushing does much more good than is generally realised. It is the first attention my lawns receive in the spring.

Lawn Feeding

Another of the essential requirements for growing healthy grass is plant food. Lawn grass, perhaps more than other plants, requires nutrients to keep it vigorous and green because its top growth is being constantly removed. If this is done to any other plant, it usually dies completely sooner or later, but grass is remarkably tough and will stand up to this kind of treatment. However, that is all it will do, unless fed properly.

There are three very important foods plants must have in fairly large quantities: nitrogen, phosphorus and potassium, and each assists them in different ways. Nitrogen helps to keep the plant green and encourages the top growth to grow well; phosphorus is good for the roots, particularly of young plants, and potassium plays a part in helping to keep the plant 'hard', rather than very sappy, and also assists in the production of good colour in flowers and fruit. From this it can be seen that for grass nitrogen is particularly important—without it the grass is likely to be thin, and yellowish, and lack of it is one of the main reasons for pale, patchy lawns.

Other plant foods are required, but only in small quantities. These are known as 'trace elements' and in the average soil there are sufficient present for a lawn's needs. It is usually only on intensively grown commercial crops that a deficiency of one of these is likely to arise, or on a soil which has an extreme defect; for instance, it may be very acid or very chalky, or consist mostly of sand.

Time of Application. On the whole, more satisfactory results are obtained if the three main plant foods are applied together, since each interacts with the other and affects the uptake of them all, and a good balance in plant growth is obtained where all are given at the same time. There are plenty of proprietary compound lawn fertilisers available, and the best time to put them on is during the spring. I like to treat my grass with a balanced fertiliser in late February-early March, and such feeding should at latest be started by early April. This can be repeated at the end of June, and if the weather is hot and dry at that time of the year, water the fertiliser in well. This will keep the grass a good colour for the rest of the summer.

In early autumn, say the middle to the end of September, a dressing of another balanced fertiliser can be given. This will strengthen the root growth and make the grass itself tougher, and so better able to stand up to the winter cold. It is best to avoid applying very nitrogenous fertilisers at this time of the year, since they will make the grass 'soft' and lush, when it can easily be damaged by extreme cold, and be badly affected by a fungus disease.

Method of Application. When applying these foods, it is important to follow the manufacturer's

A fertiliser spreader enables the task of lawn feeding to be carried out quickly and accurately. A calibrated roller spreads the fertiliser at the desired rate

instructions to the letter. For instance, applying too much fertiliser will result in scorching of the top growth so immediately producing a brown patch, and it can also do damage below ground by burning the roots.

If fertiliser is to be put on by hand, a good way of practising evenness of application is to scatter the required quantity over a square yard of paper (a square yard is the area normally specified on the manufacturer's directions) so that you can see how thinly—or thickly—the fertiliser covers the ground. Putting half on in one direction and the rest on in the opposite direction will help as well; likewise, dividing the area up into square yards with marking string or canes. For larger areas, fertiliser distributors are available, these having rollers calibrated to allow certain quantities of fertiliser to be released as required. Before using one of these it is a good idea to check it first by running it over a measured area of ground with a known quantity of fertiliser in it, to make sure it is putting the material down at the required rate. Such distributors can be bought for quite reasonable prices these days, and they can also be hired from many garden shops and garden centres.

On some soils, *i.e.*, sandy ones, it may be necessary to feed throughout the season, by putting on small quantities from about halfway through May, at four- or five-weekly intervals until about the beginning of August. The quantity applied and

A perforated hose provides a fine spray which is excellent for the grass as it approximates to natural rainfall and is easily absorbed by the soil

its frequency will depend on the soil, however, and can only be learnt by experience, or by having a full soil analysis carried out.

Watering

A lawn is practically never watered at the time when it begins to need it. It is only when the grass begins to look tired and flabby, after about 10 days without any rain, and of scorching hot sun, that it gets watered; even then it is usually only a sprinkling, and not the large quantity necessary to supply the needs of the grass. Then, because the soil has become so dry, it is difficult for it to absorb the water, and much of it runs off or evaporates. Another reason for watering before a drought has really set in is that, when the soil becomes really dry, it almost always cracks, and this tears the grass-roots and leaves them dangling in mid-air, unable to obtain food or water.

The first sign grass gives of a shortage of water is its lack of 'bounce'. When this is noticed, water should be given at the rate of about 1 gall. per sq. yd. per hour, and an average quantity at any one time should be about 4 to 5 gall. per sq. yd., but this depends very much on the type of soil, and the period of time without rain. If one remembers that the minimum depth of soil required for a healthy lawn is 4in., and that most grasses go deeper than this, then one should apply sufficient water to moisten the soil to this depth, provided it has dried out to this extent. It will mean watering once a week at least, and, in very hot weather, twice. Do not apply water in small quantities at frequent intervals. As with pot plants this does more harm than good, and will compact the surface, so encouraging the appearance of moss and flat, rosette-type weeds.

Methods of Application. There are various methods of applying water. The worst is turning the hose on and simply leaving it to flood the lawn. This wastes the water as it simply runs off the sides when the soil is dry, or pours down the cracks without soaking in first, and where it does become absorbed it tends to waterlog the top inch or so. It is much better to use a sprinkler so that water is applied as a spray which approximates as nearly as possible to natural rainfall. A perforated plastic hose laid straight across the grass is one of the best methods of doing this; it gives a fine spray which is quickly absorbed without flooding. The hose is pierced so that the spray comes out at different angles, thus enabling the whole of an area to be covered evenly on either side of the hose.

Pulsating sprinklers which water a complete circle or segments of a circle are very popular. There are also oscillating sprinklers which have a fixed central point, but which turn from side to side, fanning out the water over a rectangular area from a short tube in the centre supported on a metal stand. A more elaborate type for large lawns is the sprinkler which moves itself along the hose, and can be left on for several hours without constant attention. Before watering in very dry periods, make sure that this is allowed by your local council. Restrictions on the use of water in gardens are often made in times of drought.

Topdressing

It is often not realised what topdressing is, in relation to lawns, nor at what time of the year it should be carried out. It is thought to refer to fertiliser treatment or sometimes to lawnsand, but in fact it consists of putting on the turf a mixture of loam, peat and sand (or of leafmould, dried sludge, or well-rotted farmyard manure, mixed with or substituted for the other ingredients) in the spring or autumn to improve the structure of the top few inches of soil gradually over the years, and to prevent it deteriorating, or becoming short of humus-forming material. It also provides a good basis for new grass shoots to root into, and helps to fill in the minor hollows.

A suitable general mixture for the average soil could consist of 4 parts loam, 2 parts sand, and 1 part granulated peat. The loam should be put through a $\frac{1}{4}$ in. sieve, and the sand should be coarse river sand. Mix all these together thoroughly and then apply the mixture dry at 2lb. per sq. yd., as evenly as possible. After application, work it into the surface with a stiff broom, or the back of a rake; it will smother the grass if left on the surface and do more harm than good.

If lawns on sandy soils are being treated, the quantity of peat in the mixture can be doubled, and the sand halved; where clay soils are involved use a mixture of 2 parts loam, 4 parts sand, and 1 part peat. In every case the parts are by volume. One topdressing a year is usually enough. Sand alone can be used as a topdressing where the soil is particularly heavy, of a particle size of $\frac{1}{16}$ in. to $\frac{1}{32}$ in., spread over the grass in a thin, even layer. As a rough guide 3lb. per sq. yd. may be put on, and worked well into the surface.

Weed Control

The appearance of a lawn is marred when there are weeds growing in it, and many people seem to have considerable trouble in dealing with them. This is partly because they have not been managing the lawn correctly, with the result that the grass has become weak, and the weeds have been able to establish and spread without much competition. Another reason is that a great many of the weeds on lawns are the kind that can adapt themselves to the existing cultural conditions.
Two Main Classes. Lawn weeds can be divided

Raking a topdressing into a lawn to improve the structure of the top few inches of soil. The back of a rake or a stiff broom is suitable for working the mixture in

very roughly into two classes: one, the kind of weed with a flat, rosette habit of growth which enables it to escape the mower, such as plantains, daisies, and, to a lesser extent, docks and dandelions, which under lawn conditions tend to grow flat on the ground; the other the creeping, trailing type of weed, which roots at each leaf-joint, such as speedwell, mouse-ear chickweed, pearlwort and sea milkwort. This kind of weed often also has very small, and sometimes hairy, leaves which make it unlikely to absorb sufficient hormone weedkiller at any one time to kill it completely. Of course, there are other weeds which do not fit into these two categories, and which have their own, different methods of survival, but this does explain how weeds have adapted themselves to an unnatural environment.

Weed seeds are always liable to be introduced into a lawn by one agency or another, but when new lawns made from seed are infested with weeds it usually indicates that the preparatory work was skimped and in particular that the ground was not allowed to lie fallow long enough to allow perennial weeds to be dealt with thoroughly.

Control Methods. There are various efficient ways of dealing with weeds. If only a few are present, it is a waste to treat the whole lawn with selective weedkiller; indeed, it may not be necessary to do more than dig them out with a two-pronged fork, if they are the rosette, tap-rooting kind, or to spot-treat them with a selective weedkiller (applied either from an aerosol or a puffer pack), or with a special tool which injects weedkiller into and on to the leaves of the weed.

For bad infestations it will be necessary to treat the whole lawn, and here a word or two about the way in which the 'hormone' or selective weedkillers (which only kill certain types of plants) work is necessary. The main thing to remember is that once they have been sprayed on to the plant,

Applying a selective weedkiller to a lawn. Note that a yard-wide strip has been marked out with string and that the operator stands outside the area being treated

they are absorbed through its leaves into the sap, and are then circulated round the plant's system, reaching every part of it. Hence, they will be more effective if applied when the plant is growing fast, and the sap is at its most active in the late spring and early summer. This type of weedkiller is a solution of plant hormones, which stimulates the plant cells so that they multiply too rapidly in certain parts of the plant, particularly the growing points at the tips of the shoots, and thus the plant's metabolism is upset to a fatal extent. Another point to remember is that using as fine a spray as possible, and covering as much of the leaf and shoot surface as feasible, will provide the maximum effect. It is most important not to make up a solution at a greater concentration than is recommended in the manufacturer's directions.

A lawn made from seed must not be treated with ordinary selective weedkiller for at least three months after germination (there is a special weed-killer for such young grasses, see p. 23 for details), and the same precaution applies to treating a lawn grown from turf.

Selective weedkillers contain chemicals known by such names as 2,4-D, 2,4,5-T, MCPA, meco-prop, dichlorprop, and fenoprop. These will kill or drastically check most weeds found in lawns—and some of them are particularly effective against certain weeds. A combination of 2,4-D and 2,4,5-T is particularly good; mecoprop is especially effective against clover and pearlwort. Speedwell is one of the very few which is virtually unaffected, and requires the use of another material. In this case lawnsand might be used or a solution of tar-oil winter wash mixed with water. Especially useful are the combined selective weedkiller-fertiliser mixtures which do two jobs at once by killing the weeds and feeding the grass at the same time.

Where spot treatment can be carried out, it is sometimes better to use the weedkiller known as paraquat, which kills all plants through their green foliage. This, of course, includes grass, so if used it must be very carefully applied, to the leaves of the weeds only. One way of doing this is to 'paint' the solution on to the plants, and this could be quite useful when dealing with isolated weeds in a newly sown lawn after the first two or three weeks of germination.

I mentioned lawnsand (a mixture of sulphate of ammonia and ferrous sulphate, with sand to act as a carrier) earlier on; this is an old remedy for getting rid of weeds, and still a useful one. It works by burning the top growth of the weeds and turning it black; it may also discolour the grass, but this is only temporary. Best results are obtained if no rain follows its application for 48 hours, but after this it should be watered in. As it contains nitrogen it has the two-fold merit of feeding the grass as well. It also helps to remove dog lichen, a greyish-black leathery growth which appears on some lawns where the grass is starved.

Moss Control. Another major trouble on lawns can be moss. Although green and soft to walk on, it tends to die away during the summer if the weather is dry, and is easily scratched up by birds and so on during the winter, so that once it has spread all over the lawn and choked out most of the grass, it does not form a very good substitute. Also, weeds tend to spread to mossy areas.

The chief reasons for its invasion are: poor drainage and compaction of the soil surface; starvation of the grass; mowing the grass closely; deep shade; and, occasionally, too acid a soil, but the last-mentioned is not nearly as common a reason for the presence of moss as is generally thought. On the whole, any environmental factor, particularly poor drainage, which tends to weaken the grass will encourage moss to spread. Unless the fundamental cause is removed or modified, moss will be persistent and, while it may be removed in the first instance by the proprietary moss killers, it will return to those lawns where the growing conditions are not improved.

Lawnsand can be used to burn it out, and can be applied at any time of the year. Mercurised lawnsand can also be used and has a slightly longer-lasting effect in that it kills the spores of the moss from which new moss plants are formed, as well as burning out its top growth. Liquid moss-killers containing anthracene oil (a tar acid) can also be used. But remember to improve the drainage, eliminate compaction of the soil surface, and feed the grass, if you want to prevent moss from returning.

General Care

To make a lawn look really attractive there are various things which can be done. One of these is to finish the edges off really well. I like to see them

A very ordinary lawn can be transformed in appearance if the edges are carefully trimmed with an edging iron, as above, or some other suitable tool

the grass short and, incidentally, prevent it creeping on to flower beds.

Removing Bumps and Hollows. Another small attention which much improves the appearance of a lawn is the removal of bumps and hollows. This should not be done by rolling—this will only bring different and much worse difficulties—but by the following method, which is just as quick. Cut the turf in the shape of a capital H over the bump or hollow, roll back the two flaps of turf so formed, and add or remove soil as required until that part of the lawn is at the same level as the rest. Then put the flaps back in position and dress the cuts with sand or a topdressing mixture.

Removing Autumn Leaves. Another important point for autumn care is the removal of leaves. Letting them lie on the grass until the wind blows them away may mean the grass being covered for several weeks so that it turns yellow and becomes thin and straggly; worms and other pests congregate under the leaves and the final result is a very nasty patch of grass. Sweep the leaves up frequently, either with a besom broom or with a mechanical leaf sweeper, if the area to be swept justifies such a mechanical aid. Alternatively, use a wire-tined rake or a garden rake for this job.

Renovating Jobs. When the edges of the lawn get broken, as inevitably they do, especially when there are children in the family, treatment with an edging iron may suffice, but if the breaks are deep the only thing to do is to lift a strip along the damaged edge and turn it round. The broken edges will then be on the inside and the holes can be filled with soil and re-seeded in the spring. The first part of this operation can be done at a slack time in winter, if so desired.

Another common trouble is bare patches caused by excessive wear. An easy way to eliminate these is to prick over the surface with a fork in September to a depth of about 2in. to loosen the soil and then sprinkle some high-quality grass seed over the area. Cover the seed with a little fine soil, give a light dressing of fertiliser and the patches will disappear. Do not mow these areas, though, until the grass is well-established. The more traditional way of doing this job is to remove all the grass from such areas and either lay new turves or sow the whole area again. It is not usually necessary to go to such trouble.

Lawn Mowers

Now we come to one of the most important points of all—the choice of a mower. The size one chooses will depend on the size of the lawn. For very small lawns, of course, a hand-propelled side-wheel or cylinder mower is all that is required, but for larger areas a powered mower (electric or petrol-engined) is almost a necessity.

neatly trimmed—it makes all the difference between a first-class and an average lawn. If this is done fairly frequently before the grass gets too long there is no need to pick up the clippings as they will soon wither and mix with the soil. In some cases they can be cut at the same time as the lawn is mown. If the lawn is edged by a paved path the mower can be run partly along the path or, where there is a flower border at the side of the lawn, the mower can be used along the border's edge so that it projects over it by a few inches; but do not let it overhang too much, otherwise it will overbalance, and make sure that the edges are firm and have not been undermined by moles.

Where it is not possible to use a mower, there are various satisfactory tools that can be used for edging—the standard long-handled side shears, clippers mounted on wheels, also with a long handle, and edging tools which will tidy up crumbling edges, such as the half-moon edging iron, the spiked edging iron which is pushed along the edge and rotates as it goes, and a rectangular cutting iron mounted at an angle on a long handle.

To save time, an alternative to these is to use one of the chemical growth retardants for the grass, which is mixed with water and sprayed on to the edges. It is not really a good idea to use these on the whole of the lawn, where the grass is fine, and the appearance important, but where only a narrow strip is in question, this will keep

For lawns of moderate size a mower with a cutting cylinder 12 or 14in. in width will be sufficient.

The number of blades on the cutting cylinder is also important: the more blades the more expensive the machine and the finer the grass that can be cut and maintained in good condition.

Hand-propelled Mowers. Having decided on the size of mower, then the type should be considered. Anything over 12in. is usually powered; below these sizes the mower will be hand propelled. If a hand-propelled mower is decided on, it is a question of whether to have a side-wheel or a roller model; with side-wheels, cutting right up to the edges is not possible, but such mowers are cheaper and, without a front roller, are better able to cut and control creeping weeds since they will not have been flattened even more before the cutting blades get to them. With the roller models a better appearance is given to the lawn, and the edges can be mown at the same time.

Powered Mowers. Of the powered mowers, the battery-electric models are certainly the ones with the most appeal. They are simple to use, relatively noiseless, trouble free and easy to maintain. One must, of course, obtain a machine with sufficient capacity (in terms of mowing time between charges) to cut all the grass at one go and your local dealer will give you all the details you need in this respect. The petrol-engined mowers, too, are, of course, extremely reliable nowadays but for women at least they are not such an attractive proposition—and there is certainly more to go wrong than with the battery mowers. Mains powered electric mowers suffer the disadvantage of the trailing cable, which is always liable to get in the way and can, if one is careless, get cut. Also, its radius of action is necessarily circumscribed. Such machines can also be run off a generator which gives them more freedom of action, but they still lack the convenience of other types.

For obtaining the best possible finish the choice would be a cylinder-type mower every time, but for adaptability the popular rotary-type mower, with its ability to cut both short and long grass, wins hands down. With this type of machine the cutting blades, which revolve at very high speed, are set in the horizontal plane. They also cut the grass up very finely and although many are now provided with grass catchers they are not really needed provided the grass is cut frequently. These machines also are less expensive than their cylinder-model equivalents; they are good at cutting 'bents' and some weeds which are not touched by cylinder mowers.

The most recently introduced type of mower is that using—very successfully—the 'hovercraft' or aircushion principle. These are extremely easy to handle for they will change direction at the slightest touch, and they are particularly useful

The versatile rotary motor mower is capable of cutting both long and short grass but it will not give the fine 'finish' obtainable with a cylinder type mower

for cutting steep slopes, banks and other difficult areas which are 'out of bounds' to more conventional machines.

Routine Maintenance. Whatever type of mower one owns, however, it is essential to maintain it in good working order. A clean down and attention to any needed lubrication should be standard 'drill' each time the machine has been used. On cylinder machines, the blades should be kept really sharp and the cutting cylinder correctly positioned in relation to the bed-plate. If these two points are disregarded the grass is more likely to be torn off than cut cleanly, as it should be. The front roller should be adjusted so that it is perfectly horizontal, otherwise the grass will be cut more closely on one side than the other.

At the end of the season either overhaul your machine yourself, or get this done by a servicing company. Regrinding of the blades is a skilled job and this must be done by a trained man. The prudent gardener gets this done in winter before the spring rush begins for those who delay too long may find that their mower is not available for the first few weeks of the new season, which is certainly not to be desired.

Pests and Diseases

For details of the pests and diseases of turf which are most likely to be encountered see Chapter Thirteen (pp. 289 to 291).

used in such positions because their attributes make them eminently suitable for growing in this way) but this will depend much on the architecture of the building and whether it would have its appearance enhanced by such plantings. In small gardens in particular, walls add a new dimension to one's gardening and allow a much wider and more interesting selection of plants to be grown in the space available.

I like to see individual—or, as we say in gardening, 'specimen'—trees and shrubs planted in lawns as features in their own right. Those with attractively shaped or coloured leaves are ideal for this purpose. Flowering and berrying trees and others with coloured bark also look very handsome when grown in this way. Weeping trees, those with fully arching branches, make superb lawn specimens. Shrubs one would choose for this purpose are those with a neat, attactive habit rather than those with straggly, untidy growth.

Some trees and shrubs look well near water features and in this connection the willows immediately come to mind, also the dogwoods with coloured barks. Even a rock garden can be planted with a selection of trees and shrubs if miniature forms are chosen.

Both formal and informal hedges make pleasing garden features and provide an attractive background for other planting schemes. I like to see hedges of flowering and berrying shrubs, and hedges of evergreens with golden- or silver-variegated foliage can be very attractive indeed.

It can sometimes be difficult to find suitable plants for shady parts of the garden which are never reached by the sun but there are quite a number of smaller shrubs which are perfectly happy in shade, especially under large trees. Shade lovers which come immediately to mind are rhododendrons, hydrangeas, mahonias and skimmias. There are others which will tolerate very dry, open, sunny positions and for planting between larger trees and shrubs there are numerous low-growing, spreading shrubs from which a choice can be made. Shrubs such as these are described as ground-cover plants, for once they have formed a carpet they will to a large extent suppress weeds and make soil cultivation unnecessary.

Creating a Well-balanced Garden. To create a well-balanced garden a few trees and shrubs at least must be grown with the other plants. Some of the taller kinds should be selected to provide height where this is needed, for a flat-looking garden should be avoided at all costs. Try, too, to maintain a careful balance between the various types of trees and shrubs—evergreen and deciduous, flowering, berrying and those with decorative foliage.

The smaller the garden the more care must be

Trees and Shrubs 3

In almost every garden space can usually be found for at least one or two trees and a selection of shrubs, and there are many ways in which the latter can be used. They can be grouped in beds and borders on their own, be used in mixed borders together with perennial and other plants, or be planted near the boundaries of the garden for screening purposes. A carefully positioned tree or shrub can be used to mask off, say, the corner of a building or some other obstruction which detracts from the beauty of the garden. Trees and shrubs—particularly evergreens—may also be useful for forming windbreaks where the garden is so sited that it is exposed to the prevailing wind.

The walls of the house and the outbuildings may also be used as a support for climbers and wall shrubs (the latter is a term which includes non-climbing shrubs which benefit from the shelter provided by a wall or solid fence, or are

taken to make the right choice of trees and shrubs. Size, in particular, must be given special consideration for over-large specimens will destroy the proportions of the garden. It is all too common to see forest trees such as limes, poplars, elms and planes planted in small town gardens and, inevitably, these have to be mutilated by severe pruning to keep them within bounds. How much better in such circumstances to plant smaller-growing ornamental kinds. As will be seen from the descriptive lists which follow, there is no shortage of them. Quite apart from the mutilation I have just referred to there are two other reasons why it is unwise to plant large trees in confined places: they will rob an undue proportion of the garden of light, moisture and plant foods, and the roots may damage the foundations of the house and the drains. For small gardens, narrow, columnar trees such as *Prunus* Amanogawa (see p. 52) have special value.

Preparation for Planting

It is most important that trees and shrubs should be purchased from reputable nurserymen for then one can be sure of getting first-class specimens. A really good tree or shrub will have been transplanted in the nursery before it is sold, and this encourages a fibrous root system to develop. It is usually possible to tell when a specimen has not been transplanted as it will have a large tap root and very few fibrous roots. The more fibrous roots a plant possesses, the quicker it will become established.

Planting Times. Deciduous trees and shrubs can be planted at any time between early November and the end of March, provided the ground is not frozen or waterlogged. Evergreens, however, are best planted in September and October or during late March, April and May. As evergreens are never inactive it is especially important that they should make new roots quickly and so ensure that the loss of moisture through the leaves is made good before the plant is adversely affected. By planting them at the times recommended, the roots should quickly become established, for the soil will still be warm in the autumn and in spring it will be warming up after the winter cold.

Container-grown Plants. It is possible nowadays to buy container-grown trees and shrubs which can be planted out at any time of the year, even when they are in full leaf or flower. The plants are grown in either tin, polythene or bituminised-paper containers while they are in the nursery and build up a good fibrous root system so that they suffer very little root disturbance when they are transplanted.

Choosing a Suitable Site. Choosing a site for a tree or shrub needs careful consideration for once it is planted it will probably remain in its allotted position for many years. It must be placed where it will have enough room to grow and develop naturally, for nothing is worse than to see a specimen plant which has had to be cut back severely to keep it within bounds.

Trees, especially, need extremely careful siting as they are often used to create a focal point in the garden. A badly-placed tree can be most irritating, but one that has been sited carefully can greatly enhance the attractions of a garden. Trees should not be planted too close to the house as they may rob the rooms of much light and roots of large trees can damage foundations and drains.

Soil Requirements. Many trees and shrubs will grow on a wide range of soils but there are some which are more difficult to please. I am thinking now of the lime-hating kinds such as rhododendrons, azaleas and many of the heathers. These must have an acid soil—in other words, one that does not contain lime or chalk.

To be absolutely sure that your soil is suitable for growing lime-hating kinds I would recommend that a soil test be carried out. Soil-testing kits, together with instructions, can be obtained quite cheaply from garden sundriesmen. If the soil test shows a reading of pH 7·0 or over then you must keep to the kinds that do not mind lime. If the result is pH 6·0 or under then you will be able to grow lime-hating shrubs successfully. On a soil with a pH of between 6·0 and 7·0 you may just be able to grow some of the lime-haters if plenty of peat or leafmould is added. The ideal pH for shrubs other than lime-haters is 6·5 or 7·0.

Preparing the Soil. Once a tree or shrub has been planted it will usually remain in that place for the rest of its life. The soil must, therefore, be thoroughly prepared before planting as there is little that can be done once the plant is established. I always try to prepare the ground as far ahead as possible of the planting date—at least a month in advance and preferably more. This allows the soil to settle properly and also gives any organic matter added time to start breaking down into humus, thereby releasing its plant foods.

Single digging to the full depth of a spade is usually sufficient, but if the soil tends to hold water during wet weather, double digging may be advisable. I have described how to do this on p. 74. Whichever method is used, however, incorporate plenty of organic matter, such as well-rotted farmyard manure, garden compost, moist peat, leafmould, spent hops, shoddy or any similar material.

After digging I like to fork a slow-acting fertiliser such as bonemeal into the top few inches of soil at the rate of 2 to 4oz. per square yard. This releases the plant foods over a long period, thereby supplying the tree or shrub with the necessary nutrients. I do not use bonemeal for the lime-hating plants as it contains a certain amount of calcium.

If organic matter is not available then a good alternative is a mixture of 2 parts bonemeal and 1 part hoof and horn (all parts by volume) applied at 4 to 6oz. per square yard and forked well into the soil.

Planting and Aftercare

When trees and shrubs arrive from the nursery they have their roots well wrapped in sacking or polythene sheeting. They may also have their top growth wrapped in straw. This is to keep the plants in good condition until they are planted.

If, for any reason, you cannot plant the specimens immediately on arrival, then they may be placed in a cool shed or garage for up to one week. Remove the straw packing from around the stems to allow free circulation of air but leave the packing around the roots. Give the roots some water if they are dry. If you have to delay planting for more than a week, then the plants will be far happier if they are heeled-in temporarily in the open ground in a sheltered position.

Spacing and Arrangement of Plants. Many gardeners make the mistake of planting trees and shrubs too close together with the result that within a few years they grow into one another. Consequently, each specimen will be fighting for light and air, and this will affect its flowering potentialities. It will also mean hard cutting back in many instances, as I have already pointed out, and this is something I do not like to see. I would recommend spacing shrubs grown in beds and borders so that the distance between them is equal to two-thirds of their height when fully grown. The ultimate height of shrubs is given in most good catalogues and the spacing mentioned is a minimum one; it may certainly be increased with advantage. I usually plant two or three shrubs of the same kind in a bold group to obtain the

Above. The beautiful flowering peach, *Prunus persica* Clara Meyer, a variety well suited to small gardens for its height and width do not exceed 20ft. This variety flowers during April
Left. The brilliant berries of *Sorbus aucuparia*, the Mountain Ash or Rowan, are a familiar sight in gardens. This native tree reaches a height of 30 to 40ft. and has a spread of roughly equal proportions

37

Left. There are few lovelier shrubs
than the camellias which, given a
sheltered position to protect the
buds and blooms from frost damage,
will give a great deal of pleasure.
The variety shown here is
Inspiration, a shapely *williamsii*
hybrid with semi-double flowers
Right. The choice *Genista lydia,* a
dwarf Broom of outstanding garden
value. It has numerous garden uses
but is perhaps seen at its best
cascading over rocks, as in this
illustration
Right, below. The chaenomeles,
which many gardeners still call
by their old name of cydonia, are
among the most popular of all
spring-flowering wall shrubs.
The richly coloured *C. simonii* is
among the best of many handsome
varieties

Right. Kalmia latifolia, a beautiful ericaceous evergreen shrub from North America. This shrub grows 6 to 8ft. tall, needs similar conditions to rhododendrons (lime-free soil is essential) and flowers in June
Below. Skimmia fortunei, a very free-berrying evergreen shrub which also needs lime-free soil, bears its decorative berries for many months in winter

best effect. The spaces between the permanent shrubs can always be filled temporarily with quick-growing expendable shrubs or other plants.

Standard ornamental trees should be planted 25 to 30ft. apart and, of course, very large trees must be given proportionately more space. Small shrubs can be planted between these trees.

Planting Trees and Shrubs. It is necessary to make the planting hole of sufficient size to allow the roots to be spread out to their full extent, and it should be of such a depth that when the specimen has been planted it is at the same depth as it was in the nursery. With most trees and shrubs you will be able to see the original soil mark on the stems. Before planting, examine the roots of the tree or shrub you are putting in; any that are torn and broken should be cut back to the undamaged part, otherwise they may die back and will more easily be infected by disease.

If you are planting trees then they will almost certainly need supporting with strong wooden stakes. The stake must be inserted before the tree is planted and the top should be just below the lowest branches of the planted tree. The distance between the stake and the trunk of the tree should be about 2in.

Place the tree or shrub in the centre of the hole and spread the roots well out. Place some fine soil over the roots and gently shake the plant up and down to allow the soil to filter among the roots. Then firm the soil by treading. Add more soil gradually, at the same time treading it well until you reach the natural soil level. The soil must be in close contact with the roots and be made really firm.

Planting Container-grown Specimens. Before removing plants from their containers they must be given a good watering. Tin containers will usually be split by the nurseryman with a special tool at the time of purchase and the removal of the root ball must be done carefully so that there is no damage. Make a hole just large enough to take the root ball and insert the plant carefully, firming it in thoroughly.

Aftercare. When trees are planted these should be staked at the same time. The space between the trunk and the stake allows the insertion of a soft buffer to prevent the trunk from rubbing against the stake. I like the proprietary plastic tree ties as they are fairly cheap to buy and extremely easy to use. They also hold the tree a safe distance from the stake.

Young trees and shrubs, and particularly some of the evergreens, including conifers, may be slightly on the tender side during their first winter after planting. Therefore, I would suggest that you erect a wattle hurdle or hessian screen on the windward side of the plants, to break the force of the wind. The screen should be partially open in structure as its purpose is to filter the wind to break its force. A solid structure will cause eddying which can severely damage the plants. Make the screen slightly higher than the plants and secure it to firm stakes. It must be removed in the spring at the onset of milder weather.

During the first winter, frosts or strong winds may loosen the plants as they will not have rooted into the soil. Throughout the winter, at regular intervals, I check my newly planted trees and shrubs and refirm them if necessary.

Watering. The critical period for all newly planted trees and shrubs is April, May and June, when there are likely to be dry spells. This is the time when plenty of water should be given, and it will make all the difference between success and failure. Once the plants become dry at the roots for any length of time their growth will be seriously retarded—and in bad cases of moisture depletion they may die.

To keep evergreens fresh for the first few weeks after planting, it is a good idea to spray them overhead with plain water during the evenings. This is I consider even more beneficial if they have been planted in the spring. Alternatively, they can be covered with a polythene 'tent' for the first few weeks to keep them fresh.

Mulching. I find that newly planted trees and shrubs benefit greatly from mulching as this helps to keep moisture in the soil during dry periods. Use either well-rotted farmyard manure, garden compost, peat or spent hops and spread the material about 3 to 4in. thick around the plants. In the case of rhododendrons and azaleas, I like to mulch them every year with peat as they are far happier in a good moist soil which is at the same time well drained. I always try to put the mulch on sometime during January, February or March, when the soil is not frozen.

Pruning Trees and Shrubs

Shrubs usually need more pruning than trees and I shall start by discussing shrub pruning.

The Reasons for Pruning. The main reasons for pruning shrubs are to encourage the formation of flowers and in some cases fruits; to ensure that the plants remain healthy and vigorous, and to keep them shapely. Some shrubs do not require any pruning; for example, evergreens such as conifers, rhododendrons and azaleas. On the other hand, if those shrubs which benefit from pruning do not receive such attention they will deteriorate rapidly within a few years and will certainly not flower as well as they should do.

I do not like to see pruning carried out in a haphazard way, just for the sake of doing it. It is essential that you should become familiar with the habits of growth and flowering of your shrubs and prune them accordingly—you must know which ones need pruning and which ones do not.

Maintaining their Shape. Another important point when pruning is to try to maintain the shrub's natural shape. How often do we see shrubs being clipped hard all over by gardeners who know little or nothing about pruning, just to keep them looking neat and tidy. This invariably prevents them from flowering because in the process the flower-producing wood is removed.

When to Prune Shrubs. Deciduous shrubs can be divided into two groups for the purpose of pruning. Those that flower in the spring or early summer produce their blooms on growths formed in the previous year. These are pruned immediately after flowering has finished to give them a chance to make new growths during the summer and autumn. These new shoots will then produce flowers the following year. The second group comprises shrubs which flower during the summer and autumn. Many of these produce their blooms on growths which are formed in the current season—during the spring. These kinds are pruned very hard in March or April to encourage plenty of young shoots to form, so ensuring an abundance of large flowers later in the year.

With evergreen shrubs pruning is confined to the shortening of growth to ensure shapeliness, when this is necessary, and the removal of dead or weakly growth.

How to Prune. First, let us consider those shrubs that flower in the spring and early summer on the previous year's wood. These include chaenomeles, cytisus or brooms, deutzia, *Jasminum nudiflorum*, kerrias, philadelphus, *Prunus triloba*, *Spiraea arguta*, *S. prunifolia plena*, *S. thunbergii* and weigelas. The stems that have flowered are cut back to young shoots lower down the stems. There are usually two or three of these shoots which will develop and bear flowers in the following year. Forsythias flower on shoots which are two years old or more but these are pruned in the same way—after flowering.

When pruning brooms, all the stems with developing seed pods are cut back to young shoots. With weigelas and deutzias, cut out the old flowered wood near to the base of the plant, at the same time leaving as many new shoots as possible. They will flower much better for this harder pruning.

We come now to the shrubs which flower on the current season's wood. The main ones to include here are *Buddleia davidii*, caryopteris, deciduous varieties of ceanothus, hardy fuchsias, *Hydrangea paniculata*, and tamarix. I always cut all the old stems down to within a few inches of the ground in early spring. These shrubs are very vigorous and soon send up plenty of young shoots.

When *Buddleia davidii*, *Hydrangea paniculata*, ceanothus and tamarix are cut right down each year they do not grow quite as large as they would do naturally. While this suits owners of small gardens admirably, some gardeners like to have larger specimens. Therefore, a framework of older branches can be built up by pruning some of the shoots less severely. However, hard pruning of the flowered wood must still be carried out. The flowered shoots contained on the main framework must be cut back to within two or three buds of the main stems during March or April.

Those shrubs which are grown for the beauty of their coloured stems should be cut down to within an inch or two of the ground in early spring each year. They will then send up masses of whippy shoots. This is a good way, for example, of growing willows in a smallish garden.

The coloured-leaved elders can also be pruned hard each spring, either to within a few inches of the ground, or a framework of older wood can be

An old shoot is removed from a weigela after the flowers have faded, the cut being made just beyond two young shoots which will flower in the next year.

built up and the shoots from this cut back to within two or three buds of their base.

After pruning those shrubs which flower on the current year's wood, I usually give them a dressing of an all-purpose fertiliser. I find that this encourages them to make strong growth.

Removing Dead Flower Heads. Certain shrubs, especially some of the evergreens, require no more in the way of pruning than the removal of dead flower heads. It is best to remove them as soon as the blooms have finished, before the seed pods are properly formed, as seed production wastes the plant's energy and will mean fewer flowers in the following year. With rhododendrons and azaleas I twist off the clusters of seed pods, taking care not to damage the buds which are developing just below. I remove the old flower heads of lilacs with a pair of secateurs, cutting them off at their base.

I give my heathers and lavenders a light trim with a pair of shears after they have flowered – this also keeps them compact. I trim them just sufficiently to remove the dead flowers and never cut into the main shoots or old wood. The winter-flowering heaths I usually leave until March or April before trimming them over.

Cutting out Suckers. Some shrubs which have been budded or grafted on to a rootstock invariably throw up sucker growths from the stock. The main shrubs which come into this category include rhododendrons, azaleas and lilacs. I remove the suckers by cutting them out at their point of origin. They are easily identified as the leaves are usually slightly different from those of the main growth.

Removing Dead Wood. I always keep an eye open for any dead or diseased wood in shrubs and remove this regularly. Dead wood offers an open invitation to diseases and it is, of course, un-

previous summer's growths cut back to within a few buds of their base in February. The *alpina* and *montana* groups are trained to form a framework of branches. The side shoots from the main framework are cut back almost to their base in early August. The *florida* and *patens* groups produce their blooms on short growths from the previous year's shoots. A framework of branches must be built up. When the flowers are over they are cut off just above the buds which can be seen below them. I prefer to treat those of the *lanuginosa* group in the same way as the *jackmanii* and *viticella* groups as plenty of flowers are then produced.

Rambler roses I prune in September after flowering. These make a good deal of new growth each year and therefore must be kept thinned out annually. Shoots that have flowered are cut out completely to ground level and the new growths made in the current year are then tied back to

A dead flower head is removed from a rhododendron. If it is done as soon as the flowers fade the plant's energies are diverted from seed production

An electric hedge trimmer—here used on a quickthorn hedge—can save much effort and time. One form of power is a battery-electric mower

sightly. For both reasons it should be removed without delay. The need for the removal of diseased wood needs no explanation.

During the spring I also cast an eye over my shrub roses for dead wood. At the same time I remove a little of the very old wood, especially if the bushes are becoming congested. Apart from this, shrub roses need little attention with regard to pruning.

Pruning Climbers and Wall Shrubs. Wisteria I do not prune much until it has covered its allotted space. Then I prune the young shoots in August each year by cutting them back to within five or six buds of their base. During the winter, I cut them back farther still, to within an inch or two of the older wood.

The pruning requirements of clematis differ according to the group to which they belong. The *jackmanii* and *viticella* groups should have the

their supports. I prune my climbing roses during February but not so severely as the ramblers. Each year I cut out some of the old wood and leave as much young wood as possible.

Trimming Hedges. Most formal hedges need regular trimming to keep them shapely and well covered with foliage. Although informal ones do not need regular treatment I go over them occasionally with a pair of secateurs to remove any long or straggly growths. Flowering hedges, if trimmed too hard, will produce very few blooms. It is inadvisable to let formal hedges get out of hand before trimming them, as this means that hard cutting back will be necessary which invariably ruins their appearance. (For details of trimming particular kinds of hedging plants see pp. 71 and 72.)

A pair of garden shears or an electric trimmer are the most suitable tools for trimming the

The edges of large saw cuts on tree limbs should be pared smooth with a sharp knife and afterwards treated with a bituminous tree paint

majority of formal hedges. However, for large-leaved evergreen types, such as aucuba, rhododendrons and laurels, trimming is best carried out with secateurs so that the cuts are made on the wood. If such hedging plants are trimmed with shears the leaves will be cut and turn brown at the edges. This is very unsightly.

Pruning Trees. Trees do not need any regular pruning as most of them should be allowed to grow to their natural shapes. The only pruning I do is to remove any dead or diseased wood during the winter and possibly a badly placed or crossing branch.

Snags must not be left when branches are cut out for these will die back and may encourage diseases to gain entry. Trim the branch off flush with the trunk or main branch as appropriate. Large cuts should be made smooth with a pruning knife, and then painted with a bituminous tree paint to prevent the entry of diseases.

The pruning of ornamental plums, peaches and cherries is best carried out in the growing season after flowering—in June, July or August. At this time of the year the cuts heal quickly and there is less likelihood of them being infected by Silver-leaf Disease (see p. 293).

Removing Suckers. Trees which have been grafted, such as ornamental cherries, peaches and plums, are often troubled with suckers which emerge from the rootstock. These should be cut right back to their point of origin. They may either emerge from below ground or from the trunk of the tree below the graft union.

Propagating Trees and Shrubs

If I were asked which branch of gardening gave me most satisfaction I might well say 'Propagation and all it entails'. There is nothing quite like the thrill of seeing difficult plants successfully raised from seed, a frame full of healthy softwood cuttings prepared by one's own hands, or a graft which has 'taken' when the 'marrying together' proved rather a ticklish job. Various methods of increase can be adopted, such as cuttings, seed, layering, air layering, grafting, budding and division and I will deal with each of these in turn.

Cuttings

This is undoubtedly the most widely used method of propagating trees and shrubs. There are two main types of cuttings which can be taken, each of which requires a different technique and rooting conditions. First, there are half-ripe cuttings which are taken during the summer and, secondly, there are hard-wood cuttings which are taken during autumn.

HALF-RIPE CUTTINGS

Many of our popular garden shrubs can be propagated by half-ripe cuttings. These include abelia, buddleia, caryopteris, ceanothus, cotoneaster, deutzia, escallonia, fuchsia, hebe, heathers, hydrangea, potentilla, rosemary, santolina, senecio, viburnum and weigela.

Selecting Suitable Shoots. Half-ripe cuttings, which are taken during July, August or September, are obtained from the current year's shoots. These shoots, as the name suggests, are only half-ripe, that is, the wood has not completely hardened. You should avoid any shoots which are still rather soft; you may, therefore, have to be patient for a week or two until they have hardened and ripened slightly.

Preparing the Cuttings. I find it best to pull off the shoots with a heel of older wood attached, and the cuttings, when prepared, are 6 to 8in. long. If the tip of a shoot is very soft then I cut it off. Some gardeners like to cut off the shoots just below a leaf joint, but the method used depends upon personal preference. Heather cuttings, incidentally, are best taken with a heel and should be about 1 to 2in. in length. The leaves on the lower half of all cuttings must be cut off with a sharp knife, close to the stem.

Suitable Rooting Mediums. I prefer a mixture of equal parts by bulk of loam, peat and sand, but a mixture of equal parts peat and sand is also suitable, particularly for cuttings of lime-hating plants.

Inserting the Cuttings. The cuttings can be inserted in a cold frame, about 2in. apart each way, with half their length below the soil. Before I insert them, however, I like to dip the base first in water and then in a hormone rooting powder to assist in rapid root development. I use a piece of wood, shaped rather like a pencil (and known as a dibber), for making the planting holes. I then firm the cuttings thoroughly with this, paying particular attention to their bases. The base of each cutting must be in close contact with the soil, otherwise roots will not form.

I water the cuttings thoroughly to settle them in and then close the frame. They like a moist, close atmosphere as this encourages them to root more quickly. The frame can be kept closed for three or four weeks, opening it only for a few minutes each morning to allow condensation to drain from the underside of the glass. I also syringe the cuttings for the first week or two to keep the leaves moist and to prevent flagging. After about four weeks a little ventilation can be given, by which time many of the cuttings will be starting to form roots. Ventilation should then be gradually increased.

The rooted cuttings remain in the frame during the winter and should be given plenty of air when the weather is fine. They must be protected with a frame light, though, during periods of heavy rain, snow or severe frost.

Planting Young Specimens. The young plants may be planted out in a nursery bed during the following spring, in rows about 18in. apart with 12in. between the plants. Many of them will be of a suitable size by the following autumn for planting out in their permanent positions.

Alternative Methods of Rooting. Half-ripe cuttings may also be rooted under a cloche, each end of which is sealed with a pane of glass. Some gardeners also root them in pots—placing the cuttings round the edge of the pot—and enclose them in a polythene bag. A heated propagating case in the greenhouse will encourage such cuttings to root rapidly, especially if the case has bottom heat.

Mist Propagation. I consider mist propagation to be the quickest way of increasing shrubs at the present time. Half-ripe cuttings will root in as little as three weeks. Many of the large leaved evergreen shrubs which are normally difficult to increase will root comparatively easily under mist. The cuttings are usually inserted direct into a bed of sand in the mist unit, which is heated by electric soil-warming cables. The leaves of the cuttings are kept constantly moist by spray units which are automatically controlled. This prevents the cuttings from flagging. When the cuttings have rooted they are potted individually and put back under the mist to be gradually weaned off for about a week or so.

Cuttings which have been rooted under mist, or in a heated propagator, must be hardened off, first in a cool greenhouse and then in a cold frame, before planting them in the open ground.

HARD-WOOD CUTTINGS

These are taken in the autumn when shrubs or trees are devoid of leaves. I usually take mine during November. Again, the current year's wood is used for cuttings, but at this time of year it is really well-ripened and hard.

Many shrubs which are difficult to root from cuttings in the usual way are comparatively easy to increase under mist propagation. Such a unit is shown above

There are many shrubs which can be increased from hard-wood cuttings, including *Cornus alba, C. stolonifera*, forsythia, garrya, laburnum, philadelphus, privet, ribes, sambucus, tamarix and willows (*Salix*).

Taking the Cuttings. I remove the shoots with a heel of older wood attached, trim this smooth and cut off the tip of the shoot just above a bud to give me a cutting 9 to 10in. in length. Some gardeners prefer to cut the base of the shoot just below a leaf joint. These cuttings may be up to $\frac{1}{4}$in. thick, but avoid using any weak or spindly shoots.

Inserting the Cuttings. Hard-wood cuttings are rooted in the open ground, so try to select a sheltered situation for them. The easiest way to insert them is to make a trench (with one vertical side) sufficiently deep to allow the cuttings to be inserted to about two-thirds of their length. Unless the soil is very light and well drained it is best to place a layer of coarse sand along the bottom of the trench.

Before I insert the cuttings I dip the base of each in a hormone rooting powder. They are then placed along the trench, against the vertical side and about 2in. apart, with the base of each cutting in close contact with the sand. Then the soil is returned to the trench and the cuttings are firmed thoroughly.

If frost loosens the cuttings they should be re-firmed as soon as possible, otherwise they may not form roots. Leave them in the bed until the following autumn when they can be lifted and planted out into a nursery bed, about 1ft. apart in rows 2ft. apart.

Seeds

I find it great fun gathering the berries, fruits and various seed pods from my trees and shrubs, mainly because many of these plants grow readily from seed. Although some seeds may take a long time to germinate others will take a comparatively short time and will make flowering specimens within two or three years.

The hybrid shrubs and trees will not come true from seed, but nevertheless there is always the chance that an exciting plant may be produced. If you want hybrids to come true (that is to say, have exactly the same characteristics as the parent) then you should propagate them by one of the vegetative methods.

The shrubs which I grow from seed include berberis. Very few berberis come perfectly true but some of the variations in flower, fruit and foliage colour are most attractive. Cotoneasters are also easily raised by this method and while there may be slight variations from the parent plant in some cases I have noticed that C. horizontalis and C. simonsii come true. Brooms, Euonymus europaeus and other deciduous species of this genus, genistas, hollies and pyracanthas can all be grown from seed.

Now we come to trees, and of these I like to try my hand at acers, especially the Japanese maples (varieties of Acer palmatum). Most acers produce seed freely and while some will come fairly true to type the varieties of A. palmatum will not. However, many of the resulting seedlings are very beautiful. Amelanchier, beech, crataegus, crab apple, mountain ash, sycamore, oak, ash, elm, many conifers and silver birch can all be raised easily from seed.

Gathering the Seeds. Most seeds will be ripe and ready for harvesting during late summer and autumn. They can be stored in a dry, frost-proof place during the winter and then sown in the spring. There are exceptions to this rule, however, as all fleshy fruits and berries which contain hard-coated seeds have to be 'stratified' before sowing. This does not apply to the 'dry' seeds.

Stratifying Seeds. Fleshy berries and fruits such as those of berberis, cotoneaster, hollies, crab apples, pyracantha, hawthorn, mountain ash and Euonymus europaeus must be stratified during the winter preceding sowing, in order to soften the hard coats of the seeds. This speeds up germination in the spring. I must also include here the fruits of ornamental plums, peaches, almonds and cherries, the true species of which I like to raise from seed.

My method of stratification makes use of an old cocoa or coffee tin. First I make small holes in the lid of the tin so that air can get to the seeds. Then I place alternate layers of equal parts moist peat and sand and berries in the tin. Finally, the tin is buried in the ground for the winter. Of course, I only put one species or variety of berry in any one tin and I am careful to label each tin with the appropriate name.

Many gardeners like to use ordinary clay flower pots for stratifying seeds. Again, the berries are placed between layers of moist peat and sand and the pots are placed under a north-facing wall for the winter. It is essential, though, to cover them securely with wire netting to prevent vermin from disturbing the seeds.

Sowing. The best time to sow seeds is in early spring, about March. Those stratified should be sifted from the peat and sand mixture and separated from the remains of their fleshy covering. Then, together with the seeds which were stored 'dry', they can be sown either in seed boxes or 6-in. pots filled with John Innes seed compost. The containers must be well drained by placing a good layer of crocks in the bottom.

After sowing the seeds cover them with fine compost to a depth not exceeding the diameter of an individual seed. The containers can then be placed in a cold frame or unheated greenhouse. Keep the pots well watered as drying out will inhibit germination. To prevent rapid drying out the pots can be plunged to their rims in ashes.

If preferred, some seeds may be sown in drills in a prepared bed in the open garden. Hollies, hawthorn, cotoneasters, mountain ash and so on can be treated in this way.

Care of Seedlings. When the seedlings are large enough to handle easily they can be pricked out into seed boxes, using John Innes No. 1 Potting Compost. If they have been raised in a greenhouse they must be hardened off thoroughly before planting them out in a nursery bed the following autumn.

Layering

Layering is a good method of propagating those trees and shrubs that are difficult to raise from cuttings. There are many kinds which lend themselves to this method of increase, including azaleas, rhododendrons, all the viburnums, magnolias, kalmias, camellias, daphnes, lilacs, cotoneasters, clematis and wisteria.

The Method of Layering. Layering merely involves the rooting of shoots or branches while they are still attached to the parent plant. The best time, I find, is in the spring but it can also be done throughout the summer. Choose a branch or shoot that is near to the soil, or one that can be brought down comfortably to soil level; and always use a shoot of the previous year's growth, not an old shoot.

Before I peg down the layer I prepare the soil thoroughly by forking it over and incorporating liberal quantities of peat and sand. To prepare the shoot make a 2-in.-long slit with a sharp knife, half way through it, lengthways, and preferably through a joint. This cut should be made 9 to 12in. from the tip of the shoot.

Then make a depression in the prepared soil about 3 to 4in. deep. The part of the stem which has been cut is pegged down into this hollow, ensuring that the cut remains open. Use either a wire or wooden peg. I always tie the shoot to a cane so that it is held in an upright position. Then the part of the shoot which is pegged down is covered with soil which must be made really firm. After layering, just keep the soil moist until the shoot has rooted. Placing a large, flat stone over the layered part of the branch will help to keep it moist and assist in good root formation.

The Rooted Layers. Some layers will root during the same summer, in which case they can be severed from the parent plant in the autumn and planted in a nursery bed. Others, such as rhododendrons, azaleas and kalmias, usually take two seasons to form their roots. No layers should be severed until a good root system has been formed.

Air Layering

Air layering is also an easy way of increasing those trees or shrubs that are difficult to raise from cuttings; for example, azaleas, rhododendrons, acers, and magnolias. It is basically the same as ordinary layering except that the shoot is not rooted in the soil but in its normal position on the plant, the prepared area being enclosed in wet sphagnum moss which is held in place with a polythene sleeve. I do my air layering in May or early June.

The Method of Air Layering. Again, it is a matter of selecting a shoot of the previous year's growth and making an incision as I have just described for normal layering. The cut should be kept open with a small piece of wood and then dusted with a hormone rooting powder. There is a rooting powder on the market especially for air layering. The cut is wrapped in a large handful of wet sphagnum moss—some is pushed inside the cut—and then covered with a polythene sleeve. This must fit fairly tightly and each end is sealed to retain moisture.

You will be able to tell quite easily when the air layer has rooted as the white roots will begin to emerge through the moss. At this stage the layer may be severed from the parent plant and potted. Water it in well and place it in a propagating case or a polythene tent until it becomes established in the pot.

Grafting

Grafting is a more specialised form of propagation and is a quick method of raising plants. It is particularly suitable for hybrid kinds which you want to be true to type. Grafting involves uniting living parts of plants so that they form a permanent union. One plant supplies the root system only and is called the 'stock'. A small part of the plant of the variety required is joined to the stock and is known as the 'scion'. This eventually produces the shoots and branches.

Some fleshy berries and fruits need stratifying before sowing. The berries are placed between layers of moist peat and sand and the pots placed under a north wall for the winter

Layering a rhododendron. The shoots are tied to canes so that they are held in an upright position and the layers are covered with soil which is then made really firm

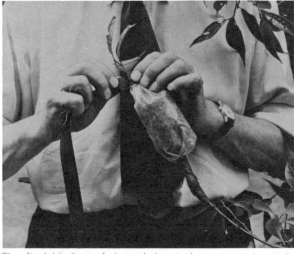

The final binding of the polythene sleeve around an air layer of *Prunus serrula*

It is essential that the stock and scion should be compatible—i.e., that they will unite or grow together. In most cases they are of the same genus. Sometimes plants of different genera can be grafted—for example, amelanchier on to *Sorbus aucuparia*.

Selection of Stocks. Varieties of holly can be grafted on to the common holly; Crab Apples on to the common Crab or apple stocks; ornamental peaches, plums and almonds on to Common Mussel plum stock; ornamental cherries on to *Prunus avium*, the Gean or wild cherry; rhododendrons on to *R. ponticum*; amelanchier and varieties of Mountain Ash on to *Sorbus aucuparia*; crataegus varieties on to the common Quickthorn and laburnums on to *L. vulgare*. You will notice that most plant varieties are grafted on to their common counterpart.

Preparation of Stocks. The stocks are bought from a nurseryman and planted in rich, well-

Tying a newly inserted bud tightly into place with raffia. The T-shaped incision in the bark of the rootstock should be completely covered, leaving the bud free to grow

prepared soil during October or November. They are then left to grow until the following March when grafting can take place. During this time it pays to rub off all buds on the stems 9 to 12in. from the ground.

Selecting the Scions. I use one-year-old shoots for grafting as these are not too thick. They can be cut from the parent plant in February, tied in bundles and heeled in under a north wall until they are required in March. Just before use they must be washed free of soil.

Whip and Tongue Grafting. This is the most popular method and is used for many trees and shrubs. For this, and any other type of grafting, always use a really sharp knife.

First, the stock is cut down to within 3 or 4in. of the ground, then a slanting upward cut, 1½in. long, is made at the top of the stock. This should remove about half the thickness of the stock.

Then a small downward cut is made in the cut surface near the top, thereby forming a 'tongue'. To prepare the scion the selected shoots are cut so that each piece contains three or four buds. Do not use the soft tips of the shoots. The base of the scion is prepared by making a slanting downward cut similar to the one on the stock, but opposite a bud. Then a tongue is cut so that it corresponds with the one on the stock.

The tongue of the scion is fitted into the one on the stock. If the cut surfaces are perfectly smooth they will fit closely together. The graft is then bound very tightly with raffia which in turn is covered with grafting wax to render it airtight and watertight. If the widths of stock and scion differ, then it is essential that one side of each should meet perfectly. This is to allow the cambium layers—seen immediately under the bark —of stock and scion to unite. Unless the cambium layers meet somewhere a union will not occur.

Once the scion is growing vigorously the raffia may be cut away, and shoots or buds which appear on the stock should be rubbed out regularly.

Budding

Budding is essentially the same as grafting, the only difference being that the scion is in actual fact a growth bud which is inserted behind the bark of the stock. It is really only another form of grafting. The shrubs and trees which can be grafted can be budded as well, but this is carried out between June and mid-August. Roses are also increased by this method. Again, the appropriate stocks must be used, in many instances the common counterpart. Roses are budded on to the common briar. Purchase the stocks from a good nurseryman and plant them during October or November in well-prepared ground. Space them 1ft. apart in rows at least 3ft. apart. They are then left to grow on until budding time the following year.

Selecting the Buds. The buds can be seen at the base of leaf stalks and when selecting them ensure that they are good plump ones and are contained on a young shoot. Just before budding commences remove a complete shoot from the plant and cut off the soft tip, as the buds in this area are not suitable. This is then called a 'bud stick'. Remove the leaves from the bud stick, leaving about ½ to 1in. of each leaf stalk, and place the sticks in a bucket of water to prevent them from drying out.

The Method of Budding. Budding is done only when the bark of the stock can be lifted easily with a knife. During dry weather this may not be possible, therefore the stocks may have to be watered two or three days before budding takes place. Incidentally, I would recommend using a proper budding knife with a flat end to the handle.

The stocks are not cut down as is the case with grafting. Rose stocks should have some soil scraped away from their base as the bud is inserted just below ground level. This soil is not replaced after budding. Other trees and shrubs are budded 3 to 4in. above the ground.

A T-shaped cut is made in the bark of the stock; the cross cut first and then a 1½-in. vertical cut to join it. The end of a budding knife is used to prise open the bark on each side of the T-shaped cut. A bud is taken from the bud stick by making a slanting cut, starting ½in. below it, drawing the knife under the bud and finishing ½in. above it. The result is a shield-shaped piece of bark with a thin sliver of wood behind it. This sliver must be removed carefully so as not to pull out the small green protuberance behind the bud, otherwise the bud will not 'take'. The base of the shield is then placed in the top of the T-shaped cut and pushed down behind the bark. Ensure that the bud is inserted the right way up. If part of the shield remains above the T-shaped cut carefully snip it off, and bind the whole length of the cut very tightly with raffia, leaving the actual bud showing. The bud is then left to grow.

Aftercare. In the case of roses the stocks are cut back to within ½in. of the bud in the following February or March. With other trees and shrubs they are cut to within 8in. of the ground. The shoots from the buds can then be tied to these snags. In September cut out the snags just above the budded area. Remove any shoots from the stocks regularly.

Training. Trees that are grown as standards, such as cherries, crabs, sorbus, laburnum and so on, must be trained after they have been budded or grafted to produce a straight, single stem. This is a matter of keeping the terminal shoot growing, at the same time removing most of the side shoots. The stem should be tied to a stake to keep it perfectly straight. When it has reached the desired height the tip can be pinched out, and it will then branch out at the top. Standard trees usually have a 6-ft. stem.

Incidentally, weeping trees and standard roses can be budded at the top of a stock of a suitable height. Sometimes two or three buds are inserted to give a really good head to the specimen (see p. 86 for further details).

Division

This is one of the easiest forms of propagation and simply involves lifting a shrub and splitting it into several parts, each with some fibrous roots attached. These portions are then replanted. The best time for this, I find, is in early spring before growth commences. Not a great many shrubs lend themselves to this method of increase, of course, but the usual ones which are divided include *Hypericum calycinum*, Butcher's Broom (*Ruscus aculeatus*), some of the spiraeas and the Snowberry (*Symphoricarpos albus*). Some gardeners increase hardy fuchsias by division.

Recommended Trees

Acer (Maple). This large family contains both small ornamental trees and huge stately specimens, most of which are deciduous. For small or medium-sized gardens I would recommend the Japanese Maples with their attractive foliage. *Acer japonicum* has good autumn colour; *A. j. aconitifolium* is noted for its crimson leaves in autumn; while *A. j. aureum* bears soft yellow foliage throughout the spring and summer. *A. palmatum* is another good Japanese Maple with spreading branches. Some excellent varieties of *A. palmatum* are *atropurpureum*, with bronze-crimson foliage throughout spring and summer; *dissectum*, with finely cut green leaves; *dissectum*

Amelanchier canadensis, a deciduous tree of modest size which is well worth a place in gardens. The white flowers appear in spring

atropurpureum, with bronze-crimson foliage; and *heptalobum* Osakazuki, with brilliant scarlet autumn foliage.

Other Maples I can recommend are *A. ginnala*, growing to about 20ft. and with crimson autumn leaves, and *A. griseum*, the Paperbark Maple, with beautiful autumn foliage. The latter also has an attractive trunk and branches—the surface flakes, curls back and exposes the orange bark beneath. It makes a good lawn specimen.

Almond, see Prunus

Amelanchier (Snowy Mespilus). The deciduous *Amelanchier canadensis* is a popular small tree, attaining about 20ft. in height. In the spring it produces masses of small white flowers on a rounded head. It is not fussy about soil.

Arbutus (Strawberry Tree). The arbutuses are among my favourite evergreen trees. They do not usually exceed 20ft. in height—10 to 15ft. is more usual—and have beautiful deep green glossy leaves. Their most decorative feature, however, is their strawberry-like fruits. They will thrive in most soils. *Arbutus unedo* is the best-known species and produces its small white flowers and fruits in the autumn.

Beech, see Fagus

Betula (Birch). These graceful deciduous trees will grow well in almost any soil. *Betula ermanii* is tall, growing to 60ft. or so with a white trunk and orange-brown branches. The Paper Birch, *B. papyrifera*, also makes a tall tree with a beautiful white trunk and good yellow foliage in the autumn. The best-known species is *B. pendula*, the common Silver Birch, again with attractive white bark and a semi-pendulous habit. This grows 50 to 60ft. tall. An erect form of this is *B. p. fastigiata*, similar in habit to a Lombardy Poplar. A smaller variety with a weeping habit is *B. p. youngii* (25 to 30ft.) which makes an excellent lawn specimen.

Catalpa (Indian Bean Tree). The white flowers, with yellow and purple spots, appear on *Catalpa bignonioides* in late summer. This excellent specimen tree attains 25 to 30ft. but has a very spreading habit. It is deciduous, having very large green leaves. In the variety *C. b. aurea* the foliage is yellow.

Cercis (Judas Tree). The clusters of rose-coloured, pea-shaped flowers of *Cercis siliquastrum* are borne all along the twigs and branches during April or May. This is a deciduous tree growing somewhere between 15 to 40ft. in height and makes an interesting lawn specimen.

Cherry, see Prunus

Crab Apple, see Malus

Crataegus (Hawthorn or May). These adaptable small trees are equally suitable for industrial areas or coastal districts and will thrive in almost any soil. They are deciduous and flower in May and June, most of them attaining over 20ft. in height. Hawthorns are usually grown as standards but some species make good hedges (see p. 71). Some of my favourites are *C. lavallei* with white flowers and orange-red berries which remain in winter; *C. monogyna pendula rosea* with weeping branches and pink flowers followed by red berries; *C. m. stricta*, of erect habit and with white flowers and red berries; *C. oxyacantha coccinea plena* with double crimson flowers and red berries; and *C. o. plena*, double white flowers and red berries. *C. prunifolia* is noted for its bright red fruits and crimson autumn foliage.

Davidia (Handkerchief or Dove Tree). *Davidia involucrata* is an unusual deciduous tree in that it is draped with pairs of large white bracts during May or June. It grows to 20 to 30ft. and is fairly adaptable regarding soils and situation. The popular name Handkerchief Tree is very apt.

Dove Tree, see Davidia

The easily pleased laburnums are among the showiest of spring-flowering trees. *Laburnum vossii* has especially long racemes of yellow flowers

Eucalyptus (Gum Tree). One of the hardiest of the eucalypts is *Eucalyptus gunnii* which reaches 60 to 70ft. in height. It is very fast growing and should be given a lime-free soil. Its beautiful evergreen foliage is blue-grey in young specimens and sage-green in more mature trees.

Fagus (Beech). The beeches are fine large deciduous trees but, of course, only spacious gardens can accommodate them satisfactorily. Beech are trees of chalk soils but they grow well in other soils as well. The common beech is also widely used as a hedging plant (see p. 71). As a specimen, *F. sylvatica*, the common beech, will attain 80ft. or more. Some good forms of this are *F. s. fastigiata*, the Dawyck Beech, of columnar habit; *F. s. cuprea*, the Copper Beech, a well known and beautiful tree; *F. s. pendula*, the Weeping Beech; *F. s. purpurea*, the Purple-leaved Beech; and *F. s. purpureo-pendula*, the Weeping Purple Beech.

Gum Tree, see Eucalyptus

Handkerchief Tree, see Davidia

Hawthorn, see Crataegus

Indian Bean Tree, see Catalpa

Judas Tree, see Cercis

Laburnum. The deciduous laburnums produce their pendulous racemes of yellow flowers during May and June. Any garden soil seems to suit them. They grow to about 20ft., or perhaps more, in height. The common laburnum, *L. anagyroides*, flowers profusely, together with the excellent

Of excellent value in smaller gardens are trees of narrow, erect habit, like the elegant *Prunus* Amanogawa with pink, semi-double flowers.

hybrid *L. vossii*, which has very long racemes of flowers.

Liquidambar (North American Sweet Gum). *Liquidambar styraciflua* is one of the very finest trees for autumn colour, its maple-like leaves turning to vivid crimson. But it is only in medium-sized and large gardens that room could be found for this 50-ft.-tall tree. It must, also, have a lime-free soil.

Magnolia. Some of my favourite magnolias, which form medium-sized trees, include *M. denudata*, about 30ft., with white cup-shaped blooms in March, April and May, and *M. kobus* with white flowers in April and May which are not produced until the tree is fairly well developed. All are deciduous and, with the exception of *M. kobus*, must have a lime-free soil.

Malus (Crab Apple). These small to medium-sized deciduous trees are almost equal to the ornamental cherries in floral beauty. Their blooms appear in spring and are followed by attractive fruits in many of the varieties.

The Japanese Crab, *Malus floribunda*, has arching branches covered in pink flowers which are red when in bud. Golden Hornet is white-flowered but the beauty lies in its crop of bright yellow fruits which last until fairly late in the year. John Downie is another good fruiting variety with beautiful yellow and red fruits borne abundantly.

One of the Siberian Crabs is *M. prunifolia*, which has red fruits. Another good Siberian Crab is *M. robusta*, together with its varieties Red Siberian and Yellow Siberian with red and yellow fruits respectively. *M. purpurea* is noted for its mass of crimson flowers and purplish-green leaves. The fruits are reddish-purple in colour. Finally, there is Wisley Crab, with large red flowers and fruits and bronzy-coloured foliage, which is another of my favourites.

Maple, see Acer
May, see Crataegus
Mountain Ash, see Sorbus
North American Sweet Gum, see Liquidambar
Paper Birch, see Betula
Parrotia. Noted for its brilliant autumn leaves of red and gold, *Parrotia persica* usually grows to 15 to 20ft. in height. Its branches are wide spreading, however, so it must be given plenty of room. It thrives in most garden soils.

Paulownia. The deciduous *Paulownia tomentosa* is a handsome flowering tree attaining 30 to 50ft. in height. The violet foxglove-like flowers are borne on erect panicles in June. The leaves are large. It is not fussy as to soil but flowers better in a sunny position.

Peach, see Prunus
Pear, see Pyrus
Plum, see Prunus
Prunus. This huge genus includes some of the most beautiful of all flowering trees, namely the ornamental cherries, peaches, plums, almonds and, my favourites, the Japanese cherries. They are smallish trees suitable for medium-sized gardens and they are a delight in spring in the full flush of their flowering. The Japanese cherries are undoubtedly the most beautiful and these are available in colours from deep pink to white and with single or fully double flowers. Ornamental cherries and almonds I like to see planted as lawn specimens. Some of the flowering cherries also have bright autumn leaf colouring.

The common almond, *P. amygdalus*, is excellent for town gardens with its pink flowers, together with *P. a. roseo-plena* with double pink blooms. *P. avium*, the well-known Gean, has white blooms while the variety *P. a. plena* has double white flowers. The Cherry Plum or Myrobalan, *P. cerasifera*, has white flowers and its very popular variety, *P. c. atropurpurea* (syn. *P. pissardii*), the Purple-leaved Plum, is noted for its beautiful reddish-purple foliage. *P. c. nigra* has much deeper purple leaves and pink flowers. The white-flowered *P. cerasus* is a very popular cherry.

The Fuji Cherry, *P. incisa*, forms a delightful small tree with pink flower buds which open white. The Japanese Apricot, *P. mume*, bears pale pink flowers early in the year, while the Bird

The lovely weeping willow *Salix alba tristis* is an unusually attractive tree for screening a corner of a garden. It is most effective of all when overhanging water

The striking foliage of the Blue Cedar, *Cedrus atlantica glauca*, makes this large spreading tree an attractive acquisition—where space is available

Aesculus parviflora, a Horse Chestnut which forms a shrub of 8 to 12ft. tall bears white flowers in July and August

Cherry, *P. padus*, has beautiful sprays of white blooms which are slightly scented. The peach, *P. persica*, is a delightful garden tree and its pink blooms are freely produced. There are some good varieties of this, my favourites being Helen Borchers, a double-flowered bright pink, Clara Meyer and Windle Weeping, also double pink.

For beautiful autumn colour I do not think you can beat *P. sargentii*, as the leaves turn to brilliant red shades at that time. Its single pink flowers are borne abundantly in early spring.

The winter-flowering cherry, *P. subhirtella autumnalis*, is an excellent small tree to bring cheer at the least colourful time of the year, for it bears its semi-double white flowers on and off from the onset of winter until the spring. These flowers are a lovely sight on the leafless branches, especially if they are highlighted by the dark background which a conifer (or conifers) could provide. I also like to see winter-flowering heathers as a carpet around this tree. A pink-flowered form is named *P. s. autumnalis rosea*.

Weeping trees are also useful for adding interest to a garden and the handsome *P. subhirtella pendula*, with pale pink flowers, can be a lovely sight in spring.

Now we come to the vast range of free-flowering and very beautiful Japanese cherries. A popular variety for a very small garden is the pink, semi-double Amanogawa with its narrowly columnar habit of growth. Others which are widely grown include: Fukubana, semi-double, rose; Kanzan, double, rose-pink; Kiku Shidare Zakura, known as Cheal's Weeping Cherry, deep pink, double, drooping branches: Shirofugen, double, white; Shirotae, semi-double, white; Tai-Haku, huge white blooms; and Ukon, semi-double, pale yellow.

Pyrus (Pear). There are a number of good ornamental pears and one of the most attractive is *Pyrus salicifolia pendula*. This is a medium-sized weeping tree with silver-grey, willow-like leaves and I like to see this planted as a lawn specimen. It is deciduous and easily grown.

Rowan, see Sorbus

Salix (Willow). Especially suitable for moist soils, the willows are most attractive deciduous trees and species can be selected for all kinds of planting schemes. Some are weeping while others are noted for their coloured shoots. *Salix alba*, the Huntingdon or White Willow, has some of the most beautiful varieties; the type has grey-green foliage and forms a fairly large tree. *S. a. tristis* is a huge weeping tree, possibly the most popular weeping willow for waterside planting. *S. vitellina britzensis* is another large willow with young shoots of orange-red colouring. For a smaller garden *S. matsudana tortuosa* is suitable and it is unusual in that it has curious contorted or twisted branches and shoots.

Silver Birch, see Betula
Snowy Mespilus, see Amelanchier
Sorbus (Mountain Ash or Rowan, and White-beam). The Mountain Ash is a good deciduous medium-sized tree noted for its display of colour-ful berries. *Sorbus aucuparia* is the species most often seen, this bearing white flowers in spring and orange-red fruits in the autumn. The Chinese Mountain Ash, *S. hupehensis*, bears large clusters of white berries in early autumn which later turn pale pink and last into the winter. The White-beam, *S. aria*, usually attains 30 to 40ft. in height. The leaves are very striking as they are bright green on the upper surface and white on the undersides. The fruits are scarlet. This is a good tree for town gardens.
Strawberry Tree, see Arbutus
Sweet Gum, North American, see Liquid-ambar
Whitebeam, see Sorbus

Selected Conifers

Cedrus (Cedar). The Blue Cedar, *Cedrus atlan-tica glauca*, is a magnificent 80- to 100-ft. tree with spreading branches and makes a good speci-men for a really large expanse of lawn. The glaucous foliage is very striking. The Cedar of Lebanon, *C. libani*, is slower growing than the Blue Cedar but eventually makes a splendid lawn specimen.
Chamaecyparis. There are some excellent varie-ties of Lawson's Cypress, *C. lawsoniana*, all of which are easily grown. The most popular is *C. l. allumii*, 30ft. or so high, which forms a narrow pyramid of bluish-green foliage. *C. l. fletcheri* also has glaucous foliage which is feathery and forms a pyramid about 15ft. in height. Kilmacurragh is bright green and columnar in habit, while *C. l. lutea* has golden-yellow foliage (its popular name is Golden Lawson's Cypress) and is narrowly cone shaped. Silver Queen is noted for its silvery-white tipped young foliage in spring and early summer. *C. obtusa crippsii* has golden-yellow foliage and forms a broad pyramid. These are all suitable for the medium-sized garden. *C. pisifera squarrosa* forms a large tree with glaucous foliage of a soft texture.
Cryptomeria (Japanese Cedar). *Cryptomeria japonica elegans* forms a tall, bushy, broad pyra-mid which turns a bronzy colour in the autumn. This is easily cultivated and suitable for a medium-sized garden as its maximum height is about 30ft.
Dawn Redwood, see Metasequoia
Ginkgo (Maidenhair Tree). *Ginkgo biloba* is a good conifer for town gardens but eventually reaches a height of 60 to 80ft. It is deciduous and the attractive two-lobed, fan-shaped leaves turn yellow in the autumn.
Japanese Cedar, see Cryptomeria
Juniperus (Juniper). My favourite juniper is *Juniperus communis*, a British native which thrives on chalky soils. It reaches a height of about 30ft. The Irish Juniper, *J. c. hibernica*, is a columnar variety with glaucous foliage and attains a height of 10 to 15ft. *J. media aurea* is a particularly good yellow-foliaged variety. The leaves turn bronzy-yellow and green during the winter.
Maidenhair Tree, see Ginkgo
Metasequoia (Dawn Redwood). *Metasequoia glyptostroboides*, a lovely fossil-age deciduous conifer thought to be extinct until a Chinese botanist rediscovered it in Szechwan Province in 1945, is very fast-growing and the indications are that it will make a very large tree (the largest plant seen in the wild was 110ft. tall). It is a columnar tree with distinctive feathery foliage which turns to lovely shades of reddish-brown in the autumn. It likes a moist but well-drained soil. A good specimen tree for a lawn.
Picea (Spruce). The best-known picea is *P. abies* which is used extensively as a Christmas tree. This is the Common or Norway Spruce and is easily grown in a reasonably deep soil. It makes a tall, broad, pyramidal tree. *P. pungens glauca*, the Blue Spruce, is of medium height with beautiful grey-blue foliage and I like to see this as a lawn specimen.
Pinus (Pine). Pines are fairly adaptable trees when it comes to soils but cannot be recom-mended for industrial towns as they dislike a polluted atmosphere. Our native Scots Pine, *P. sylvestris*, forms a very tall, flat-topped tree and is picturesque when planted in a woodland setting.
Spruce, see Picea

Recommended Shrubs

Abelia. Related to the honeysuckle, abelias flower freely during summer and autumn and can be recommended for a small garden. *Abelia flori-bunda* has funnel-shaped, rose-red flowers and grows to a height of about 4ft. This evergreen species is slightly tender and therefore more suitable for milder localities. A semi-evergreen, *A. grandiflora* carries pale pink flowers on its slightly pendulous branches and attains about 5ft. in height.
Aesculus (Horse Chestnut). Although most of the Horse Chestnuts are large trees, *Aesculus parviflora* is a bushy shrub 8 to 12ft. in height. The huge white flowers are borne freely in July and August. It is deciduous and will tolerate most soils.
Artemisia. I like artemisias for their delightful greyish or silvery aromatic foliage. *Artemisia abrotanum*, the Southernwood or Lad's Love, has soft, greyish-green fern-like leaves and reaches $3\frac{1}{2}$ to 4ft. in height. The beautiful silvery foliage of *A. arborescens* is also attractive.
Arundinaria (Bamboo). One of the most useful

bamboos for a small garden is *Arundinaria angustifolia*. This has slender canes about 5ft. in height and attractive bright green leaves. It is non-rampant and, like all bamboos, is evergreen.

Atriplex. For coastal districts especially *Atriplex halimus* is an excellent shrub, growing to about 6ft. in height with semi-evergreen, silvery-grey leaves.

Aucuba. The evergreen aucubas were much favoured by Victorian gardeners, but the variegated varieties are very popular today, are invaluable for sunless positions and will grow in any soil. They form dense, rounded bushes 6 to 10ft. in height. There are some variegated kinds of *Aucuba japonica* including *A. j. crotonoides*, with gold-mottled leaves; *A. j. sulphurea*, with yellow-edged leaves; and *A. j. variegata*, with cream-variegated leaves, which is the common form.

Azalea, see Rhododendron

Bamboos, see Arundinaria, Pleioblastus, Pseudosasa and Sinarundinaria

Barberry, see Berberis

Bay, see Laurus

Berberis (Barberry). These are some of my favourite shrubs, with their masses of colourful flowers early in the year and often brilliant berries in autumn. There are both evergreen and deciduous species and varieties, dwarf and tall kinds, and they thrive in a good range of soils.

Of the deciduous barberries my favourite is *B. thunbergii*, which makes a compact bush about 4ft. tall. The pale yellow flowers appear in spring but the real display comes in autumn when the scarlet berries appear and the foliage turns to brilliant shades of red. I am also fond of its variety, *B. t. atropurpurea*, which has purplish-red foliage throughout the spring and summer, turning to deep red in autumn. It is similar in other respects to the type, although possibly more vigorous. I must also mention *B. wilsonae* as an autumn attraction, for the small leaves at that time turn to brilliant shades and coral-red berries are borne freely. It makes a spreading bush some 3ft. tall.

Other deciduous berberis I am fond of are *B. aggregata*, a dense, 4 to 5ft. bush with clusters of red berries and rich autumn foliage; Buccaneer, with large red berries; Cherry Ripe, with brilliant crops of red berries, 6ft. tall; *B. thunbergii atropurpurea nana*, a dwarf with purple foliage throughout the season; *B. yunnanensis*, with colourful leaves in the autumn and glowing red berries; and some of the hybrids of Chinese origin, noted for brilliant autumn foliage and berries.

Some excellent evergreen kinds are *B. buxifolia*, early flowering, with purplish-blue berries, 5 to 6ft.; *B. b. nana*, which forms a compact, 18-in. mound; *B. darwinii*, with masses of orange-yellow flowers during May, 6ft.; *B. irwinii*, with

Camellia williamsii Donation, a variety of exceptional beauty with semi-double, clear pink flowers which appear in early spring

masses of yellow flowers, 3ft., compact; *B. i. corallina*, with yellow blooms, dwarf; *B. i. corallina compacta*, very dwarf; *B. stenophylla*, with masses of deep yellow blooms in April, 10ft.; *B. s. autumnalis*, a dwarf variety; and *B. wilsonae stapfiana*, with yellow blooms followed by coral-red berries, semi-evergreen.

Bladder Senna, see Colutea

Broom, see Cytisus, Genista and Spartium

Buddleia. The buddleias grow in a wide range of soils and flower profusely during the summer. *Buddleia alternifolia* has fragrant, lilac-coloured flowers borne on long, arching branches. It attains a height of about 12ft. and can be trained into a small tree, in which form it looks best of all. The Orange Ball Tree, *B. globosa*, is one I would not be without for, during June, this tall shrub (up to 15ft.) is covered in orange, ball-shaped flowers.

The Butterfly Bush, *B. davidii*, hardly needs an introduction as it is so widely grown. There are many good varieties (up to 12ft. or so tall), but I like Black Knight, deep violet; Charming, a good pink; Dubonnet, deep purple; Royal Red, red-purple; and White Cloud.

Callicarpa. I like *Callicarpa japonica* for its masses of conspicuous, bead-like, violet-coloured fruits. To obtain these several plants should be grown together. The leaves turn to a beautiful rose shade in the autumn, which is an added attraction of this neat, 5-ft.-tall shrub.

Camellia. These beautiful evergreen shrubs are well-worth growing if you have a moist, lime-free soil; they make their flowering display from March onwards. The varieties of *Camellia japonica* are hardy, but they must be grown in a sheltered position to protect the flower buds from frosts. They are very suitable for a north-facing wall. The flower colours range from red through shades of pink to white and these show up well against the glossy, dark green foliage.

The Winter Sweet, *Chimonanthus praecox*, is one of the joys of the winter garden. The pale yellow flowers of wax-like texture are delightfully fragrant

Some of my favourite *C. japonica* varieties are Adolphe Audusson, deep red, semi-double; *C. j. alba grandiflora*, semi-double, white; Arejishii, double red; Augusto Pinto, deep pink, double; Lady Clare, soft pink, semi-double; and *C. j. nobilissima*, a pure white double.

Camellia reticulata may be too tender for cold or exposed districts but it is a beautiful plant. Some good varieties include Captain Rawes, semi-double, rose-crimson; Lion Head, semi-double, red, variegated with white; Shot Silk, semi-double, pink; and Trewithen Pink, single, soft pink.

Camellia williamsii has several good varieties and my favourite is Donation with semi-double, clear pink blooms, fully 3 to 4in. across. Also, I am successful with the *williamsii* variety Mary Christian, with bright pink single blooms.

Calluna, see p. 73

Caryopteris. These small, verbena-like shrubs like a sunny situation where the bright blue flowers, set against grey, aromatic leaves, will be freely produced. My favourite is Heavenly Blue; *Caryopteris clandonensis* can also be recommended. Its bright blue flowers appear in August and September and are set off handsomely by the small grey-green leaves.

Ceanothus. There are both evergreen and deciduous species of these blue-flowered medium-sized shrubs. My favourite evergreen ones are A. T. Johnson with deep blue flowers in spring and autumn, and *C. burkwoodii*, a rounded 6- to 8-ft. shrub flowering in summer and autumn. Of the deciduous varieties, all 6 to 8ft. tall, I grow Gloire de Versailles; Indigo, indigo-blue; and Topaz, a lighter blue, and I would recommend the deciduous kinds rather than the evergreens, beautiful as these are, because they are hardier and easier to grow.

Ceratostigma (Plumbago). Growing to 3ft. in height, *Ceratostigma willmottianum* bears blue flowers during the summer and autumn and is a good front-of-the-border shrub.

Chaenomeles (Ornamental Quince). These spring-flowering shrubs are easily grown in almost any soil. The spreading, 6-ft.-tall *Chaenomeles speciosa* (syn. *Cydonia japonica*) has some attractive varieties including *C. s. cardinalis*, salmon-pink; Knap Hill Scarlet, orange-scarlet; *C. s. nivalis*, white; and *simonii*, red, low-growing, crimson. *C. japonica* (syn. *Cydonia maulei*) has some good varieties such as the apricot-coloured Boule de Feu and the bright pink Hever Castle.

Cherry, see Prunus

Chimonanthus (Winter Sweet). The winter-flowering *Chimonanthus praecox*, which I so value in my garden, bears pale yellow blooms on leafless stems from mid-winter until March. The common name comes from its delightful fragrance. It is 8 to 10ft. tall when fully grown.

Choisya (Mexican Orange Blossom). *Choisya ternata* is a 4 to 6ft.-tall, bushy but shapely evergreen shrub. The clusters of white, highly fragrant orange-blossom-like blooms are freely produced during late spring and early summer. It needs shelter from cold winds.

Cinquefoil, Shrubby, see Potentilla

Cistus (Rock Rose). The Rock Roses are bushy plants, mostly between 2 to 5ft. in height, which like a really sunny position and a dry soil. They are evergreen and some of them are too tender for exposed positions. The hardiest kinds include Silver Pink; *Cistus crispus*, rose-mauve; *C. laurifolius*, white, to 8ft.; and *C. populifolius lasiocalyx*, white, yellow-stained flowers, to 6ft. Other good ones, but not so hardy, are *C. purpureus*, reddish-purple; and *C. ladaniferus*, white, with maroon blotches. The single rose-like flowers are produced freely in June and July.

Colutea (Bladder Senna). The deciduous *Colutea arborescens* is a quick-growing shrub reaching about 10ft. in height. It has attractive, small, yellow, pea-shaped flowers during the summer, but it is the fat, inflated seed pods which fascinate me. It is extremely tolerant with regard to soil conditions.

Cornel, see Cornus

Cornus (Dogwood or Cornel). The Dogwoods are attractive garden plants, providing displays at different times of the year. Those species with coloured bark are worthy of a place in the garden for the delightful effect they create in winter. They form thickets of whippy shoots which look well growing by water. The young shoots of *Cornus alba*, the red-barked Dogwood, are deep red; those of *C. a. sibirica*, the Westonbirt Dogwood, are brilliant crimson; and *C. stolonifera flaviramea*, the yellow-barked Dogwood, has attractive yellow stems. There are coloured-leaved varieties of *C. alba*; *C. a. spaethii* has golden-variegated leaves and red winter stems, and a good white variegated form is *C. a. variegata*.

A much taller deciduous shrub is *C. kousa chinensis* which bears its handsome white bracts in June. This variety also colours well in autumn, the leaves turning to shades of crimson. *C. florida rubra* has rose-pink bracts and also has good autumn colour.

The Cornelian Cherry, *C. mas*, which makes a tree of shrub-like habit up to 20ft. tall, carries small yellow flowers on the bare branches in February and can be recommended for any garden.

Corylopsis. Flowering in early spring, *Corylopsis pauciflora* is a dense, 4-ft.-tall deciduous shrub. The freely produced cup-shaped blooms are pale yellow and pleasantly scented.

Corylus (Hazel). This group of plants contains some strong-growing species which are good for screening. All are deciduous. *Corylus avellana aurea*, which is a variety of the wild Cobnut or Hazel of our hedgerows, has attractive yellow leaves, which contrast beautifully with those of *C. maxima purpurea*, the Purple-leaved Filbert.

Cotinus (Smoke Tree). *C. coggygria* is, in fact, a tall (14 to 15ft.) shrub with two periods of beauty: in summer it bears a profusion of fawn, plume-like flowers which turn pink and finally smoky grey by the end of the season; then, in autumn, the small, round leaves turn to beautiful shades of red and orange. Several varieties of rather lesser height are also excellent. These include Flame, with orange-red autumn leaves: *foliis purpureis*, with young purple-coloured foliage which turns to light red in autumn; and Royal Purple with deep purple foliage.

Cotoneaster. My favourite cotoneasters are *Cotoneaster horizontalis*, known as the Herring-bone Cotoneaster because of its flat, spreading branches which form a herringbone pattern; *C. simonsii*, *C. salicifolius* and *C. cornubia*.

Cotoneaster horizontalis bears white flowers in early summer but its most colourful display is in autumn when the deep crimson berries are produced. At that time, too, the leaves turn to brilliant shades of orange and red. The erect-growing, semi-evergreen *C. simonsii* is ablaze with scarlet berries in autumn and this makes a large shrub of 10ft. or more in height and almost as much through. The evergreen *C. salicifolius* is a tall shrub of elegant, weeping habit which carries masses of bright red berries in autumn. *C. cornubia* is even taller growing, often reaching a height of 20ft. or so. The showy red berries are borne in great profusion.

Other cotoneasters which I would recommend are *C. conspicuus*, 4ft. tall with arching branches covered in red berries in autumn and into the winter; *C. hybridus pendulus*, 8ft., with glossy green leaves and long prostrate branches which carry red fruits (this makes a small weeping tree if grown on a stem); and *C. franchetii*, 10ft. tall with orange berries. These three are all evergreens. Another useful deciduous species is *C. frigidus* which makes a very large bush and bears red berries in autumn and winter.

Currant, Flowering, see Ribes

Cytisus (Broom). This colourful family is not fussy as regards soil—in fact, the plants like a fairly dry soil in a sunny position. They flower during spring and early summer. *Cytisus albus* is the white Portugal Broom and attains 8ft. in height. A tall shrub and suitable for a wall is *C. battandieri*, with silky grey laburnum-like leaves and conical clusters of yellow, pineapple-scented flowers, borne in July. *C. beanii*, golden-yellow, is semi-prostrate while *C. kewensis* is completely prostrate and bears masses of creamy flowers. The last two are ideal for a rock garden.

Other favourites of mine, which are fairly tall, are Cornish Cream, an ivory or pale yellow hybrid; Dorothy Walpole, crimson; Lady Moore, rich red and buff; *C. praecox*, cream; *C. purpureus*, purple; and *C. scoparius*, the common yellow broom, and its varieties Firefly, bronze-crimson, Golden Sunlight, Andreanus, yellow and brownish-red, and *C. s. sulphureus*, deep cream.

Daisy Bush, see Olearia

Daphne. These shrubs are usually fragrant and need a good soil which is moist but at the same time freely drained. The Mezereon, *Daphne mezereum*, is perhaps my favourite early-flowering shrub. What a joy the very sweetly scented purple flowers are in late winter and early spring when they clothe the bare branches of this 4-ft. shrub. There is a white form of this species named *alba* which also has its attractions.

Other good kinds are *D. burkwoodii*, a 3-ft.-tall semi-evergreen with pale pink, scented flowers in spring; *D. odora*, a very fragrant winter- and early spring-flowering evergreen, about 3 to 4ft.; and *D. o. aurea-marginata* with creamy or yellowish margins to the leaves. Much the same height as the type, *aureo-marginata* is more spreading.

Deutzia. The deutzias are easily grown shrubs, and those I name are about 4 to 6ft. in height and flower in June. The variety Avalanche bears an abundance of white flowers; Mont Rose bears clusters of rose-pink blooms; *D. rosea* is a compact, pink-flowered hybrid: and *D. r. grandiflora* has white, flushed pink blooms.

Dogwood, see Cornus

Elaeagnus. There are both evergreen and deciduous species of these attractive, medium to large foliage shrubs. *Elaeagnus angustifolia* is deciduous and has attractive silvery leaves, like *E. commutata*, which is probably more intensely silver. The varieties of *E. pungens*, which are evergreen, are some of the most attractive variegated shrubs.

Hibiscus syriacus Woodbridge. a variety of Tree Hollyhock with flowers of especially fine substance

Left. The delightfully scented
flowers of the lilacs are evocative
of all the joys of the garden in May
and early June. Shown here is
Syringa vulgaris Clarke's Giant, a
single flowered variety of
outstanding charm and quality
Right. The glorious *Rhododendron*
Pink Pearl, one of the most
celebrated of the hardy hybrids.
This easily pleased variety flowers
in May. All rhododendrons need,
of course, a lime-free soil
Left, below. The distinctive
Cystisus battandieri which makes
an admirable wall shrub. The
flowers, borne in July, are
pineapple scented

Clematis Nelly Moser – a wall shrub with excellent decorative qualities

E. p. maculata has leaves with a yellow blotch, while the leaves of *E. p. variegata* have cream-yellow margins.

Elder, see Sambucus

Erica, see p. 73

Escallonia. For gardeners living on the coast, the species and varieties of escallonia are invaluable. All are evergreen shrubs which flower during summer and early autumn but some are not hardy inland. Most kinds have attractive glossy foliage.

The following kinds are considered to be hardy, are easily grown in most soils and make medium to large shrubs: Apple Blossom, pink and white flowers; *E. bellidiflora*, small white blooms; C. F. Ball, crimson, tall, good for coastal planting; *E. langleyensis*, rose-crimson flowers on arching branches, 5 to 6ft.; and *E. macrantha*, a vigorous rose-crimson species, good for coastal planting. The Donard varieties are excellent and include Donard Beauty, rose-red; Donard Gem, large pink flowers; Donard Seedling, pale pink buds opening white; Glory of Donard, carmine; and Slieve Donard, apple-blossom pink.

Euonymus (Spindle Tree). Differing greatly in habit from one species to another, the euonymus are very easy to grow and will thrive in a wide range of soils. There are both evergreen and deciduous kinds. The deciduous species are undoubtedly the most decorative with their colourful berries and autumn foliage. *Euonymus alatus* grows to about 6ft. and is noted for its brilliant autumn colour in shades of scarlet. Our native Spindle Tree, *E. europaeus*, attains a height of 10 to 12ft. or more and is noted for its attractive autumn fruits which are reddish-pink with orange seeds. *E. e. albus* is white-fruited while *E. e. atropurpureus* has leaves which are purple during spring and summer and turn vivid red in autumn. *E. latifolius* has large scarlet fruits and is more brilliant in autumn colour than *E. europaeus*. (See also p. 73.)

Exochorda. The exochordas are deciduous shrubs which bloom freely in April and May and have white flowers about 1½in. across. They grow well in most soils and reach about 10 to 12ft. in height. *Exochorda racemosa* is my favourite but this species is not happy in chalky soils.

Fatsia. *Fatsia japonica* is an impressive evergreen shrub, with leathery leaves over 1ft. across. It thrives in shade, has thick, sturdy stems and grows to about 10ft. in height. It carries spikes of ivory flowers in autumn and gives a delightful tropical effect to the garden.

Firethorn, see Pyracantha

Forsythia. Perhaps the best known and most widely grown of all the forsythias—which do so much to lift our spirits in March and early April —are *Forsythia intermedia spectabilis*, which bears its golden-yellow flowers on its bare branches with especial freedom, and the semi-pendulous *F. suspensa*, which makes a splendid shrub for planting against a north wall where colour will be particularly welcome. Both of these forsythias grow to about 10 to 12ft. in height.

Others I would mention are the fairly new *F. intermedia* Arnold Giant with large, rich yellow flowers, and *F. i.* Lynwood which has rich yellow blooms of even larger size. For small gardens *F. ovata* is ideal as it attains a height of only 4 to 5ft.

Fuchsia. If hardy fuchsias are carefully sited so that their lovely flowers are seen to best advantage, they can be an especially attractive feature during summer and autumn. Some varieties suitable for outdoor planting may be cut to ground level by severe frosts but they will make strong growth again in the following spring. They are excellent for planting in small gardens as they vary in height from about 2½ to 6ft. *Fuchsia magellanica* and its varieties are probably the most widely grown. *F. m. alba* is a good white, and *F. m. riccartonii* has crimson and purple flowers. *F. gracilis* is of upright habit, with red and purple flowers, while its variety *F. g. variegata* has silvery leaves flushed with pink. Madame Cornelissen produces large, carmine and white, semi-double blooms and Mrs Popple, also with large flowers, is carmine and violet. Two charming dwarf varieties are *F. magellanica pumila*, with red and purple flowers, and the red and violet Tom Thumb.

Furze, see Ulex

Garrya. *Garrya elliptica* is a magnificent evergreen, 8 to 12ft. tall, with 9-in.-long silvery-green catkins in January and February. The male plants should be grown, as the catkins on female plants are much shorter and less impressive.

Genista (Broom or Gorse). These shrubs are related to cytisus and need similar conditions. They include both dwarf and tall shrubs, and bear their yellow, pea-shaped flowers in spring or summer. The Mt. Etna Broom, *Genista aethnensis*, grows to 12ft. or so and blooms in July. An excellent dwarf species is *G. hispanica*, the Spanish Gorse, which flowers profusely in May and June. It forms 2-ft.-high hummocks and is excellent for hot dry banks. *G. lydia* is another dwarf species, flowering in May and June, and the delightful 2-ft.-tall *G. tinctoria* is a late-flowering kind.

Gorse, Common, see Ulex

Gorse, Spanish, see Genista

Hamamelis (Witch-hazel). Where room can be found for large winter-flowering shrubs the hamamelis are an excellent proposition, for their distinctive flowers with strap-shaped petals, which are borne on bare branches in mid-winter, are a delight at that least colourful time of the year. The best-known species is *Hamamelis mollis*, which makes a shrub or small tree of upright habit some 20ft. tall and bears its scented yellow flowers in January and February. Even more desirable, perhaps, is its variety *pallida*, with sulphur-yellow flowers of large size. *H. japonica*, the other species which is widely grown, has a more spreading habit and reaches much the same height. The rich yellow flowers appear between January and March.

Hazel, see Corylus

Hebe. Many of these evergreen shrubs are on the tender side but they grow well in maritime districts. They will thrive in most soils including almost solid chalk, and like full sun. The hebes were once included under the generic name *Veronica* and may still be listed as such in some catalogues. Many are suitable for small gardens as they grow to 2 to 4ft. in height and can be had in flower from spring to autumn.

The following kinds are considered to be hardy: *Hebe albicans*, white; Autumn Glory, violet; *H. brachysiphon*, white; Hielan Lassie, blue-violet; *H. franciscana latifolia*, mauve, and Mrs E. Tennant, light violet. Some excellent, but unfortunately tender, kinds are Great Orme, bright pink; *H. hulkeana*, pale lavender; Midsummer Beauty, lavender; Purple Queen; and Simon Delaux, crimson.

Hibiscus (Tree Hollyhock). A sunny position is needed for the beautiful varieties of *Hibiscus syriacus*. All are deciduous and attain about 6ft. in height. The large blooms are trumpet shaped and appear in August or September. The following are some of my favourites: Blue Bird, single blue; Duc de Brabant, double red; *H. s. monstrosus plenus*, double white, purple centre; *H. s. violaceus plenus*, double purple; and Woodbridge, rose-pink, with maroon central patches.

Hippophaë (Sea Buckthorn). A good seaside shrub, though rather tall (it can be well over 20ft.), is *Hippophaë rhamnoides* which is easily grown in almost any soil. The willow-like leaves are silvery and female plants bear orange berries. Plant both male and female kinds in groups to ensure good crops of berries.

Holly, see Ilex

Honeysuckle, Shrubby, see Lonicera

Horse Chestnut, see Aesculus

Hydrangea. The hydrangeas, which make such a colourful display in summer and autumn, are splendid plants for moist soils and partial shade. The varieties of the common hydrangea, *H. macrophylla*, can be recommended. A particularly good variety is Blue Wave, a lacecap type which is rich blue in colour on acid soils. Another of my favourites is the lacecap variety *H. m. mariesii* with large, flat heads of mauve-pink flowers which turn blue in acid conditions. There are also the many fine varieties of *H. macrophylla hortensia*, with large, globular flower heads. These include: Altona, rose; Blue Prince, rose-red; Générale Vicomtesse de Vibraye, rose-red; Goliath, pink;

Hydrangea Blue Wave, a lacecap variety of rich blue colouring. The hydrangeas are plants for moist soils and partial shade

King George, light red; Madame E. Mouillière, white; and Maréchal Foch, red.

On an acid soil the above varieties will turn blue naturally—except the white one—but on slightly chalky or alkaline soils red varieties of hydrangeas can be treated to change their colour. The soil around each plant can be dressed in early spring with alum (aluminium sulphate) and at intervals thereafter. Proprietary blueing powder can also be applied at intervals, as recommended by the manufacturers. All the above varieties form rounded bushes 3 to 5ft. in height.

Two more of my favourite hydrangeas are *H. paniculata grandiflora*, with massive cone-shaped heads of white flowers, and *H. sargentiana* with velvety leaves and blue and white flowers. The latter usually needs a sheltered position.

Hypericum (St John's Wort). Summer and autumn flowering, hypericums thrive in any soil which is well drained. They are all yellow-flowered and bloom very freely. *Hypericum patulum* Hidcote has really large blooms and attains 5ft. in height. Rowallane Hybrid is evergreen and needs a sheltered position but is one of the most beautiful hypericums, reaching about 8ft. It bears large golden-yellow blooms. (See also p. 73.)

Ilex (Holly). I am sure that hollies need hardly any description, except possibly the delightful variegated kinds which are my favourites. These

Kerria japonica pleniflora, an easily pleased shrub of some 6ft. in height. The double, yellow flowers appear in spring and form a delightful mass of colour

Lavandula spica, the much-loved Old English Lavender, is one of the finest fragrant shrubs and makes a delightful edging for a path or lawn

make ideal specimen shrubs, if kept clipped, especially on a lawn. Most hollies are evergreen and tolerate a wide range of soils. They will also withstand a polluted atmosphere and are happy in sun or shade.

The common English Holly, *Ilex aquifolium*, has some delightful varieties; for example, *I. a. argenteo-marginata*, silver-variegated, berrying; *I. a. argenteo-marginata pendula*, a silver-variegated weeping form, berries well: *I. a. aureomarginata*, yellow edges to the leaves; Golden King, broad yellow margins to the leaves, berries freely; Golden Milkmaid, gold leaves with narrow green margins; *I. a. pyramidalis fructuluteo*, a pyramidal green-leaved variety with bright yellow berries; and Silver Queen, creamy-white margins to the leaves.

Kalmia. Similar to a rhododendron in its foliage and requirements, the handsome *Kalmia latifolia* must have a lime-free soil. It attains about 6 to 8ft. in height, is evergreen, and in June the small, bright pink flowers appear quite freely.

Kerria. The single yellow flowers of *Kerria japonica* appear on the arching branches in the spring. It attains about 6ft. in height and is deciduous. I prefer, however, *K. j. pleniflora* which is similar in habit but has double flowers.

Lad's Love, see Artemisia

Laurus (Bay). The evergreen *Laurus nobilis* is an aromatic, dense pyramidal shrub which is useful as a lawn or tub specimen. It will withstand clipping exceptionally well. The leaves are a dull green.

Laurustinus, see Viburnum

Lavandula (Lavender). Lavender has been grown by generations of gardeners and is undoubtedly one of the best of all fragrant shrubs. I always have a few plants and I also like seeing it as a dwarf hedge. My favourite is the Old English Lavender (*Lavandula spica*), with grey leaves and spikes of blue-mauve flowers. This delightfully fragrant shrub will grow in a wide range of soils. Grappenhall Variety reaches about 3ft. in height but Munstead Dwarf is very compact. An attractive variety with strongly coloured flowers is Twickle Purple. The French Lavender (*L. stoechas*) is only about 1ft. tall and has bright green foliage and purple flowers.

Lavender Cotton, see Santolina

Leycesteria. Growing to about 6ft. in height, *L. formosa* is a deciduous shrub with unusual green stems and white flowers with reddish-coloured bracts. These appear throughout the summer months.

Lilac, see Syringa

Lonicera (Shrubby Honeysuckle). My favourite shrubby honeysuckles are *Lonicera fragrantissima*, a 6-ft.-tall, partly evergreen shrub with very fragrant cream flowers in winter, and *L. standishii* which is almost identical. The latter bears red

The small (6- to 10-ft. tall), white-flowered *Magnolia stellata*, aptly named the Star Magnolia, is a good choice for modestly-sized gardens

berries in early summer. They are both easy to cultivate.

Magnolia. Some magnolias are classed as shrubs while other species are trees (for details of two of the latter see p. 51). They all like a deep, moist soil, and while most of them need lime-free conditions there are some which will tolerate chalk. They make excellent lawn specimens.

Magnolia highdownensis is a deciduous, 8-ft. shrub that will grow in a chalky soil. The large white flowers appear in early summer. *M. liliflora nigra* blooms in April and May and has large cup-shaped flowers, pale purple inside and a much deeper shade on the outside. It is slow growing, eventually attaining about 10ft., and is deciduous. Another very slow-growing species, also deciduous, is *M. sieboldii* which may attain anything between 8 and 15ft. It bears large, white flowers with crimson stamens from May until August. *M. soulangiana*—the most commonly seen magnolia, in one or other of its numerous forms—may attain a height of 25 to 30ft. and looks well against a large wall. Its wide-spreading branches carry large, white, purple-stained flowers. Finally, for a small garden, there is the distinctive *M. stellata*, the Star Magnolia, which grows to 6 to 10ft. The white, scented, star-shaped flowers appear in great abundance before the leaves in March and April.

Mahonia. The mahonias are most attractive evergreen shrubs, suitable for planting in gardens of all sizes. The racemes of flowers are in various shades of yellow. The most distinguished species is *Mahonia japonica*, 5 to 6ft. tall, a very beautiful plant indeed with its long, pendulous racemes (sprays) of fragrant lemon-yellow flowers for which the deep green, large, pinnate leaves provide a perfect foil. The flowers appear in February and March. Very similar but with upright racemes of flowers of lesser size is *M. bealei*. An

extremely useful and tough species is *M. aquifolium*, the Holly-leaved Berberis as it is sometimes called. This has golden-yellow flowers in spring and thrives in sun or shade. The leaves turn red in autumn and are an additional attraction. It grows about 4ft. tall.

Mexican Orange Blossom, see Choisya
Mezereon, see Daphne
Mock Orange, see Philadelphus
Nut, Hazel, see Corylus
Olearia (Daisy Bush). The olearias, which are evergreen shrubs, like a position in full sun, grow well by the sea, tolerate a wide range of soils and vary in height from 4 to 8ft. The daisy-like flowers are cream or white in most species. *Olearia haastii* flowers in July and August, the blooms being scented. This is considered to be the hardiest species, the others being more suitable for coastal planting. *O. semidentata* has silvery leaves and large lilac flowers with a purple centre.

Ornamental Quince, see Chaenomeles
Osmanthus, see Siphonosmanthus
Paeonia (Tree Peony). *Paeonia delavayi* bears the most beautiful crimson flowers on 5-ft. stems while *P. lutea ludlowii*, of about the same height, has golden-yellow blooms about 4in. across. They are reasonably easy to grow and thrive in a good deep loamy soil.

Pernettya. Pernettyas are evergreen berrying shrubs which require a lime-free soil and prefer it to be rather moist. They are good companions for rhododendrons and azaleas. *P. mucronata* and its varieties are the most popular and grow to about 3ft. in height. The large berries of *P. mucronata* appear in late summer and persist through the winter. The colour of these may be white, various shades of pink, lavender, purple or red.

Philadelphus (Mock Orange). These early-summer-flowering deciduous shrubs are invaluable for their white, fragrant blossoms and will thrive in almost any soil. These shrubs are often erroneously called Syringa, which is confusing because this is the botanical name of Lilac. The correct common name of philadelphus is Mock Orange, in reference to the scent of the white flowers which is reminiscent of orange blossom. The varieties which I particularly like are Avalanche, with masses of small flowers on arching stems; Beauclerk, with $2\frac{1}{2}$-in.-wide flowers; Belle Etoile, with blooms 2in. across, flushed with maroon in the centre; and Virginal, a double variety. All the foregoing attain about 6 to 8ft. in height, but for a very small garden *P. microphyllus* is superb. It forms a compact bush about 3ft. high, has very small leaves and wonderfully fragrant blooms.

Pieris. This genus of attractive evergreens requires the same conditions as rhododendrons; in other words, a moist, lime-free soil. The most beautiful kind is undoubtedly *Pieris formosa forrestii* which grows to about 8ft. tall. It has long panicles of white, fragrant, waxy blooms, each one similar to a lily-of-the-valley flower. These appear in spring, at which time the young growths are a brilliant shade of red.

Pleioblastus (Bamboo). Some of our most useful bamboos are included in this genus, the following species being non-rampant. *Pleioblastus variegatus* (syn. *Arundinaria fortunei*) has slender stems, 2 to 4ft. high, with white-striped leaves, while *P. viridi-striatus* (syn. *Arundinaria auricoma*) has 2 to 4ft. purplish-green stems and yellow-striped leaves. Both are evergreen.

Plumbago, see Ceratostigma

Potentilla (Shrubby Cinquefoil). These dwarf summer and autumn-flowering plants are invaluable for a small garden. Their small, single, rose-like flowers are produced in great profusion. The best-known varieties belong to *Potentilla fruticosa* and attain 2½ to 4ft. in height. They are suitable for almost any soil and situation. *P. f. beanii* has white flowers; Elizabeth, canary yellow; Katherine Dykes, primrose-yellow; *P. f. parvifolia*, deep yellow; and Tangerine, copper-yellow. They flower more or less continuously throughout the summer and autumn.

Prunus (Cherry). A good selection of *Prunus* will be found on pp. 51 and 52 and p. 72, the latter reference being to those suitable for hedging purposes. However, there are one or two dwarf shrubs in this family which I can recommend for a small garden. *P. besseyi*, the Sand Cherry, grows to 4ft. and produces clusters of small white flowers along its branches during the spring. The glaucous foliage colours well in autumn. *P. cistena*, the Purple-leaved Sand Cherry, has beautiful red leaves and white blossoms. *P. prostrata* forms a rounded bush about 2½ft. in height and produces bright rosy-red blossoms. Finally, *P. triloba multiplex* is a dwarf almond with double, pink flowers borne all along its numerous stems in early spring.

Pseudosasa (Bamboo). One of the most common bamboos but a very useful evergreen for backgrounds and screening is *Pseudosasa japonica* (syn. *Arundinaria japonica*). This has 12-ft. canes.

Pyracantha (Firethorn). I consider pyracanthas to be the finest of the evergreen shrubs with their colourful autumn and winter berries. They make good wall specimens, when they will attain 10 to 12ft. in height, but they are equally effective in a border and are exceptionally easy to grow. My favourites are *P. coccinea lalandii* with small white flowers in May and June and orange-scarlet berries in August which last until January or February, and *P. crenulata flava* (syn. *P. rogersiana flava*) with bright yellow berries.

Rhododendron (including Azalea). Azaleas are classified as rhododendrons and are grouped in

The canary-yellow *Potentilla* Elizabeth, which grows to 4ft. tall. This splendid variety flowers from early summer to autumn

the azalea series of this genus. Both rhododendrons and azaleas require a moist, lime-free soil —preferably peaty or medium-light loams—and range from dwarf rock-garden shrubs to very large shrubs and tree-like specimens. There are both evergreen and deciduous kinds and their main flowering period is spring and early summer.

There is now such a wide range of good species and varieties that it is very difficult to make a small representative selection. However, I have picked out some of my favourites which I consider to be good garden plants.

I will start with some species of rhododendron —all are evergreen. *R. augustinii*, beautiful blue flowers, 10ft.; *R. barbatum*, crimson-scarlet, 10 to 15ft.; *R. calostrotum*, grey foliage, crimson flowers, 1ft.; *R. campylocarpum*, yellow, 5 to 8ft.; *R. haematodes*, crimson, 3 to 4ft.; *R. impeditum*, purple-blue, low growing; *R. moupinense*, white, 4ft.; *R. orbiculare*, pink, 4 to 6ft.; *R. saluenense*, purplish-crimson, 2ft.; *R. scintillans*, lavender-blue, 2 to 3ft.; *R. thomsonii*, deep red, 8 to 12ft.; and *R. yunnanense*, white or pale pink, 8ft.

Some good hardy evergreen rhododendron hybrids are as follows, most of them being of moderate size while some are suitable for small gardens: Bagshot Ruby, ruby-red; Beauty of Littleworth, white, spotted crimson; Blue Tit, lavender-blue, 2½ to 3ft.; Bo-Peep, primrose-yellow; Bric-a-Brac, pure white, 2½ft.; Britannia, crimson-scarlet; Dairymaid, pale yellow; Goldsworth Yellow, primrose-yellow; Lady Chamberlain, reddish, shading to orange-buff; Lady Clementine Mitford, peach; Loder's White; Pink Pearl, rose-pink; *R. praecox*, lavender-pink, early, semi-evergreen; Purple Splendour; *R. shilsonii*, blood-red, tall; and Tessa, pink, early, 3ft.

Of the azaleas, one of my favourites is the common yellow *A. pontica* (now more correctly *Rhododendron luteum*), with fragrant flowers and richly coloured autumn foliage. The deciduous Mollis azaleas flower prolifically in May, usually before the leaves appear, and attain about 5ft. in height. Some good varieties include Anthony Koster, deep yellow, flushed orange; Comte de Gomer, pink; Hugo Koster, salmon-orange; Koster's Brilliant Red; Salmon Queen, salmon-apricot; and Snowdrift, white, yellow markings.

Finally we come to the evergreen azaleas which bloom exceptionally freely in May and grow to 2 to 3ft. in height. The following varieties are known as Kurume hybrids: Addy Wery, vermilion-red; Daphne, white; Hinomayo, clear pink; Hoo (syn. Apple Blossom), white, tinged pink; Salmon Beauty, salmon-pink; Suga-no-ito, clear pink; and Surprise, pale orange-red.

Rhus. The Stag's Horn Sumach, *Rhus typhina*, 12 to 15ft. tall, is a good foliage shrub, having pinnate leaves 2 to 3ft. in length which turn orange-red in the autumn. Female plants also bear a crop of showy crimson fruits. *R. t. laciniata* has deeply cut leaves and really good autumn colour. These are good town shrubs.

Ribes (Flowering Currant). The easily grown ribes are invaluable for their early spring display. *R. sanguineum* provides the popular varieties such as *carneum*, with deep pink blooms, and King Edward VII, brilliant crimson. They are deciduous and make bushy plants about 8 to 10ft. in height.

Rock Rose, see Cistus

Rosa (Rose). The shrub roses and rose species are superb plants either grown as single specimens or associated in a border with other shrubs. They need little pruning. Some of the old shrub roses are delightfully scented and the modern shrub roses include many varieties of merit. They flower between June and September, and many are perpetual-flowering. The rose species are superb flowering shrubs, many having additional attributes in the form of attractive foliage, which sometimes colours well in autumn, and brightly coloured heps. Nearly all have single blooms and produce only one flush of flowers during the summer.

There are so many shrub roses that it is rather difficult to compile a very short list. However, some of my favourites include Bonn, rose-red, attaining a height of about 6ft.; Golden Wings, large, single, canary-yellow, 5ft.; Will Scarlet, semi-double, 6ft.; Canary Bird, single yellow, early flowering, 7ft.; Nevada, creamy-white, 8ft.; Frühlingsgold, creamy-yellow, semi-double, early flowering, 10ft.; and Frühlingsmorgen, single pink with yellow centre, 10ft.

Two of my favourite rose species, and possibly the most popular, are *R. moyesii* with single, deep crimson flowers followed by flask-shaped scarlet heps, 9 to 10ft.; and *R. rubrifolia* with attractive blue-grey foliage tinged with pink, and masses of crimson hips after the small pink flowers, 7 to 8ft.

Rosmarinus (Rosemary). The common Rosemary is *R. officinalis* and it has highly aromatic foliage. This shrub reaches 6 to 7ft. in height and is very bushy. Blue flowers appear in spring. Severn Seas is a good variety and is a smaller, more compact plant than the type, with deeper blue flowers.

Ruscus, see p. 73
St John's Wort, see Hypericum

Sambucus (Elder). The varieties of the common Elder, *Sambucus nigra*, are suited to town gardens and will grow in most soils. They are fairly vigorous, tall, deciduous shrubs. *S. n. albo-variegata* has white-edged leaves, and *S. n. aurea* is the Golden Elder, with bright yellow foliage.

Santolina (Lavender Cotton). The santolinas are dwarf, aromatic, evergreen shrubs making fairly dense bushes. *S. chamaecyparissus* has silvery foliage with small yellow flowers in summer. This

The evergreen *Senecio laxifolius* with silvery-grey foliage and yellow flowers is an accommodating shrub which associates happily with many other plants

attains 1½ to 2ft. in height, while *S. c. nana* reaches only 1ft. *S. virens* makes a good companion with its bright green foliage and yellow flowers and grows to about 2ft.

Sarcococca, see p. 73
Sea Buckthorn, see Hippophaë

Senecio. The senecios are good evergreen shrubs for coastal planting and will withstand most soil conditions. *S. laxifolius* is the commonly grown species and it has silvery-grey foliage, with yellow 'daisy' flowers in summer. It forms a rounded bush 3 to 4ft. in height.

Siphonosmanthus. The shrub we used to know as *Osmanthus delavayi* is now *Siphonosmanthus delavayi*; it is an evergreen which I find very attractive. It grows to a height of about 6ft. and is much the same in diameter. The white, scented flowers, which appear in spring, look very decorative against the small green leaves.

Sinarundinaria (Bamboo). A good bamboo for a reasonably moist soil is *Sinarundinaria murieliae* (syn. *Arundinaria murieliae*) with 8 to 12ft. canes and beautiful green foliage. Excellent for a background to a planting scheme.

Skimmia. I like these attractive evergreen shrubs for their masses of red berries which are held throughout the winter. They are also useful in that they can be grown in semi-shady positions. *Skimmia japonica* makes a dome-shaped bush about 4ft. in height. It is necessary to grow male and female plants together for berries to be produced. *S. fortunei* is hermaphrodite and bears berries if planted on its own.

Smoke Tree, see Cotinus
Snowberry, see Symphoricarpos
Southernwood, see Artemisia
Spartium (Spanish Broom). *Spartium junceum* is an adaptable 8-ft. shrub which thrives almost anywhere. It has beautiful, very fragrant, deep

autumn and winter, *Symphoricarpos albus laevigatus* is a very adaptable deciduous shrub. It will grow almost anywhere and is extremely useful for shady situations. The fairly new variety Constance Spry has, I believe, even more striking berries.

Syringa (Lilac). The varieties of the common lilac, *Syringa vulgaris*, must undoubtedly be among the most widely grown of all flowering shrubs—and little wonder, for their large, delightfully scented blooms in many colours add great beauty to the garden in May and early June. They attain a height of about 10ft. and are of easy cultivation. Some good doubles include Charles Joly, deep red; Edith Cavell, white; Katherine Havemeyer, purple-lavender; and President Grevy, lilac-blue. Single lilacs include Congo, red; Lavanensis, pale pink; Maréchal Foch, carmine; Prodige, purple; Souvenir de Louis Spath, deep red; Clarke's Giant, lavendar, and Vestale, white.

The Snowberry, *Symphoricarpos albus laevigatus*, is an especially useful shrub for shady positions. The white berries are carried in autumn and winter

Spiraea arguta, the Foam of May. The masses of white flowers are borne in spring and have a special attraction, as the bushes are graceful

yellow pea-like blooms which are carried in summer from about July onwards.

Spindle Tree, see Euonymus
Spiraea. The spiraeas are very graceful shrubs with attractive foliage and dainty blooms, usually of white or cream colouring, which appear either in spring or summer. They are easy to grow and their average height is about 5ft. I particularly like *S. arguta*—or Foam of May—with masses of flowers in the spring and very dainty leaves. *S. japonica* has pink flowers borne in flat heads which appear in the summer. *S. sargentiana* is slightly taller than the others and has creamy-white flowers in early summer, while *S. vanhouttei* attains a height of 8ft. and blooms during the spring.

Stag's Horn Sumach, see Rhus
Symphoricarpos (Snowberry). Producing its large, round white berries in profusion during

Tamarix (Tamarisk). Tamarix are deciduous shrubs usually planted in coastal districts where they make excellent windbreaks of up to about 10ft. high. Their feathery pink flowers and delicate green foliage on slender, graceful stems add distinction to the garden. They like to be in a sunny spot and most soils seem to suit them. *T. tetrandra* blooms in spring, while *T. pentandra* flowers in August.

Tree Hollyhock, see Hibiscus
Tree Peony, see Paeonia
Ulex (Furze or Gorse). Our beautiful native Gorse, *U. europaeus*, so commonly seen on heathlands and downlands, has given us a good garden shrub in the form of *U. e. plenus*, the double form. This also has the deep yellow flowers of the type. It is excellent for growing on dry banks or in coastal districts and reaches a height of 4 to 5ft. The flowers appear in April and May.

Viburnum. All the viburnums, evergreen and deciduous, are easy to grow. Two which I admire for spring flowering are the evergreen *V. burkwoodii*, which bears heads of blush pink, fragrant blooms, and the deciduous *V. carlesii* with fragrant white flowers which are a delightful pink shade in the bud stage. The first makes a shrub of about 8ft., the second little more than half this height. Following these two in flowering, in May and June, is *V. tomentosum mariesii* which is 10ft. or so in height. This deciduous variety is very handsome indeed with its tiered, wide-spreading branches and large white flowers. The Snowball Tree, *V. opulus sterile*, also deciduous, of similar height to the last, carries its round, white flowers in June on a rounded shrub up to 15ft. tall.

The 8-ft.-tall *V. bodnantense* is a fairly new hybrid of considerable merit which bears its fragrant, pink flowers on the bare branches in mid-winter. I also like the variety *V. b.* Dawn, with its arching branches of lovely pink blooms. Better known winter flowerers are the Laurustinus, *V. tinus*, and *V. fragrans*, whose fragrant, pink-tinged white blooms I like to cut and bring indoors. These flowers are borne on leafless stems intermittently from November to March. The Laurustinus bears its quite large flower heads on and off from late autumn to early spring, and its evergreen leaves provide a deep green background for the white flowers when they appear.

The spring-flowering evergreen *Viburnum burkwoodii* which bears blush-pink fragrant blooms. This species grows to about 8ft. tall

Vinca, see p. 73

Weigela. The weigelas, formerly known as diervillas, are easily grown deciduous shrubs, forming fairly neat bushes about 5 to 6ft. high. They flower freely in May and June, the blooms being bell-shaped. Although I prefer the garden hybrids, I do like the pale pink *W. florida variegata*, the leaves of which are margined with cream. Some good large-flowered hybrids are Abel Carrière, deep pink; Bristol Ruby, ruby-red; Espérance, pale pink; Eva Rathke, crimson; and Mont Blanc, white.

Winter Sweet, see Chimonanthus
Witch-hazel, see Hamamelis

Yucca. These peculiar evergreen shrubs have stiff, erect, spear-like leaves formed in rosettes. I like them for the sub-tropical effect they create in the garden. *Yucca filamentosa*, known as Adam's Needle, is most usually grown. It is stemless and has 3- to 6-ft.-tall spikes of cream, bell-shaped, lily-like flowers in July. This species blooms when quite young and flowers very freely.

Conifers

These are all evergreen shrubs which will grow in most garden soils. The dwarf ones are excellent for the rock garden.

Arbor-vitae, see Thuya

Cephalotaxus (Plum Yew). *Cephalotaxus harringtonia drupacea* has leaves similar to a yew only much larger and makes a spreading bush about 10ft. in height. It is noted for its peculiar plum-like fruits.

Chamaecyparis. Two delightful dwarf forms of *Chamaecyparis lawsoniana* are *C. l. minima*, which is 3 to 4ft. tall, and *C. l. minima aurea*, with beautiful golden leaves, which reaches a similar height. Another one which I especially like on a rock garden is *C. l. pygmaea argentea* with deep, bluish-green foliage and white-tipped shoots. *C. obtusa pygmaea* attains a height of 2 to $2\frac{1}{2}$ft. and has fan-shaped sprays of green foliage which takes on a bronzy hue when mature.

Cryptomeria. Growing to $2\frac{1}{2}$ to 3ft., *Cryptomeria japonica elegans compacta* is of globose habit with dense foliage which turns reddish-bronze in the autumn and winter.

Juniperus (Juniper). The delightful *Juniperus communis compressa* is certainly one of the best of all dwarf shrubs for a rock garden. The foliage is bluish-grey and very dense and the plant forms a neat little column about 1 to $1\frac{1}{2}$ft. in height. It is very slow growing.

Picea (Spruce). One of my favourite rock-garden conifers is *Picea albertiana conica* which is slow-growing, eventually attaining a height of about 3ft. It is perfectly conical and has beautiful green foliage.

Plum Yew, see Cephalotaxus
Spruce, see Picea

Thuya (Arbor-vitae). *Thuya occidentalis* Rheingold forms a broadly conical, dense bush about 3 to 4ft. in height. The foliage is golden during summer but turns to bronze and deep gold shades during the autumn and winter.

Climbers and Wall Plants

Actinidia. The beautiful *Actinidia kolomikta* has white, pink and green variegated leaves. To encourage this variegation it should be grown on a south or west wall in full sun. It is deciduous.

Ampelopsis. These deciduous climbers usually bear attractive bluish grapes. *Ampelopsis brevipedunculata* is fairly vigorous, has hop-like leaves and produces a good crop of grapes after a fine summer.

Ceanothus. These shrubs can be trained against a wall most successfully (see p. 55 for varieties).

Chaenomeles (Ornamental Quince). Chaenomeles also make good wall shrubs if they are trained (see p. 55 for descriptions of species and varieties).

Clematis. The species and large-flowered hybrids of clematis are indispensable for wall or pergola culture. They should be grown in full sun but with their roots in cool, moist soil. A large flat piece of stone placed over the root area will provide these conditions. Their average height is 8 to 10ft.

Some of my favourite species are: *C. macropetala* with violet-blue flowers in spring and summer; *C. montana*, masses of white flowers in May; *C. m. rubens*, a pink variety; and *C. tangutica* with deep yellow blooms in autumn.

The large-flowered hybrids which I like are Crimson King, rose-red; Ernest Markham, red; Lasurstern, deep purplish-blue; *jackmanii*, violet-purple; Nelly Moser, pale mauvish-pink with a carmine bar; and Perle d'Azur, pale blue.

Cotoneaster. *Cotoneaster horizontalis* makes an excellent wall shrub (see p. 56 for description).

Forsythia. *Forsythia suspensa* is another fine wall plant (see p. 61 for description).

Hedera (Ivy). These evergreen self-clinging climbers are extremely well known and will grow in any soil and in any aspect. The Canary Island Ivy, *H. canariensis*, has large, rounded, bright green leaves while its variety *H. c. variegata* has green, grey and white variegated foliage. The common ivy, *H. helix*, has some excellent varieties, including *H. h. aureo-variegata*, the leaves of which are suffused with yellow; Chicago, small leaves blotched with purple; *H. h. congesta*, small, grey-green, triangular leaves; *H. h. marginata* (syn. Silver Queen), leaves edged with white; and *H. h. tricolor*, leaves greyish-green, edged with white, which turns reddish in autumn.

Honeysuckle, see Lonicera

Hydrangea. *Hydrangea petiolaris* is a deciduous, self-clinging climber well suited to growing on a north wall. It has large, deep green leaves and bears white flowers in June.

Ivy, see Hedera

Jasminum (Jasmine). *Jasminum nudiflorum* is, perhaps, the best-known of all winter-flowering shrubs, bearing an abundance of bright yellow flowers on leafless stems from November or December onwards. It is an excellent wall shrub, will grow in almost any soil and reaches 10 to 12ft. in height. It is ideal for use in flower arrangements. The Common White Jasmine, *J. officinale*, is a semi-evergreen twiner and produces fragrant white blooms in the summer.

Lonicera (Honeysuckle). The honeysuckles look well growing up large trees or rambling over tree stumps, but they are equally effective on a house wall where their fragrance can be appreciated. The wild Woodbine, *Lonicera periclymenum*, produces very sweet cream and red flowers from June to September. The blooms of the Early Dutch Honeysuckle, *L. p. belgica*, are flushed with red on the outside and appear in May and June and sometimes again in late summer. The Late Dutch Honeysuckle, *L. p. serotina*, is reddish-purple on the outside of the flowers and blooms from July to October.

Magnolia. *Magnolia soulangiana* is a good wall shrub if a large wall is available to take its spreading branches (see description on p. 64).

Less often grown than it deserves to be is the deciduous, self-clinging climber *Hydrangea petiolaris*. It will cover large areas of wall, fence or tree

The rampant-growing *Polygonum baldschuanicum*, the Russian Vine, is one of the most useful shrubs for clothing walls, outbuildings and so on

Vitis coignetiae with its bold leaves—these turn orange and crimson in autumn—provides an effective cover for walls and other surfaces

The violet-blue-flowered *Wisteria floribunda*, a well-known member of this decorative genus

Ornamental Quince, see Chaenomeles

Parthenocissus (Virginia Creeper). These are usually vigorous climbers, self-clinging and are noted for their brilliant autumn foliage. *Parthenocissus quinquefolia* (syn. *Vitis quinquefolia*) is very tall and has bright green leaves which change, in autumn, to scarlet and orange shades. There is, in my opinion, no creeper to beat the intense autumn colouring of *P. tricuspidata veitchii* (syn. *Ampelopsis veitchii*). It is a very vigorous species.

Polygonum (Russian Vine). The fastest-growing twining shrub is, without question, *Polygonum baldschuanicum*. It should not be grown where space is limited but it is invaluable for covering unsightly objects—which it does in a very short time. The masses of white blooms are borne during summer and autumn. It is deciduous.

Pyracantha (Firethorn). The pyracanthas I have described on p. 65, and these are ideal for training on a north or east wall.

Rosa (Rose). Climbing and rambling roses always make a colourful display during the summer. Among the many climbers, my favourites are the new Golden Showers, with large, fragrant yellow flowers, 9 to 10ft. tall; Danse du Feu, orange-scarlet, 8ft.; and Mme. Gregoire Staechelin, carmine buds opening to large pink blooms, very vigorous. Some good ramblers are: Alberic Barbier, creamy-white, very vigorous; Albertine, large, coppery-pink, vigorous; American Pillar, single rose-pink and white blooms, vigorous; and Paul's Scarlet Climber, scarlet, moderately vigorous.

Russian Vine, see Polygonum

Vine, see Vitis

Virginia Creeper, see Parthenocissus

Vitis (Ornamental Vines). These are excellent climbers for foliage colour and my two favourites are *Vitis coignetiae* with huge leaves which turn orange and crimson in autumn, and *V. vinifera purpurea*, the Teinturier Grape, with claret-red leaves which later turn deep purple. Give them plenty of wall space.

Wisteria. These are magnificent climbing deciduous shrubs which produce their long, pendulous racemes of blue, mauve or white flowers in May and June. They do best in a sunny position and fairly rich soil. *Wisteria floribunda macrobotrys* has lilac flowers borne in exceedingly long racemes. The type has violet-blue flowers. The most popular species is *W. sinensis* with mauve flowers. This has two good varieties: *W. s. alba*, white, and *W. s. plena* with double mauve flowers. *W. venusta* has pure white blooms.

Hedging Plants

Atriplex. *Atriplex halimus* (described on p. 54) makes a first-class hedge in coastal districts. Plant about 1½ft. apart for a good thick hedge.

Aucuba. The varieties of *Aucuba japonica* (see p. 54), will make a good informal hedge if specimens are planted about 2 to 3ft. apart. Can be trimmed if desired in the spring.

Azalea, see Rhododendron

Barberry, see Berberis

Bay, see Laurus

Beech, see Fagus

Berberis (Barberry). The barberries mentioned on p. 54 form very thick hedges. The dwarf ones are excellent if a very low hedge is required. The most popular ones, however, seem to be *Berberis stenophylla* and *B. darwinii* which make excellent evergreen flowering hedges. *B. thunbergii* makes a good deciduous hedge. Set the plants 1 to 2ft. apart. Do not trim too hard otherwise flowering will be affected.

Buxus (Box). In my opinion there is nothing better for a formal evergreen hedge than *Buxus sempervirens*, the Common Box. It stands clipping exceptionally well and makes a really dense hedge if the plants are set 1½ft. apart. The small, thick leaves are deep green and it does well on almost any soil, particularly chalk. It can be trimmed when necessary in summer. A hedge up to about 15ft. in height can be obtained. There is a form with golden blotches on the leaves which I particularly like, called *B. s. aurea maculata*. The Edging Box, *B. s. suffruiticosa*, forms a 2 to 3ft. evergreen hedge.

Carpinus (Hornbeam). The Common Hornbeam, *Carpinus betulus*, suitable for clay or chalk soils, is used extensively as a hedging plant and makes a good thick screen if the plants are set 1 to 1½ft. apart. Double staggered rows may also be planted with 8 to 10in. between the rows. Hornbeam can be trained as a formal hedge and is best clipped in late summer. The plants must not be trimmed for the first two years after planting.

Crataegus (Hawthorn). The species *Crataegus monogyna* and *C. oxyacantha* form fairly dense deciduous hedges. They stand very hard clipping during the summer. Hawthorns are very often mixed with hornbeam or beech and make a good boundary hedge. The plants are set 1ft. apart, and a 20-ft.-tall hedge can be easily obtained, of either a formal or informal nature.

Escallonia. The escallonias are very popular flowering evergreen hedges, especially in coastal districts. The varieties I mentioned on p. 61 will all make good hedges if the plants are set 1½ft. apart. They are trimmed after flowering, removing the old flowered wood.

Euonymus. For a tough evergreen hedge in either a town or coastal garden *Euonymus japonicus* is ideal. The leaves are deep green and glossy. It will thrive in almost any soil and can be trimmed during summer. It makes a good formal hedge and will attain 10 to 12ft. in height, if allowed.

Plant about 1½ft. apart. There are some good coloured-leaved varieties also, including *E. j. ovatus aureus* with yellow-blotched foliage, and *E. j. macrophyllus albus* with broad, silver margins to the leaves.

Fagus (Beech). The common beech, *Fagus sylvatica*, is widely used as a formal hedge as it stands clipping well and makes good backgrounds for borders and so on. It is interesting to note that when planted as a hedge the leaves persist throughout the autumn and winter—a beautiful golden-brown colour. A hedge can be grown to any reasonable height and the planting distance is 1 to 1½ft. apart. Double staggered rows should be spaced 8 to 10in. apart, with 1 to 1½ft. between the plants in the rows. Do not clip for the first two years after planting but thereafter trim during summer or autumn.

Firethorn, see Pyracantha

Forsythia. *Forsythia intermedia spectabilis* (see p. 61) makes a beautiful flowering hedge and the plants are best set 2ft. apart. Trim after flowering. *F. ovata* is also useful if a low-growing hedge is required.

Fuchsia. The fuchsias I mentioned on p. 61—except the very dwarf ones—are all suitable for informal hedges. *F. magellanica* is the most commonly used for this purpose, however, especially on the West Coast. Plant specimens 1½ft. apart. They may be cut down to 'ground level in early spring or, alternatively, trimmed lightly.

Furze, see Ulex

Gorse, see Ulex

Hawthorn, see Crataegus

Holly, see Ilex

Honeysuckle, Shrubby, see Lonicera

Hornbeam, see Carpinus

Ilex (Holly). For a dense, impenetrable evergreen hedge the hollies are superb and those I described on p. 62, except the weeping one, would all be suitable. They may be clipped in August or September and make a perfect formal hedge, up to 20ft. tall, particularly suitable for backgrounds to borders. Plant them 1½ to 2ft. apart.

Laurel, Common, see Prunus

Laurel, Portugal, see Prunus

Laurus (Bay). *Laurus nobilis* (see p. 63), makes a good formal hedge, fairly tall if required, and should be clipped during the summer. Plant 1½ to 2ft. apart.

Lavandula (Lavender). I have described the lavenders on p. 63, all of which make excellent, low-growing hedges. Plant them 1 to 2ft. apart, and do any trimming needed in early spring. Remove dead flower heads.

Ligustrum (Privet). The Oval-leaved Privet, *Ligustrum ovalifolium*, is almost certainly the most widely planted shrub for formal hedges. It is semi-evergreen and has deep green leaves. A popular form of this is *L. o. aureo-marginatum*, the Golden Privet, with bright yellow leaves. The privets will tolerate almost any soil and are particularly suitable for town conditions. Clipping can be carried out when required during the summer. Plant 1ft. apart. It is usually best not to let a hedge grow above 5 to 6ft. in height, as it will then keep a better shape.

Lonicera (Shrubby Honeysuckle). The Chinese Honeysuckle, *Lonicera nitida*, is an evergreen shrub with small, deep green, box-like leaves. It forms a really thick hedge and is usually grown formally. It is fairly fast growing and can be clipped when necessary during the summer. Space the plants 1ft. apart.

Privet, see Ligustrum

Prunus. An attractive informal purple hedge can be made with *Prunus cerasifera atropurpurea* and *P. c. nigra* which I described on p. 51. These are best planted 1½ to 2ft. apart and should be trimmed back hard each year in early spring.

The common laurel, *P. laurocerasus*, is ideal for a boundary hedge or windscreen. It thrives in almost any soil. The laurel is evergreen with large, deep green leaves, and should be planted 1½ to 2ft. apart. Clipping may be carried out in the spring and a hedge 15 to 20ft. in height may be obtained. There are numerous varieties but *P. l. rotundifolia* has lighter green leaves while *P. l. schipkaensis* has narrow leaves and grows to only 5ft.

The Portugal Laurel, *P. lusitanica*, is also evergreen with glossy, deep green leaves, and is dense in habit. Space the plants 1½ to 2ft. apart and trim in spring if necessary. It makes a good formal hedge up to 20ft. A decorative silver-variegated variety is *P. l. variegata*.

Pyracantha (Firethorn). The pyracanthas (see p. 65) make good berrying informal or semi-informal hedges. If they are clipped too hard, though, the berries will not form. Plant 1½ to 2ft. apart.

Rhododendron (including Azalea). Rhododendrons and azaleas make good informal flowering hedges and many of those I mentioned on p. 65 would be suitable, except the dwarf spreading types. Plant them about 2ft. apart. The common mauve-flowered *R. ponticum* makes a good boundary hedge or wind shelter.

Ulex (Gorse or Furze). This makes a really dense flowering hedge and also a good windbreak. I described *Ulex europaeus* and its double variety on p. 67. Plant 1½ft. apart and clip if necessary after flowering. A 5ft. hedge can be obtained.

Conifers for Hedging Purposes

Arbor-vitae, see Thuya

Chamaecyparis. Lawson's Cypress, *Chamaecyparis lawsoniana*, is used extensively for hedging purposes and has evergreen foliage in a good shade of green. It makes an excellent formal hedge up

Heathers and Heaths (callunas and ericas respectively) are among the best of all ground-cover plants. Planted with conifers they are even more effective

to 15ft. and should be clipped in the spring. Plant 1½ or 2ft. apart. A good deep green variety which I like is Green Hedger, although some of the other varieties of Lawson's Cypress can also be used effectively.

Cupressocyparis. *Cupressocyparis leylandii* (a cross between *Cupressus macrocarpa* and *Chamaecyparis nootkatensis*) is the fastest growing conifer in this country and it has beautiful deep green foliage. It will make a formal hedge of from 8 to 25ft. Plant 1½ to 2ft. apart and clip in late summer.

Cupressus (Cypress). The Monterey Cypress, *Cupressus macrocarpa*, is also fairly fast growing and the evergreen foliage is bright green. This is not always hardy in cold districts but does well on the coast. Plant 1½ to 2ft. apart and clip in the spring. It can be grown as a formal hedge and will reach 15ft. in height grown in this way.

Metasequoia (Dawn Redwood). *Metasequoia glyptostroboides*, described on p. 53, has proved to be an excellent hedging plant, making a screen 4 to 10ft. tall. Plant 2ft. apart.

Taxus (Yew). The yews are very adaptable evergreen plants and the Common Yew, *Taxus baccata*, is used extensively as a formal hedge. It is very often used as a background for shrub or herbaceous borders and will grow well on chalk as well as many other well-drained soils. Plant 1½ to 2ft. apart and trim in late summer. A tall hedge can be obtained of up to 15ft.

Thuya (Arbor-vitae). There are two good species of this evergreen conifer for formal or informal hedges—*Thuya occidentalis*, with bronzy winter foliage, and *T. plicata* (syn. *T. lobbii*), with deep green, glossy leaves. They should be planted 1½ to 2ft. apart, may be clipped in late summer and fairly tall hedges can be obtained. They make good windbreaks.

Yew, see Taxus

The St John's Wort or Rose of Sharon, *Hypericum calycinum*, is another ground-cover plant of especial value for it will succeed under the most adverse conditions

Useful Ground-cover Plants

Bamboo, see Pleioblastus
Butcher's Broom, see Ruscus
Calluna (Ling or Heather). Many of the varieties of *Calluna vulgaris* have beautifully coloured foliage and flower from July to November. They must, however, have a lime-free soil and a sunny position. Good compact bushes are formed if trimmed after flowering, and they cover the ground well if planted 1½ft. apart.

Some of my favourites are *C. v. alba minor*, white flowers, bright green foliage, 6in.; *C. v. alportii*, crimson, 2ft.; County Wicklow, double pink, 9in.; Gold Haze, bright yellow foliage, white flowers, 2ft.; H. E. Beale, soft pink, 2ft.; J. H. Hamilton, double bright pink, 9in.; *C. v. multicolor*, orange, yellow and bronze foliage, 6in.; and *C. v. nana*, small pink flowers in profusion, 6in.

Erica (Heath). There are many dwarf, spreading heaths which are used extensively as ground cover—and a very attractive display they make, too. All ericas are evergreen. The numerous varieties of *E. carnea*, the Winter-flowering Heath, will grow in an alkaline soil provided a good quantity of peat is added. The flowering period is from October until April and they look best planted in drifts. The more of the same variety that are planted together the better the effect. Plant them 1½ft. apart. Trim off dead flower heads after flowering. Some good varieties of *E. carnea* are Eileen Porter, rich carmine, 6 to 9in.; King George, rose-pink, 6 to 9in.; Springwood Pink, bright pink, 6 to 9in.; and Springwood, pure white, 6 to 9in.—the last two are the very best carpeters.

The Grey Heath, *E. cinerea*, flowers from June to September. Plant it in a sunny position in a peaty, acid soil, setting the plants about 1½ft. apart.

Trim off the dead blooms after flowering. Some well-known varieties include *E. c. alba minor*, white, compact, 6in.; *E. c. atrorubens*, ruby red, 9in.; Frances, cerise-pink, bronzy foliage, 12in.; and Golden Drop, golden-copper foliage in summer which turns reddish in winter, 4in.

Finally, the Cornish Heath, *E. vagans*, which blooms from August to October and should be given a lime-free soil. Plant about 1½ft. apart. I particularly like the following varieties: *E. v. kevernensis*, rose-pink, 1ft.; Lyonesse, white with conspicuous brown anthers, 1½ft.; Mrs D. F. Maxwell, deep cerise, 1½ft.; and *E. v. rubra*, purple-red, 1½ft.

Euonymus. The evergreen *Euonymus fortunei* has a prostrate habit of growth and makes good ground cover in sun or shade. Its variety Silver Queen has silver-variegated foliage. It will thrive on any soil.

Heath, see Erica
Heather, see Calluna
Hypericum (St John's Wort or Rose of Sharon). The evergreen *Hypericum calycinum* is very useful for covering banks and shady places as it does not mind a dry soil. It forms a dense, spreading mat about 1 to 1½ft. high, and from June to August produces many large bright yellow flowers. Plant 1½ft. apart and keep an eye on it as it may quickly spread from its allotted space.

Ling, see Calluna
Periwinkle, see Vinca
Pleioblastus (Bamboo). A creeping bamboo which makes good ground cover is *Pleioblastus viridi-striatus vagans* (syn. *Arundinaria vagans*). This will thrive in almost any soil.

Rose of Sharon, see Hypericum
Ruscus (Butcher's Broom). *Ruscus aculeatus* is a low, dense, spiny bush which makes good ground cover if planted closely. It is an excellent evergreen for a shady position—especially under trees.

St John's Wort, see Hypericum
Sarcococca. The evergreen *Sarcococca humilis* has small, deep green glossy leaves and small white flowers which appear in late winter. It is excellent for a shady spot and makes dense cover; about 1 to 1½ft. in height. Plant in a fairly good soil about 1½ft. apart.

Vinca (Periwinkle). These are trailing evergreens suitable for both shady and sunny positions—they are often used for covering banks. The flowering period is from April to September. *Vinca major* has large blue flowers and is very vigorous. *V. minor* is smaller in habit and not quite so rampant; it also has blue flowers. The variety *V. minor alba* is white-flowered and *V. m. aureo-variegata* has yellow-blotched leaves.

Pests and Diseases

For details of pests and diseases of trees and shrubs see pp. 291 to 293.

light and air, the better. If the soil is naturally light and quick draining it must have sufficient organic matter added to ensure that during dry spells it will not become parched. Organic matter is equally useful on clay soils to improve their texture and prevent them cracking in hot weather. While, of course, partly rotted organic materials provide the basis of nearly all natural plant food taken up by the roots, they also act as a sponge, holding on to soil moisture which would otherwise be lost. At the same time soil texture is improved enormously by the air spaces left as the material breaks down further into humus and it is from this that clay soils particularly benefit.

Humus. This term, in my opinion, is used far too loosely. Humus itself is the end product of the complex breaking-down of organic materials added to the soil. It takes some time to reach this state and, because of this, I try to give my roses regular dressings of organic matter each year, knowing that this breaking-down process is going on all the time.

Humus can be provided in a variety of ways, the best being as well-rotted farmyard or stable manure. Unfortunately, this is not often readily obtained near a town, and haulage over long distances may make the price prohibitive.

Any decayed vegetable matter may be used with advantage if well worked in. There must be tons of kitchen vegetable trimmings put into dustbins each year which could, and should, be added to the garden compost heap.

Lawn mowings and other garden refuse, stacked for a few months and turned occasionally, will rot down into good manure. Spent hops offer another excellent means of supplying humus, and turf can frequently be obtained quite cheaply from building sites. It should be secured and stacked whenever possible for there is no better foundation for the rose beds than good top-spit soil full of the fibrous roots of grasses and other plants. Peat and leafmould are other very useful humus-forming materials.

Cultivation. Roses, like other shrubs, are usually a permanent feature once planted and I always feel that thorough groundwork before planting is well worth while.

Double digging is the most satisfactory preparation. This means thoroughly breaking up the soil to a depth of about 2ft. The work is commenced by taking out a 2-ft.-wide trench across the plot to a depth of 1ft. and wheeling the soil removed from this to the other end of the patch. Then the gardener gets into the trench and, with a strong digging fork, breaks up the bottom soil as deeply as possible. The work proceeds by opening an adjacent trench 2ft. in width and turning the soil on to the broken soil lying in the previous trench. Now the bottom of this new trench is forked.

Roses 4

Roses, ideally, like a deep, good quality loam, not waterlogged or sour, but well supplied with plant foods and stiff enough to allow the roots to find a congenial cool run. Yet many successful rose growers whom I know have to produce their plants and blooms under quite contrary conditions to these. They have had to make the most of what they possessed and their success should be an encouragement for anyone who feels he has little chance of excelling because his garden is not composed of just the sort of soil which he has been given to understand is essential. From the results I have seen in all parts of the country I believe that, with good cultivation and the proper use of manure, almost any garden in the British Isles may be made to produce quite satisfactory roses.

Roses do not like to be dried out, yet they appreciate enough sun to ensure thorough ripening of the wood. The more open the beds are to

So the work goes on, 2ft. strip by 2ft. strip, until the far end of the bed is reached, when the heap of soil removed from the first trench will be used to fill up the last trench.

As the work of double digging proceeds, well rotted manure or any of the other humus-forming substances I have mentioned should be incorporated with the soil which is dug over at the bottom of each trench. Manure is not wanted at first in the top soil because it may check early root development in newly planted roses, but there is no reason why peat or rotted turf should not be used to improve the texture of the top soil.

Early Preparation. Try to prepare the site as long as possible before the roses will be planted, so that the ground may have a chance of settling and being broken up by the weather. Leave the surface rough until a few days before planting commences and then go over the plot with a fork and break the surface soil down to as fine and crumbly a condition as possible, choosing a fine day for the work.

To complete the pre-planting preparations it is a good plan to dress the soil with bonemeal applied at the rate of about 4oz. per square yard. This is lightly forked into the surface of the soil and will benefit the roses over an extended period.

Planting Procedure

Roses can be planted in the open ground at any time from late October to late March, but from past experience I have found that November is probably the most favourable month. Nowadays many nurserymen grow some roses in old food cans or large treated paper pots and from these they can be transplanted at any time, even in mid-summer, provided moist soil is kept around the roots.

Faults To Avoid In Planting. Roses are not difficult to plant and they will probably survive rough treatment better than most plants, but a little care and understanding will help to give them a good start. There are, in particular, three things to avoid as each of them may check the plants severely. They are: allowing the roots to become dry before or during the time of planting; planting too deeply, and doubling up some of the roots so that their ends are pointing upwards instead of outwards or downwards as they should.

To these I would add insufficient firming of the soil but I do not regard this as quite so serious as the other three, particularly if roses are planted in autumn, because the amount of rain we usually get then soon consolidates the soil even if it has not been well trodden down in the first place. All the same it is safer not to trust to the weather but to firm properly directly the soil has been returned around the rose roots.

Arrival From The Nursery. Roses, if properly

packed, should arrive from the nursery with their roots reasonably moist but, if they do appear to be dry, do not hesitate to stand them in a bucket of water for a few minutes to get a thorough soaking. Then either dig a hole big enough to accommodate all the roots in the bundle of roses and cover them immediately with soil, or wrap damp sacking or straw around them. It may take several hours to plant a large number properly and in that time roots can be seriously damaged by drying out if left exposed to the sun or a drying wind.

Planting. The planting holes should be of ample size—a bit too big rather than a little too small. A hole 10 to 12in. deep and about 15in. in diameter is usually about right for a well-grown bush rose.

Look carefully at each bush before it is planted and, if some of the root ends are broken or damaged, cut them back a little with sharp secateurs. Then hold the bush in the middle of the hole with one hand and with the other spread the roots out as evenly as possible. None should curl upwards. If they do it may be necessary to make the hole a little larger.

The bush should be held so that the point where the branches join the main root is just below the natural soil level—say ½in. but no more. Then return the soil a little at a time, working it round and between the roots. Whenever possible I try to have an assistant—one pair of hands to hold the bush and work in the soil, another pair to wield a spade to throw the soil. But it can be done alone with a trowel instead of a spade. When nearly all the soil has been returned tread it down firmly. Firming with the hands is not sufficient for roses. Finally the remaining soil can be scattered over the surface to obliterate foot marks and leave a clean, level finish.

Feeding and Pruning. I know that many gardeners are tempted to add manure to the soil as it is returned around the roots, thinking that they will give the plants an added boost in their first season, but it should not be used at this stage. Its proper use is in the preparation of the rose bed a month or so before planting and as a mulch the following spring. The roses will benefit, however, from the addition of really coarse bonemeal well mixed with the planting soil—a handful of bonemeal for each rose bush unless, of course, this has already been applied when the rose beds were being prepared. If roses are planted in the autumn they are left unpruned, or at most are only cut back a little, but in late winter or early spring all my newly planted roses are pruned really hard.

Spacing. The correct distance apart for the plants will depend on the vigour of the variety. This information is usually given in the catalogue descriptions of each variety from leading growers and I always check before planting. Most floribunda roses make a good deal of growth and should be spaced 2ft. apart or even 3ft. for the very vigorous, newer kinds such as Queen Elizabeth. There are some less vigorous or upright growing hybrid tea varieties that can be spaced as closely as 18in. All these distances are for bush plants. Standards and half-standards, i.e., roses growing on main stems which may be anything from 3ft. to 6ft. in height, should be given a good deal more room, 4ft. being about the minimum.

Ties and Labels. Standard and half-standard roses must be provided with good strong stakes. These should be of at least 1-in.-square wood sufficiently long to be driven 18in. into the soil and to come at least to the top of the main stem of the rose. Patent adjustable ties are available or strong tarred twine can be used but in an emergency I find old nylon stockings excellent—strong, long lasting and not liable to chafe the bark of the rose. But all ties, whatever they are made of, should be examined from time to time to make certain they are not cutting into the stems. All roses, or groups of the same variety of rose, should be labelled, and here individual taste must be allowed to determine what is most acceptable. There are plenty of excellent alternatives from which to choose. Here again, however, it is necessary to warn that labels tied to branches may in time cut into them.

Pruning

Pruning is one of the tasks which the rose grower must master thoroughly. There is little point in paying good money for roses and taking care and time to plant them properly if they will be ruined later on by incorrect pruning. Nor can all types of roses be pruned in the same manner. The treatment most suitable for wichuraiana ramblers of the Dorothy Perkins type, which bear the best blooms on young growth, would be ruinous for many climbers, which flower well on older wood. Similarly, the hard pruning used when roses are grown for exhibition would spoil the effect of roses used for massed effect in the garden.

Reasons for Pruning. The principal reasons for pruning at all are to concentrate growth upon a limited number of selected shoots, each of which is capable of producing blooms of first-rate quality, and to get rid of old, worn out growth. Some roses, if left to themselves, would make numerous thin, spindly shoots too weak to bear any but small flowers, but this is less true of modern varieties than of some of the older ones. The growth of all roses tends to loose vigour after two, three or four years and needs to be replaced with young growth. This is the natural habit of roses and not a weakness of cultivated varieties.

In a general way, the more severely a branch is cut back the more vigorous will be the new shoots starting from it. The reason for this is that the sap forced up from the roots, which are not pruned, is concentrated upon a smaller number of shoots. Consequently, each gets more nourishment. Other reasons for pruning are that branches must be evenly spaced so that light and air can reach the leaves freely and that all dead, diseased or damaged shoots must be removed.

Cutting Back After Planting

The first cutting back, in the spring after planting, of almost all the many classes of roses is the simplest of all pruning operations. The aim is to build up a good foundation of stems, so the plants should be cut back hard while young.

Bush Roses. To achieve this I cut each sturdy growth of a bush rose to within about three buds of its base. These buds, or 'eyes' as they are sometimes called, are the points from which new

The hybrid tea rose Piccadilly, a bicolor of outstanding value

Above. The floribunda rose
Lilli Marlene—a first-rate variety
by any standard
Right. The shrub rose
Frühlingsmorgen, a variety of
exceptional beauty

Above. Roses make a splendid informal screen if grown, as in this case, on tripods. Here, *Clematis jackmannii superba* clambers over a specimen of Climbing Michèle Meilland which bears delicate pink flowers in early summer, before the clematis

Left. Albertine, one of the best and most widely grown of all rambler roses, needs plenty of space to accommodate its growth, which is very vigorous. The flowers are sweetly scented

Roses can be mixed effectively with other garden plants, as this very decorative border indicates. The objective should always be to contrast bold pockets and sweeps of colour one with another so that the eye is led from one object of interest to the next

shoots will presently sprout, and they may easily be detected, situated at more or less even distances along every growth and usually marked by the scar of an old leaf stalk.

The buds grow in all directions on the stems and, when pruning, it is always best to make each cut immediately above a bud that is pointing away from the centre of the plant, the object being to encourage the formation of an open-centred, roughly goblet-shaped bush. The shoot growing from any bud will tend to follow the direction in which that bud was originally pointing, and it is the bud to which a branch is pruned that will usually produce the new replacement growth, this time growing in the direction chosen by the pruner.

Standards. Standard roses are treated just as though they were bushes perched on the top of a tall stick. Take no account of the standard stem, which does not require any treatment beyond the complete removal of any sucker growths which may come from it. The head of branches is the portion that must be cut back, and each shoot should be pruned hard in exactly the same manner as I have described for bushes.

Ramblers and Climbers. Ramblers and vigorous climbers are even simpler to deal with and it is not always necessary to make all cuts to outward pointing buds, the shoots in any case having to be trained and tied to supports. Still, it will help matters if the pruner does follow the general policy already outlined and make as many cuts as possible to buds that point in the direction he wishes the resultant shoots to take. Roughly speaking, each strong growth on freshly planted ramblers and vigorous climbers should be cut to within 1ft. of the ground, while weaker shoots may be shortened to 6 or 9in.

Climbing Sports. These are the only kind of climbers which are exceptions to this simple rule of general hard pruning in the first season after planting. They are very vigorous forms, termed sports, of bush roses. Their strong-growing habit and large, well-formed flowers render them suitable for covering the walls of houses or clothing pillars and arches. But on account of their peculiar origin they have a tendency to revert to the former bush habit and this may be intensified if they are severely pruned. For this reason they must be pruned with caution, and I usually do no more than remove the ends of the shoots which have been damaged in the process of transplanting or by frost, and shorten any weak growths. The climbing sports always carry the prefix 'Climbing' before their names in catalogues—e.g., Climbing Étoile de Hollande, Climbing Madame Butterfly —and so are easily picked out.

In districts which have a normal climate, with no extreme weather conditions at the end of March, all the pruning I have described for newly planted roses may be done during March. In exceptionally exposed gardens, the first pruning should be delayed until early April.

Pruning in Later Years

Treatment of old roses is not quite so simple. Different methods must be employed for the various types and classes, while the severity with which bushes and standards are cut back will depend on whether blooms are required for exhibition or for garden display.

Hybrid Teas. Moderate pruning is needed for most hybrid tea roses grown for display. This means that each strong main growth will be shortened to within four, five or six buds of its base, and each sturdy side growth to one or two buds. Very old, weak or diseased growths must be cut right out. I begin by taking out these and also such old growth as can be removed without sacrificing good stems growing from it. By tackling the job this way I find it simplifies the task of pruning the remaining younger growth.

Some very vigorous roses, and particularly several hybrid perpetuals, well-known examples of which are Frau Karl Druschki and Hugh Dickson, must be treated even more lightly. After removing old, worn-out and diseased wood, the strongest young growths should be left about 3ft. in length, more weakly shoots being cut back a little further and laterals shortened to a few inches.

Hard pruning of bush roses is almost entirely confined to hybrid tea varieties grown for exhibition. Here the aim is to produce a few blooms of the finest quality, and so the whole energy of each plant is concentrated on a very limited number of shoots. Each strong growth is pruned to within two or three buds of its base and practically everything else is cut out entirely.

Where bush roses are going to be cut back hard in March I have found no harm in cutting them back by about half in November. Admittedly, this means going over the bushes twice but I believe it prevents a lot of wind damage and root disturbance, particularly in exposed areas.

Floribundas. These roses like fairly light pruning. After cutting out all old, weak, diseased or dead wood the best of the younger stems are shortened by about half, a few being cut back rather more severely, especially if the plant is not growing very strongly. The object is to make a fairly big, well-balanced bush with plenty of strong young growth to bear an abundance of flowers.

Pegging Down. A difficulty with some very vigorous bush roses is that they soon attain such a height that they look ungainly. Pegging down offers a simple and effective means of overcoming this.

When the roses are pruned only four or five of the strongest shoots are retained, all other

Top. Standard rose after pruning. The same technique is employed as with ordinary bush roses. This floribunda variety has been lightly pruned, and all weak and crossing growths have been removed
Bottom. A weeping standard rose after pruning. Note the even spacing of the growths to provide a well-balanced head. All weak growth has been removed
Far right. A correctly trained and pruned climbing rose of Wichuraiana type. These do not produce new growths too freely and pruning is limited to thinning out the main stems and removing the oldest shoots

growth being cut out. Then these long shoots are shortened slightly, so as to get rid of the un-ripened tips, and are bent down and fastened to pegs driven into the soil. In this way a large area of ground is covered. Side shoots grow freely from the 'eyes' along the length of each shoot, and bear flowers profusely. Only the number of shoots required to form the next season's plant are allowed to grow from the base of the plant. Then the following spring all growths that have flowered are cut out, and the young stems retained for the purpose are pegged down in their place.

Many climbing roses may also be treated in this way, and, so grown, make a charming cover-ing for steep banks. Lady Waterlow and Zéphirine Drouhin are two that respond well to this treat-ment. The second mentioned is a very good friend, being one of the few thornless varieties and so easy to handle for this method of training. Rather more shoots may be retained according to the strength of the plant. Some moss roses are also suitable for the pegging-down method.

Standards. Standard roses are treated on similar lines to bushes. The main stems must be kept clear of all growth, but the branches forming the head are pruned each year like those of bush roses of the same kind.

Ramblers. Rambler roses of the wichuraiana and multiflora types should be pruned in autumn, or even in late summer after the flowers have faded. These vigorous roses bear their best flowers on strong growths made the previous year. My aim in pruning these is to get rid of as much as possible of the old growth, including that which has just flowered, but to retain all the strong young stems that have not yet flowered. With some varieties, such as Dorothy Perkins, Excelsa and American Pillar, it may actually be possible to do just that every year, for when growing well they make a lot of new growth right from the base. Other varieties, such as Alberic Barbier and Albertine, almost always make some good new growth from the old stem, sometimes many feet above soil level, so it would be im-possible to cut out all old growth without at the same time sacrificing most of the new stems. With these one must compromise and keep as much of the old wood as is necessary to bear plenty of young growth too.

The reason for pruning ramblers early is that the wood retained is exposed in this way to all the autumn sun and air, which ripens it well to with-stand the winter.

Weeping Standards. Weeping standard roses have a tall main stem which may be as much as 7ft. high and on this a very vigorous rose, usually of the rambler type, is grown so that it forms a shower of growth reaching nearly to the ground. Rambler roses grown in this way must be pruned in a similar manner to other ramblers but even

more care must be taken in order to space the young shoots, after pruning, at regular distances on the umbrella-like wire trainer which is commonly used to shape these roses. There is no object in letting them hang too far down at first, as in summer the heavy flower trusses will hang down even further and look unsightly if stained with mud splashings.

Climbing Sports. I have already referred to the peculiarities of the class of roses known as climbing sports. Even when established, such types require a somewhat specialised form of pruning. Some of these roses are inclined to make several main stems from which strong shoots grow, but not particularly freely towards the base. The best of this new growth should be retained at practically full length but occasionally I either bring a main stem to the horizontal or cut one back quite severely, to encourage more strong growth from low down and so prevent the bottom of the plant becoming bare.

Other Climbers. In addition to the climbing sports there are various other climbing roses which thrive on moderate thinning. The aim should be to remove some of the older stems but to retain plenty of young growth, either at full length or shortened a little but not by more than a third.

Shrub Roses. The hybrid musk roses and many of the vigorous hybrids now lumped together as 'shrub roses' require little pruning. It is only necessary to thin out in March a little of the older wood to make room for new growth. Every third or fourth year I find it pays to give a more drastic thinning out, particularly if the bushes tend to fill up with a good deal of thin growth. Much the same applies to the wild roses, or species, which require little or no pruning but may be thinned out if growth is too crowded or if they are exceeding the space available for them.

Cultural Routine

Thinning New Shoots. The great majority of modern roses are naturally free growing. Practically every eye left on the plants after pruning will 'break', i.e. start into growth, and for that reason I find it often necessary to supplement pruning with judicious thinning of the new shoots towards the end of April. Surplus shoots, and particularly those growing inwards and so threatening to crowd the centre of the plant, may be pinched out between finger and thumb. Sometimes the bud to which the shoot has been pruned goes blind and refuses to grow. When this occurs the shoot should be pruned again to the first live shoot or bud, otherwise the stem above this will die. Such 'snags' of dead wood often provide an entry point for diseases such as canker and dieback.

An all-purpose fertiliser being spread around roses in late March, after pruning. A mulch is applied afterwards to conserve moisture, supress weeds and provide food

Hoeing, Mulching and Feeding. Hoeing should be done fairly frequently when the soil surface is reasonably dry as it keeps down weeds and aerates the soil. The regular dressings of organic material to which I referred on p. 74 are best applied as a mulch spread over the surface of the rose beds after pruning. This reduces the loss of moisture from the soil, and besides rotting down to provide food for the roses, it acts as an excellent weed barrier, enough to last until the plants are in full leaf and creating heavy shadow to suppress weed growth among them. Horse manure containing a good deal of litter is the best material, but horticultural peat or spent hops will serve well, especially if supplemented by a scattering of a good all-purpose fertiliser or, better still, a specially blended rose fertiliser.

Watering. Whether roses are watered in dry weather must be decided largely by the available water supply. Established plants are capable of surviving a fairly prolonged drought but growth and flowering will be improved if the soil can be kept reasonably moist right through. If water is applied sufficient must be given to soak right down into the soil, and once started it should be continued at regular intervals until sufficient rain falls to replenish the soil. It is a mistake to water roses for a while and then stop while the weather remains hot and dry because roots will have tended to grow upwards towards the water and they will suffer severely from drought. Without any watering the roots would have tended to grow downwards out of harm's way.

It is surprising how spraying the bushes with clean water in the cool of the evening will counteract the influence of parching sunshine.

Disbudding. If large individual blooms are required disbudding must be practised. This is usually confined to hybrid tea varieties, most of which produce buds in clusters but not all in a cluster opening at the same time as in floribunda roses. To obtain quality flowers each cluster must be reduced to a single flower. For preference the central bud, which will also be the most advanced, should be retained, and the others removed in their very early stage.

Spraying. This should be a matter of routine and it is dealt with in detail in the chapter on pests and diseases and their control (see pp. 293 and 294).

Faded flowers. These should be removed without delay as, if left, the later cropping may be reduced. At first the flowers may simply be snipped or broken off, but later, as the first flush of blossom declines and all the flowers on a stem have faded, the whole stem should be cut back a full 8 or 9in. to encourage the growth of strong new shoots to provide a second crop of bloom. I always like to have some roses in the house and find that, by regularly cutting blooms with good long stems, this summer pruning is automatically achieved on the bushes used for this purpose and they are all the better for it.

Suckers. I keep a close lookout throughout the season for suckers growing from the roots below the union of the rose and the rootstock. Despite the fact that there are several distinctive features by which suckers may be recognised, I often see them allowed to reach considerable size before being removed. Briar suckers have small pale green leaves, and those of the rugosa stock can be identified by their rough crinkled leaves and dense stem covering of prickle-like spines. Suckers of *Rosa laxa* and *R. polyantha*, two other popular rose stocks, are both fairly distinct from the growth of garden roses, but whenever I am in doubt I trace the stem back to its point of origin. If this is below the point of budding, recognisable on bushes, climbers and ramblers as a slight swelling or irregularity in the main stem just above the roots and usually below soil level, I then know that the shoot is a sucker. Standard roses are worked high up on the main stem of the stock and so all shoots below the head of branches are suckers and must be removed as soon as these are noticed.

Autumn Care. The practice which I mentioned of cutting back by half, in the autumn, hybrid teas to be hard pruned in March, is equally sound for any other taller growing or large bush variety. Weakly growths can be cut right back and very long stems can also be shortened by a third or even a half to reduce wind resistance and so prevent the roses being pulled about by winter gales.

All fallen leaves, petals and prunings must be carefully gathered and burnt. Left lying on the ground they are a menace, as the winter spores of several diseases may be carried on them and re-infect the trees the following spring.

Feeding

Roses are strong-growing plants, making a heavy demand on the food reserves of the soil. A regular programme of manuring must be followed, and it must take the form of a properly balanced diet.

Bulky Organic Manures. Natural manure, such as dung from farmyard or stable, should not be used in a fresh state, but must be stored until the straw is well rotted and the manure has lost rankness. As a rule, 1 large barrow load of manure to 12 square yards of bed is an ample dressing and the best time to apply this is immediately after pruning. It is best spread as a mulch over the surface of the beds.

Properly made garden compost, peat and leaf-mould are useful substitutes if dung is not available. Although their nutrient properties are nothing like so great, they are of equal value in increasing the humus content of the soil and their deficiency in actual food can be made good by means of fertilisers.

Slow-acting Fertilisers. Only in a few cases, where the more fortunate gardeners are on highly fertile soils, will natural manures be available in sufficient quantity to provide enrichment great enough to promote and maintain sturdy growth. The great majority of soils will require, in addition, a dressing of a fertiliser of a long lasting nature, such as bonemeal, or better still, meat and bonemeal which provides nitrogen as well as phosphates, two of the most essential rose foods. This fertiliser will surrender its feeding properties little by little over a period of two years or more, and may be used at rates of 4 to 6oz. per square yard according to the estimated need of the soil, and can be applied early, in February or even, if it is more convenient, in September.

Inorganic Fertilisers. Application of quick-acting fertilisers prepared with superphosphate of lime, sulphate of ammonia, sulphate of potash and similar inorganic chemicals should be left until after pruning and further small topdressings can be given at intervals of three or four weeks during the spring and summer. While rose enthusiasts have many varied home-made formulas which each claims is best, much time is taken in their preparation and mixing. It is really better to use a proprietary fertiliser, either a general fertiliser or one specially blended for roses. In either case manufacturer's instructions regarding the rate of application should be followed.

Propagation

Roses can be increased by various means such as budding, grafting, cuttings, layering and from seeds, but budding is the method most commonly employed for the majority of garden varieties. These methods of increase are described, stage by stage, on the pages which follow.

The best mulch of all for roses is horse manure containing plenty of litter. However, horticultural peat or spent hops serve just as well

Budding

Budding is really a form of grafting and enables the grower to unite a garden rose with a root system or 'stock' obtained from a wild or vigorous rose. The strength of the stock is of service to the more weakly and highly bred garden hybrids. Not only is this so, but on a commercial scale, the rose grower can make many more plants from the buds on the 'bud sticks' which would otherwise be single cuttings.

Stock for Budding. The common briar, which is frequently used as a stock for budding, is native to Britain. It matters little whether the briar is raised from a cutting or from a seed. The latter method has the advantage of cheapness, but when a few stocks are required by the amateur it is often most convenient to raise them by cuttings which will root quite readily during the autumn outdoors. Briars for standard stems are usually gathered from hedgerows, suitable strong young canes being chopped out with as much root as possible in the autumn, trimmed with a pair of secateurs, and planted in the garden for budding the following summer.

For standards, *R. rugosa* is often employed. The stock is easy to produce from cuttings and equally easy to work, but is very liable to produce suckers.

The *simplex* stock is a variety of *R. multiflora* and is much in evidence in nurseries where rose trees are grown on light, sandy soils.

The *laxa* is a rose stock of northern origin. Cultivators in Scotland like it and rear fine trees budded to it.

Time of Budding. Budding can begin as soon as the bark of both the stock and the garden variety or 'scion' separates readily from the wood. This is usually towards the end of June. From then the work may be continued through July and August, and sometimes into September.

Stocks planted during the autumn and winter will be fit for budding the following summer. The only tools and materials necessary are a sharp budding knife, a trowel and a supply of soft raffia.

Suitable Buds. Suitable buds can only be obtained from half-ripened rose shoots of the current year's growth. A simple test which I make to decide whether any shoot is in the right condition is to attempt to break off the thorns by pressing them sideways with the thumb. If they refuse to snap off cleanly but tear away, the growth is immature, but if when removed they leave a dry, hard looking scar, the shoot is over-ripe. The thorns on ideal 'bud sticks' will break off readily, exposing green, juicy-looking tissue. Stems that have just produced flowers are usually ideal.

Too many growths must not be removed at a time, as they quickly dry out if left exposed to the air. A good way of keeping them fresh is to place them upright in a jam jar containing $\frac{1}{2}$in. of water. This can be carried round as the work of budding proceeds.

Preparation for Insertion. To prepare a bud for insertion the leaves are first of all removed from one of the selected stems, but the leaf stalks are left. I have always left the lower two leaflets on the stalk, believing that these are easier to handle but I have not come across many others who do it. The knife is then inserted about $\frac{1}{2}$in. below one of these leaf stalks (a dormant growth bud or 'eye' is situated in the axil of each of these, where it joins the stem) and is drawn upwards so as to come out again, on the same side of the stem, about $\frac{1}{2}$in. above the bud.

The budder will then have a small, shield-shaped portion of green rind with a dormant bud and a leaf stalk by which to hold it. If he has made the cut properly he will find, at the back of the shield, a narrow sliver of wood. Grasping the leaf stalk firmly between the thumb and first finger of the left hand, the lower end of the sliver of wood is lifted with the point of the knife, and is pulled away from the rind and discarded.

Insertion. The prepared 'bud' (it is really a strip of rind containing a bud) is now ready for insertion under the bark of the stock. To enable this to be done, a T-shaped incision is made in the bark of the stock. The flaps of bark on either side of the downstroke of the T are then gently prised up with the thin bone scalpel which forms the handle of the budding knife. It is then a simple matter to slip the shield bearing the dormant bud downward into this incision, so that its inner surface lies snugly against the exposed tissue of the stock. If there is a small tail, formed by the upper end of the shield still protruding from the top of the incision, the blade can be pressed into the already-made horizontal cut at the top of the T to trim off the tail flush with it.

The bud is bound in position with raffia which should cover the incision from top to bottom. When there are a number of stocks to be budded and tied, it is as well to prepare the raffia beforehand by cutting it into lengths and then soaking it in water. It is less springy and easier to use when wet; the skein can be slipped under the budder's belt for easy access as each piece is wanted.

The top of the stock is not removed immediately as it is required to encourage a flow of sap past the bud.

Position of Buds. The preparation and insertion of the bud is exactly the same for all forms of roses, but the position in which it is placed varies. Bushes and climbing roses are budded as low down as possible on the main stem of young stocks. It is an advantage if the bud can be inserted below the ground level, and, with this end in view, the soil is scraped away from around the stock with a trowel, just before it is budded.

A T-shaped incision is made in the bark of a briar in preparation for budding. Briar stocks can be raised from seed or cuttings

Standards on the Dog Rose stock (*R. canina*) are budded on side shoots of the current year's growth selected at the right height to form a head. One of the advantages of *R. rugosa* as a stock for standards is that buds can be inserted directly in the main stem at any convenient height. It is usual to insert three on each stem. whether of *R. canina* or *R. rugosa*, so as to form a head of branches as quickly as possible.

Amateur gardeners after budding, particularly for the first time, are always anxious to know whether the buds have taken. I wait for three or four weeks until I inspect mine, when I look at the inserted shields of rind. If the buds on them are still fresh and plump they have probably taken. Should they have a dry, shrivelled appearance, they have failed, and the stocks can be worked again. This is most easily done by inserting another bud on the opposite side of the stem, or if there is room I sometimes make a new incision a little lower down.

At the end of the following March the stocks should be cut back to within ½in. of the top of the T-shaped incisions made for the reception of the buds.

Grafting

Grafting is chiefly employed by trade growers for the rapid increase of new varieties. The advantage in this case is that the work can be carried out in a warm greenhouse in January or February. Then buds obtained from the resultant plants can be used for budding in July; in other words, two generations are obtained in one season. However, grafting is also used to increase miniature roses which make such thin stems that it would be difficult to cut buds from them.

Stocks. There are several methods of grafting, but the simplest is splicing, which consists of

but still dormant. A thin slip of bark is removed from the lower part of the scion to expose the wood and form a tongue. Each scion need have no more than two joints, one on the portion shaved to form the tongue to attach to the stock, and one above to break into growth. Some good propagators do not bother to have the eyes on the tongue at all, but novices will be more sure of success if the scions used have the two.

First head back the stock, leaving just sufficient stem above the soil to take the tongue of the scion, very little more than an inch is required for this. A shallow strip is shaved from the side of the briar stock. This must exactly correspond in length and breadth with the tongue cut on the scion, so that the two will fit perfectly together with no gaps, cavities or overlapping edges. Union of the two is only possible when the two barks fit exactly. Bind the scion firmly and finish the job by sealing with grafting wax.

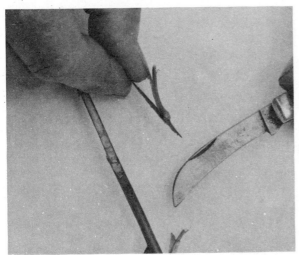

Removing a bud shield for insertion in the stock. The sliver of wood at the back of the shield is carefully removed before insertion

Inserting the bud in the T-shaped incision made in the stock. The bud is bound firmly in place with raffia, which should completely cover the incision

matching the tapered edge of the scion with precision to the cut edges of the bark of the stock. Even the best of professional knifesmen always practise their cuts before beginning the real job. Amateurs would be well advised to follow this habit by first practising on pieces of rose stems taken from unpruned bushes outside. Manettii or common briar stocks should be obtained; there are nurserymen who raise stocks for sale to amateurs, the former being easiest to work. Procure these in autumn and pot singly into reasonably small pots as soon as they arrive. Plunge them out of doors for a couple of months or so before bringing into the greenhouse. These stocks will be ready for grafting just as new growth commences.

Scions. Scions of the variety to be grafted are made from shoots of firm, but not too stout, wood in which the eyes or buds are nicely plump

Alternatively the scion can be cut with a taper and a vertical incision made in the bark at the top of the stock, which is first beheaded. The bark is then prized open, as in budding, and the tapered part of the scion is slipped down under the bark so that exposed tissue lies closely against tissue. Then all is bound with raffia.

The pots are plunged in sand or peat in a close frame in a warm greenhouse. The soil in the pots must not be allowed to get dry, and the atmosphere needs to be kept humid by use of the syringe when needed. As the eyes break into growth ventilate the frames increasingly to acclimatise the plants to cooler conditions.

Cuttings

Some varieties of rose also grow well from cuttings prepared from ripened growth in September and October and rooted outdoors in a sheltered border. This is a method specially suitable for the increase of rambler roses, shrub roses and the various species. It can sometimes be used successfully with other types, including some of the vigorous hybrid tea and floribunda roses.

Well-ripened firm growths, 9 to 12in. in length, should be selected. These can either be pulled off with a 'heel' or small slip of older wood which is then trimmed neatly with a sharp knife, or else each shoot must be cut cleanly through immediately below a joint. The lower leaves should be trimmed off and the cuttings lined out in trenches 4in. deep with a scattering of sharp sand along the bottom of each. The soil is then returned around the cuttings and made thoroughly firm with the foot. The rooted cuttings should be ready for transplanting to the flower garden by the following autumn.

Layering

This is a very useful method of propagating rambler roses, as it enables the work of increase to go on without the risk of removing shoots from the parent plant before they have any roots of their own. It can be done at the same time as cuttings are taken.

Young shoots of the current year's growth are most suitable. Those selected should be so placed that they can readily be bent down to and secured at ground level.

Select a joint on the shoot which can be buried, and insert the blade of a sharp knife just below it. The knife is then drawn upwards through the joint without actually severing the stem. Next the cut is opened, but again without breaking the stem, and is bent downwards so that it can be buried in the soil. The use of the proper strength rooting powder on the cut surface before it is buried will often make just the difference between success or failure with a shy-rooting variety and this is something which is well worth bearing in mind. The same applies to cuttings.

The 'layer', as this sliced portion of stem is termed, is held in position with a forked stick or a length of galvanised wire bent to the shape of a hairpin. Having already gone to this much trouble I like to tie the stem to a stake, otherwise it may blow about and prevent the layer from forming roots.

What the gardener has virtually done is to make and insert a cutting without entirely detaching it from the parent. The shoot continues to draw nourishment from the old plant until it makes roots of its own. It is then detached and, when a suitable time arrives, is lifted and replanted on its own. Rooting usually takes place fairly quickly, but it is often wise to leave the layers attached to the parent plant for 12 months before severing the connection and planting them on their own.

Seeds

Roses can be readily raised from seeds, but this is a method of propagation which should only be used for stocks and species, or for definite crosses between garden roses which have been made with the intention of producing new varieties. It is not worth while raising roses haphazard from seed as the results are likely to be most disappointing.

Cleaning of Seed. In all cases the procedure is the same. The fruits, or heps, are gathered in the autumn when ripe and are placed in layers between sand or peat litter in boxes. These should be stood out of doors in a position exposed to frost, as this will encourage the rapid decay of the fleshy part of the heps. To facilitate this decay the sand or peat should be kept moist. Do take precautions against mice by encasing the boxes in fine wire netting. It is surprising what trouble they will go to in order to reach such a ready-made winter store of seeds! By the following

A rambler rose being layered. A shoot is pegged down into sand after being slit on its lower side. Treating the cut surface with hormone rooting powder is helpful

March it should be possible to clean the pulp from the seeds. It is a help to place the whole mass in a fine flour sieve and hold this under a tap, so forcing the pulp through the sieve but leaving the seeds in it. Another simple method I find useful is to squeeze the pulp through muslin by twisting it tightly.

Sowing. The seeds should be sown in shallow drills in an open but sheltered position out of doors. The drills must be 15in. apart and the seeds sown evenly and covered with $\frac{1}{2}$in. of fine soil. Germination is sometimes irregular and a seed bed should not be destroyed for at least a year. The seedlings may be transplanted to nursery beds as they attain a size which makes handling easy. They will need some nursing in the way of shading, sheltering from wind, and watering until they gain strength and become well hardened.

Seed of crosses between garden roses specially made with the intention of producing new varieties is usually germinated in a frame or greenhouse, in beds of sandy soil if a large number of seeds is being sown or in pots or pans if there are only a few seeds.

Exhibiting

The general cultural routine already described will apply to roses grown for exhibition, but it is usually necessary to adopt a more drastic system of shoot thinning and disbudding than is required for garden display. The very finest blooms are often obtained from one-year-old trees and so really keen exhibitors sometimes bud their own stocks annually so as to have a constant supply of young rose bushes.

Every known means of encouraging the development of bigger and yet more perfect blooms

Six exhibition blooms arranged in a regulation show box. The stems rest in metal tubes containing water. Other such boxes hold 9, 12, 18 and 24 blooms respectively

must be adopted, but bear in mind that no amount of attention immediately prior to a show can make up for a bad start earlier in the year. Pruning and thinning must be intelligently carried out from the outset. Disbudding will demand constant attention from May to September, each cluster of buds being eventually thinned to the centre bud only, but for a start it is well to leave one side bud in case the central one should be damaged.

Protection and Selection. Flower shades will prove useful, for they serve a dual purpose, to keep off excessive sun and to provide shelter from heavy rain. Many of the brighter reds and delicate pink shades only show their most pleasing tints when partially shaded.

Look over the plants carefully in the early morning and evening prior to the day of the show and decide whether each bloom will be just right or too forward by next day. If a particularly fine

young bloom seems likely to get too forward tie up its centre with a smooth piece of white wool, not in a knot, but just a double turn pulled so that it grips the centre of the flower, yet can easily be slackened and removed at the proper moment.

Should the show be within easy distance, the blooms are best cut in the morning, very early, before the sun is on them. Have some receptacle filled with water, so that the blooms, as cut, can go straight into it.

Varieties. It is as well to label each flower before it is taken from the plant, otherwise varieties are apt to get mixed or misnamed. It is also a point of strength to have as many spare varieties as possible in reserve. Exhibition blooms in boxes should be presented to the judges in 'the most perfect phase of their possible beauty'—that is, briefly, the blooms should possess form, freshness of colour and size. Always try to aim at quality rather than over-size. A very large flower, if it has a split or muddled centre, or is dull in colour, will not carry as many points as a medium-sized bloom, well finished and bright in colour.

Even with boxes of specimen blooms some pains should be taken to contrast the colours of the different varieties as pleasingly as possible. Arrange the flowers in alternate dark and light colours with the larger and heavier blooms at the back, but keep all as even in size as possible.

As a rule, regulation boxes made in sizes to display 6, 9, 12, 18 or 24 blooms must be used. These can be purchased from any seedsman or rose specialist. The boxes have holes to take metal tubes which are filled with water and in which the flower stems are placed, one bloom to a tube.

Dressing. Most blooms will require a little dressing if they are to be shown in really perfect condition. It is a practice which is sometimes condemned, but not to do so is surely a mistake. Faking in any form is certainly malpractice, but merely blowing open a slow-opening flower or pressing the petals lightly back with a camel-hair brush is only assisting a natural process.

Different Classes. Besides the exhibition blooms shown in boxes there are also classes for vases, baskets and bowls of roses. The chief thing I look for in these classes, in addition to quality and freshness, is lightness of arrangement. Avoid crowding in too many flowers or sprays of flowers as the case may be. Sprays should be trimmed with a pair of sharp-pointed scissors, cutting out all blooms that have gone off colour or are overblown. This will lighten them considerably and give them a clean, bright appearance.

Classes of Roses

The rose has had a very long history as a garden plant having been cultivated since the earliest days of civilisation. The Persians knew and grew it and so did the Romans. It has been a favourite flower in Britain for more than three hundred years. Today it is the most popular of all garden plants in this country and supports a society, The Royal National Rose Society, which is by far the largest specialist horticultural society in the world. It has become commercially a very important plant not only in Britain but in France, Germany, Holland, Denmark, America and many other countries. Rose breeding has become quite big business, attracting some of the best brains in horticulture and protected in many countries by laws which give the raiser certain financial rights in his country for a number of years.

It is not surprising, therefore, that the rose has been remarkably developed and has produced all manner of shapes, colours and even scents not associated with it as a wild plant. Species has been crossed with species, hybrids themselves have been bred one with another, until a vast number

Thus the old Polyantha (or Polyantha Pompon) rose has virtually disappeared, its place taken by the Floribunda rose which was started by crossing hybrid tea roses with polyantha roses. At first the floribunda was quite different. The flowers were larger than those of the polyantha roses but smaller and far less shapely than those of the hybrid teas. They were produced in big clusters with all the flowers in a cluster open at practically the same time so that the overall effect was greater than that of the hybrid tea roses though individually the flowers were far less interesting.

But the plant breeders were not satisfied with that and they went on trying to 'improve' the floribunda rose which really means that they tried to make it more and more like the hybrid tea rose with the one exception of opening more flowers at the same time. In this they have now succeeded so well that it is really difficult to decide into which class some new roses should go. In America a

All the perfection of the rose in the mind's eye is embodied in a shapely hybrid tea bloom. The variety shown here is Wendy Cussons

Floribunda varieties, with their excellent full-flowering qualities, are especially popular with gardeners. Shown above is the variety Shepherd's Delight

of separate inheritance units, what the scientific breeder calls genes, has been combined to produce the rose in all its complex variety as we know it today.

Hybrid Teas and Floribundas. Years ago it was not too difficult to separate garden roses into certain well defined groups or classes. Thus rose growers knew the Hybrid Tea as a rose with large shapely blooms produced more or less continuously from June to September. It was a totally different rose from the Polyantha which produced much smaller flowers in much larger clusters, or the Ramblers which threw out great stems 10 or 15ft. in length and flowered only once each year in July or August.

For convenience we still try to keep these or similar classifications but between some of them it becomes more difficult each year to draw any really rigid and satisfactory dividing lines.

new name, Grandiflora, has been used to describe some of the large-flowered floribunda varieties but this has not been accepted by The Royal National Rose Society which, if it distinguishes at all between small- and large-flowered floribunda varieties, calls the latter Hybrid Tea-type Floribundas.

Shrub Roses. We used to know (or at least think we knew) what a Hybrid Musk rose was. This was a name given to a rose first produced by an Essex clergyman, the Rev. T. H. Pemberton of Havering-atte-Bower, who claimed to have crossed the Musk Rose (*Rosa moschata*) with some garden hybrids. The varieties he produced, among which the most famous were (and still are) Penelope, Cornelia and Felicia, made quite big bushes and had a long flowering season. But now a great many roses which certainly owe nothing to the musk rose have similar qualities and there is a growing tendency to lump them all together as Shrub Roses. It is difficult to give a hard and fast definition of a shrub rose but by and large it is a rose that is vigorous and well branched so that it makes a good bush not in need of a great deal of pruning. The flowers may be large or small, produced in big clusters or only in twos and threes but usually more or less perpetual flowering from June to September. There is not a lot to distinguish some shrub from some floribunda roses except that the floribunda varieties are less vigorous and therefore more suitable for massing in beds, whereas the shrub roses often look their best when planted individually or mixed with other kinds of shrubs.

Ramblers and Other Climbers. Rambler roses have not changed greatly, largely because not many new varieties have been raised. Essentially the rambler is a vigorous climber bearing small flowers in big clusters. As a rule it flowers once only but a few varieties are repeat-flowering.

because of their peculiar origin it is seldom wise to prune climbing sports hard, for they have a tendency to revert to their original bush forms and hard pruning aggravates this. I believe that because of this weakness climbing sports are a dying class and that their place will be taken by what one might call 'natural born climbers'.

Miniatures. The Miniatures have been known for a very long time but it is only in recent years that rose growers have begun to take much interest in them. They are miniature in every way with little flowers borne on little bushes which may be no more than a foot high when fully grown. Most of them look rather like floribunda roses seen through the wrong end of a pair of field glasses. They are useful little roses for edging rose beds in paved terraces so that they almost seem to be growing in the crevices between the slabs of paving.

Old-fashioned Roses. Some rose growers also

Shrub roses, such as the splendid Nevada shown above, are best planted as individual specimens or with other kinds of shrubs

Climbing roses can add enormous interest to a garden by providing colour at different levels. Shown above is the rambler Chaplin's Pink Climber

Climbers other than ramblers can be conveniently divided into those that were natural climbers from the moment they were first raised from seed and those that started life as bush roses and then changed (or sported) to the climbing habit. Outwardly there is often little or nothing to distinguish the two types; indeed the climbers are a very mixed lot with little to link them together save their long growth which can be trained on walls, fences, screens, etc. Some have large shapely flowers, others have smaller flowers (though not so small or clustered as those of the ramblers). Some have difficulty in 'climbing' to 8ft., others will easily reach 20ft.

The reason for the importance in the distinction between climbers and Climbing Sports resides in none of these things but simply in the fact, which I have already referred to in my recommendations on pruning (see p. 81), that

refer to Old-fashioned Roses as though they were a class in themselves, which they obviously are not because even in olden times there were many different classes of roses: the Moss roses with a curious almost moss-like outgrowth on buds and young stems, the Bourbon roses with fine shapely flowers, the Cabbage roses with very full many-petalled flowers, the Damask roses, Provence roses, Noisette roses and still more. Moreover, not all the roses that pass as old-fashioned today are really old at all. Some have been raised in the present century but they have an old-fashioned look about them such as flat quartered blossoms or an outgrowth of 'moss'. A great many of these roses look best when grown in isolation as I have suggested for shrub roses and quite a lot of them, like shrub roses, do not need a great deal of pruning. But it is really unwise to generalise too much about such a very mixed lot of roses.

Species. Finally, there are the Species; the wild roses, some of which are parents of the garden hybrids I have just been describing. There are hundreds of these wild roses though by no means all of them are grown in gardens. They nearly all have single flowers, and bloom only once a year.

Sometimes the heps or fruits which follow the flowers are very handsome and the plants are worth growing for this feature only. *R. moyesii* and *R. pomifera* are two examples of this. Sometimes the foliage is attractively coloured or turns colour in autumn. *R. alba, R. rubrifolia* and *R. nitida* are examples. Some species are prostrate, e.g., *R. multiflora* and *R. filipes*. Most commercial rose growers stock a few species and some make a feature of them. In the garden their place is in the shrub border or as a background to the rose garden rather than in the rose beds themselves. Correctly sited these roses can give a great deal of pleasure

Fragrant Cloud, a free-flowering coral-red hybrid tea variety. The large blooms have outstanding fragrance. It makes a medium sized bush

A Select List of Varieties

Hybrid Tea Roses

PINK

Ballet. This has large shapely bluey-pink flowers of good shape. It is of medium height and gives a fine display.

Gavotte. This medium-sized variety has shapely rose-pink flowers with a paler reverse. It is quite easy to place in planting schemes.

Grace de Monaco. The big globular flowers are pale pink which varies little whether they are young or old. Of medium vigour, it branches well and the flowers have good scent.

My Choice. The buds of this fine rose are gold shaded with crimson and they open to pink with a buff reverse. Vigorous in growth, it is well worth having for its scent alone.

Percy Thrower. I can recommend this rose as an excellent variety for the garden and cutting. The flowers are a delightful rose-pink shade and are carried freely.

Silver Lining. A rose whose colour is not difficult to blend with others, the flowers being soft rose paling to almost white at the base. The blooms are large, almost always of perfect shape, and the scent is exceptionally good.

Stella. A fine rose, cherry red on the outside petals and pinkish-cream or ivory in the centre. The flowers are long and shapely and the plants are of good habit.

ROSE-PINK AND OTHER VARIATIONS

Eden Rose. One of the most popular rose-pink hybrid teas, the deepest coloured petals matching the lightest coloured of Wendy Cussons. The blooms are magnificent but are not produced too plentifully. Strongly scented.

Rose Gaujard. Although the magenta colour with white reverse of this rose does not appeal to everyone, it is an excellent variety for garden display. The blooms have scent and stand up well to wet weather.

Wendy Cussons. A rose with beautifully shaped flowers although the rather hard cerise colour makes it a poor mixer and it is best planted well separated from reds and yellows. It has a good branching habit with glossy foliage and has won The Royal National Rose Society's Clay Cup for scent.

RED

Champs Elysées. This is a deep crimson rose with a velvety bloom and slight fragrance. It is moderately strong growing, and is very effective massed in a bed.

Christian Dior. This is a fine scarlet rose with beautifully shaped flowers, but it is rather prone to mildew and is well above average in height. The young foliage is an attractive coppery colour.

Ena Harkness. One of the best and most popular red roses. The flowers have a tendency to hang their heads.

Fragrant Cloud. This splendid rose has flowers of coral-red colouring, large in size and with outstanding fragrance, as the name implies. It makes a medium-sized bush.

Papa Meilland. An extremely attractive rich red colour, blooms of good shape and a rich scent are attributes which have brought this variety to the fore. It is of medium height.

Uncle Walter. A variety with very large red flowers of classical hybrid tea shape with little tendency to blueing. Growth is extremely vigorous.

Westminster. A brilliant bicolor rose, vivid deep cherry red with yellow reverse. Growth is tall and very vigorous. Strongly scented.

Apricot Silk. This is one of the newest roses, and one of the best. The name well describes the deep orange-yellow of the pointed blooms on long flower stems. It is most effective when massed.

Beauté. A pleasing apricot-yellow with well-formed flowers which stand up well to wet weather. The leathery foliage resists disease.

Doreen. A moderately vigorous rose with deep yellow to orange flowers with red shading. It is scented and flowers very freely, but has an unfortunate tendency to black spot.

Gold Crown. A tall grower, upright in habit. The colour of the well-formed blooms is a clear, deep yellow but turns paler with age. It has dark, disease-resistant foliage.

Grandpa Dickson. A very good yellow, fading a little with age, with a little pink on the margins of the petals. The dark green shining leaves offset the blooms well, and the growth is upright and strong.

King's Ransom. A shapely medium-sized yellow. It bears its flowers freely and is good for bedding. About medium size.

Lady Belper. An excellent garden rose with beautifully formed blooms of apricot warming to near orange at the heart. It flowers well and has spreading growth of medium height.

Peace. One of the finest roses ever produced. The large flowers vary from pale to deep yellow with variable cerise-pink shading at the edges of the petals. It is very vigorous and needs plenty of space.

Piccadilly. An orange-scarlet and yellow bicolor which pales with age to pink and soft gold. The foliage, reddish when young, changes to dark green with age. It grows about 3ft. tall.

FLAME AND VERMILION

Mischief. The medium-sized flowers are deep pink with orange on the outside of the petals. The coppery coloured foliage, resistant to disease, is good and bushes cover the ground well.

Mrs Sam McGredy. An old rose still outstanding for its bright coppery-orange and red flowers and reddish-bronze foliage. The blooms are produced plentifully and they stand up to wet weather well.

Super Star. Although this is an outstanding rose, the unusual light vermilion colour of the blooms makes it difficult to associate with other colours. The medium-sized blooms are well formed.

BLUE

Blue Moon. This well-scented lilac-blue rose is the best choice, perhaps, for those wanting the nearest approach yet to a truly blue rose but it is not really good when grown outdoors—it is better in a cool or unheated greenhouse where the blooms

are protected. Of medium height, it has a branching habit and is slightly stronger, it seems, than Sterling Silver.

Sterling Silver. This lavender-lilac variety has been a flower arranger's favourite since it was introduced in the late 1950s. The flowers have good shape and fragrance and, carefully placed, it is a good garden rose.

WHITE

Frau Karl Druschki. We still lack a really good white hybrid tea and I still plant this old and very vigorous variety even though its large dead-white flowers do produce traces of pink on the outside.

Floribunda Roses

PINK

Chanelle. The buds are a warm amber outside, opening to China pink and cream, but the flowers tend to become wishy-washy with age. It is a

Super Star, an eye-catching hybrid tea with light vermilion flowers. Because of its colour, this variety is best planted in beds on its own

vigorous variety that covers the ground well and seldom suffers from diseases.

Dearest. A gold medal rose with very pretty pink flowers. The buds are shapely but open to flattish flowers with golden centres; the blooms are carried in large trusses but they do not like wet weather.

Pink Parfait. The pale pink and orange flowers are produced very freely on compact, vigorous bushes. The colour is pleasing but may be too delicate for some people. Fragrance is not one of its strong points.

Queen Elizabeth. The soft pink flowers of this fine and very popular floribunda rose are almost of hybrid tea quality but they are produced on very tall bushes.

Vera Dalton. A variety that does well in my own garden; it could be described as a shorter version of Queen Elizabeth. The flowers are a lovely deep

Iceberg, a white floribunda rose of superb quality. It is a very free-flowering variety and has attractive glossy dark green foliage

pink, bluer in shade than Dearest and more pointed in bud, although they open cup-shaped. Compact in growth, the foliage of this variety is dark green and disease resistant.

Violet Carson. Shrimp pink with a paler reverse and fading a little with age, this is a promising new pink floribunda. The long, pointed flowers are of hybrid tea shape and they are said to stand up to wet weather well. The reddish-bronze foliage enhances the flowers.

CARMINE

Daily Sketch. A carmine-pink and silver bicolor, very like Ideal Home, but starting to flower a little earlier. The flowers are large and effective. Medium in height, and with bushy growth.

Paddy McGredy. Described as a hybrid tea-type floribunda it is certainly very easy to mistake the flowers for those of a hybrid tea rose, although it has the free-flowering of a floribunda. The cerise colour is not to everyone's liking, but the flowers are scented and produced on bushy plants of medium height.

Rosemary Rose. A charming variety with rosette-shaped flowers resembling those of a double camellia. The colour is carmine-crimson and the bronze-red foliage is also handsome. Growth is vigorous and the flowers are produced freely throughout the season. Unfortunately the foliage is prone to mildew.

RED

Border Coral. This variety bears coral-salmon blooms in large trusses; these blooms are still soft enough in colour, however, to tone with reds and the bright deep pinks. The growth is strong but spreading and the plants cover the ground well.

Europeana. A deep crimson floribunda with fully double flowers very freely produced.

Evelyn Fison. The colour is a little lighter than the crimson-scarlet Lilli Marlene, almost a geranium red. The effect is very good and the slight frilling of the petals is pleasing. The foliage is a clean dark green and the growth covers the ground well.

Lilli Marlene. First-rate in every way, this floribunda has velvety crimson-scarlet flowers, fully double and carried in close clusters. Growth is compact and bushy covering the ground well.

Marlena. This rich red floribunda is free-flowering and the dwarfest of all I grow. A very useful variety.

Meteor. A variety of dwarf, bushy habit. The semi-double flowers are a bright vermilion and they are produced in good trusses but there is little scent.

Paprika. This has very bright scarlet semi-double flowers, lightening a little towards the centre. It flowers freely and has shining dark green foliage.

Strawberry Fair. The strawberry-red, semi-double flowers of this variety do not fade as they mature. It flowers freely and is of medium height.

YELLOW TO ORANGE

Allgold. Outstanding for its golden-yellow colour which does not fade in the sun or rain. It grows only 2ft. tall and is resistant to black spot and mildew.

Chinatown. This variety is exceptionally vigorous, growing at least 5ft. tall in good soil. The large double blooms of hybrid tea shape, produced in small clusters, are a rich yellow. It has healthy, glossy foliage and there is some scent.

Circus. Yellow with pink and salmon tints, this is a fascinating rose, its colours deepening as the blooms age. It is fragrant and double, and its growth is vigorous, although it does not reach any great height.

Elizabeth of Glamis. This fine orange-salmon variety won The Royal National Rose Society's President's International Trophy for the best new rose of the year and the Clay Cup for the best new scented rose of the year in 1963—this last a most unusual achievement for a floribunda. Also a gold medal winner. Of medium height.

Jiminy Cricket. The colour of this charming rose varies between salmon and pink. It always makes a good show, even in wet weather, and has pleasing reddish foliage.

Woburn Abbey. This is a very showy rose with deep orange flowers of good size. It has pleasing fragrance.

YELLOW AND CARMINE

Paint Box. The colouring is red and yellow gradually turning to deep red—at first glance one could imagine that two varieties had got mixed together. Growth is vigorous.

Shepherd's Delight. This is another mixture of colours, this time with yellow, orange and flame predominating. The semi-double flowers with frilled edges are freely produced. It does well in the autumn but is rather prone to black spot.

FLAME AND VERMILION

Anna Wheatcroft. The large single flowers are of reddish-salmon colouring which fades a little with age. The yellow stamens show prominently and liven the effect. Of medium height, growth is bushy and the foliage is dark and healthy.

Firecracker. This rose is unique because of the contrast of its bright cherry red colour with yellow centre and golden anthers. It is immensely free-flowering and a deservedly popular rose.

Highlight. A good orange-scarlet floribunda with flowers in large trusses. It grows vigorously and resists disease. The old petals are inclined to hang on the trusses when the flowers wither creating an untidy effect.

Orangeade. A popular variety because of its deep orange flowers, but I find that they age to an unattractive puce. The growth of this variety tends to be a little thin.

Orange Sensation. This is such a hot orange-red that it is difficult to blend with other colours, although it should be happy with a white floribunda such as Iceberg. The flowers are double and produced in large clusters on vigorous, well-branched plants.

Orange Silk. A brilliant orange vermilion rose, introduced in 1968, this has large double flowers. It is a vigorous grower with a neat appearance and makes an outstanding display.

LAVENDER SHADES

Lavender Lassie. The lilac-pink, rosette-shaped flowers of this tall-growing rose are very handsome and have good fragrance.

Lavender Pinocchio. The blooms of this variety are greyish-lavender in colour and have pleasant fragrance. It flowers freely.

Magenta. This strong-growing rose has rosy-magenta flowers which it bears freely over an extended period. It also has good fragrance.

WHITE

Iceberg. The best white floribunda for garden display. Growth is rather tall but covers the ground well.

Mermaid, a climbing rose of unusual charm. The golden-yellow flowers have prominent yellow stamens, and the foliage is dark green and glossy

Climbing Roses

Autumn Sunlight. One of the newest varieties, with orange-vermilion flowers in clusters. It flowers profusely, and gives a second display in the autumn.

Danse du Feu. A really bright rose with orange-scarlet flowers of medium size. They are produced in succession throughout most of the summer. It grows to about 8ft. tall.

Golden Showers. This is a really beautiful climbing rose, with large, fragrant yellow flowers produced from June to September. Final height is about 9 to 10ft.

Clg. Mme. Caroline Testout. An old variety, but as vigorous as it has always been. The flowers are very large and full, round, and a delightful pink. It is scented and grows to 12 to 15ft.

Mme. Gregoire Staechelin. Another old rose, coral pink with crimson shadings. The flower is large and strongly scented and the growth is very strong; it will cover a large area of wall, and grows to about 10ft.

Mermaid. One of the most delightful of the climbing roses, this has large, saucer-shaped, golden-yellow, single flowers, with prominent yellow stamens, backed by glossy, dark green foliage. It may take some years to settle down, before it really starts to grow, but when it does it is extremely vigorous and needs much space. The flowers are produced in great profusion.

Paul's Lemon Pillar. This has beautifully shaped double flowers, lemon-yellow to white, and scented, but it is summer flowering only, between June and July. It is a strong grower.

Pink Perpetué. This variety, introduced a few years ago, has Danse du Feu as one of its parents. The flower is pink on the inside, and carmine on the outside; it is slightly scented and very free-flowering, producing a second crop later in the summer. The dark green glossy leaves provide an agreeable background.

Zéphirine Drouhin. A very old rose indeed, first recorded in 1868, which bears bright carmine-pink flowers in July. They are nearly double, very fragrant, and there is a repeat crop later. It grows to a height of about 10 to 12ft. if unrestricted, and the stems are thornless.

Rambler Roses

Albéric Barbier. One of the most vigorous of all rambler roses with coppery-purple young growth, shining dark green foliage which is almost ever-green in a mild winter, and creamy-white double flowers in July. It will grow well on a north-facing wall.

Albertine. A very vigorous and rampant rose, the reddish to coppery-pink flowers are produced in great profusion in June and July and are delightfully scented. It needs plenty of room to display its beautiful flowers and accommodate its growth.

American Pillar. This has single rose-pink flowers with white centres and golden stamens, produced in large clusters in summer. It is strong growing and needs plenty of room.

Chaplin's Pink Climber. Semi-double, bright pink flowers in summer and glossy dark green foliage make this a most attractive variety. It flowers in summer.

Dorothy Perkins. Probably one of the most well-known of the ramblers, the small bright pink double flowers are produced in clusters in great profusion in July, particularly if pruned correctly. It makes a good weeping standard and is excellent for growing over arches or pergolas.

Excelsa. A rose similar to Dorothy Perkins but with bright crimson flowers, borne in July and August, but with no repeat in the autumn.

Paul's Scarlet Climber. The flowers are large for a rambler but they have no fragrance; they are very freely produced. They are a vivid scarlet-crimson and are produced in small clusters.

Shrub Roses

Bonn. Originally this was classified as a Hybrid Musk, though it lacks the musk rose fragrance. It is a vigorous shrub rose with rose-red blooms suffused with vermilion, freely produced. It grows to about 6ft.

Canary Bird. More correctly, though less frequently, known as *Rosa xanthina spontanea*, this variety has masses of single yellow flowers in May and June on a bush about 7ft. high.

Elmshorn. A Hybrid Musk variety with deep carmine, double blooms, borne over a very long flowering period, but it is only slightly fragrant. It grows to about 6ft. tall.

Felicia. One of the original Hybrid Musk roses, this is a beautiful variety with double, well-scented blooms of warm pink colouring, on a bush about 5 to 6ft. high.

Friedrich Heyer. Officially this is a floribunda rose, but it is so vigorous it is better regarded as a small shrub. It has semi-double, rich scarlet flowers borne in large clusters and these have a good foil in the very dark green foliage.

Frühlingsgold. A beautiful variety, with large golden-yellow to cream, semi-double flowers in May and June. It produces its arching stems to about 10ft.

Frühlingsmorgen. Similar to Frühlingsgold, but with single flowers, pink in colour with yellow centres and maroon-coloured stamens.

Golden Wings. A beautiful rose making long arching growths up to 5ft. high. It bears very large, single flowers in a delicate shade of pale canary yellow.

Heidelberg. An attractive red-flowered variety carrying its blooms over a long period. It is strong growing.

Kassel. A shrub rose of Hybrid Musk origin, whose scented flowers are deep cherry red and are carried in large clusters. The flowers continue in succession for a long time, on a bush 5 to 6ft. high.

Moyesii. This is a species rose with deep red single flowers having cream stamens. The flowers are dotted all over the bush like butterflies, and are followed by large bottle-shaped red heps; the foliage is delicate and attractive, and the bush reaches about 10ft.

Nevada. An outstanding rose, up to 8ft. tall and with arching branches wreathed in June, and sometimes also in August, with large creamy-white flowers with golden anthers. It has few thorns, but it has the drawback of being susceptible to black spot.

Rubrifolia. Besides having purplish-pink blooms in June and July, this species has most attractive fern-like, greyish-blue leaves, tinted with pink. The heps are round and dark red, and the bush reaches a height of about 6ft.

Scarlet Fire. A very showy variety with single scarlet-crimson blooms which are borne freely in mid-summer. The large red heps which appear later are also attractive. It makes a bush some 6ft. tall and as much through.

Will Scarlet. Sometimes regarded as a Hybrid Musk, although it has little fragrance, this variety has brilliant scarlet, semi-double blooms. It makes a bush of up to 6ft.

Planning herbaceous borders to form pleasing shapes and patterns and delightful colour blendings means calling on all the artistic resources at one's command. But, as in every other aspect of gardening, careful attention to detail in the early stages pays handsome dividends later. Above all, in border planning, avoid the pitfall of uniformity while at the same time giving the overall planting scheme form and unity

Right. The doronicums, which make such a gay showing from early spring to June, are excellent cut flowers. Another useful attribute is that they grow well in sun or semi-shade

Right, below. The hostas, or Plantain Lilies, are foliage plants of great decorative value, with attractive flower spikes—in lilac, lavender, mauve or white—as a secondary adornment. Shown here is the delightful *Hosta crispula,* a species from Japan

Left. Dicentra spectabilis, the Dutchman's Breeches or Bleeding Heart, is one of the most beautiful of all herbaceous plants. It flowers in May and June

Left, below. The geums, available in a range of bright colours, are easily-grown border plants for summer and autumn display. They grow best in poorish soils which are really free-draining, especially in winter

Sweet Williams are flowers with their
own special kind of beauty. A
delightful range of these biennials is
now available, some being in self
colours and others having contrasting
shades in their petals

Perennials Annuals and Biennials

5

Where colour and interest are needed quickly, as for instance around a new home where the garden has been neglected or is as yet unmade, perennials, annuals and biennials, like numerous bulbous flowering plants, come into their own. They can be used to 'hold the fort' until long-term plans have time to mature, as well as being included in the permanent scheme of things.

Colourful and very pleasing displays can be created with these flowers under widely differing conditions. Their versatility is such as to make them useful to both the town gardener with little space to play with and his suburban and country counterpart who can contemplate more ambitious planting schemes.

The Mixed Border. A modern development well suited to gardens of modest size (and a good illustration, too, of how gardening does keep in step with the times) is the 'mixed border' in which perennials, annuals, biennials, bulbs and selected flowering and foliage shrubs are grown together. This has the obvious advantage of saving space, for all the above-mentioned groups of plants are grown in an area that one of them would normally occupy; and it has the merit of giving—with skilful planting—a longer period of interest and colour. I am all for adopting measures like this where they have positive, practical advantages to offer.

When making a mixed or conventional herbaceous border I consider it a mistake to give it a straight edge bordered by a path. I find this harsh and unsympathetic. Much more pleasing is to have grass running up to the border and to scallop its edge, once or more depending on the border's length. This brings the plants right out into the grass.

Annuals. The main advantage of annuals, as I see it, is that they will give one a quicker and cheaper display of colour than do any other flowers. Also, they will thrive in almost any soil

Annuals are superb plants for providing quick colour. They are best grown in poor soils so that excessive foliage is not produced at the expense of flowers

Biennials, like the pale yellow *Verbascum bombyciferum* shown above, can be used most effectively as 'fillers' while longer-term plants mature

provided it is worked down to a fine tilth. In fact, they must not have rich soil or a lot of leafy growth will be made at the expense of flowers. A well-drained, light soil with plenty of sunshine is what they really appreciate.

One difficulty in gardens where spring bedding plants like wallflowers, polyanthus and forget-me-nots are grown is that, quite often, seeds of hardy annuals cannot be sown in March or April where they are wanted to flower. I get over this problem by raising my annuals in soil blocks. I stand the soil blocks in boxes, sow two or three seeds on the top of each block and thin the seedlings out when they are about 1 inch high. The young plants then go out when the bedding plants are removed. They begin flowering in June and if the flowers are removed after they have faded I get a show of colour which lasts until September.

For filling in odd gaps in the herbaceous border annuals are also ideal. Any space noticed in April or May can be dealt with in this way, either by sowing seeds directly in the border or by planting out young plants from soil blocks following the system described above. Their use in mixed borders has already been mentioned.

Biennials. Although, compared with herbaceous perennials and annuals, biennials are a small group of plants, they are still invaluable in the garden for set-piece displays or for 'filling in' while longer term plans mature. Most of us would want to find space for such favourites as the Canterbury Bells, Wallflowers, Foxgloves and Sweet Williams.

Planting and Aftercare

The best aspect for a plant border is south or west; I would place east next with north a poor fourth, for it is very restrictive in that one is only able to grow plants which are happy with very little sun. The background to the border is rarely given the consideration it deserves. A stone wall looks superb, but it is unlikely that one will be where it is wanted. When one has to create a background there is not much to beat a cupressus, thuya or beech hedge. One or other of these will act as an excellent foil for the colours of the flowers. Fencing does not serve the same purpose, and where it does form the background for a border I would always endeavour to mask it with tall-growing plants. A fence is not sympathetic to plants in the same way as a close-knit hedge. A tall fence can be masked with rambler or climbing roses, *Clematis montana* or *Polygonum baldschuanicum*. I also like to erect posts here and there in the border itself and use these as supports for roses like Paul's Scarlet Climber or Dorothy Perkins.

Arrangement of Plants. When working out planting schemes it is more important to place plants so that they form pleasing shapes and patterns than it is to get perfect colour blendings. I always try to arrange things, too, so that early-flowering border plants which may be dull or even unsightly for the rest of the season are reasonably well masked by later-flowering plants.

It is a rule, and a good one, with perennials, annuals and biennials, always to make bold plantings of each individual species or variety. It is also true that varying the shape of the groups so that there is no uniformity, and varying the heights to carry the eye from one plant to another, adds immensely to the attractions of a border. Broadly speaking, one wants the tall plants at the back, the medium-sized plants in the middle and the short plants in the front—but I do not like to keep too closely to this formula. Bring some of the medium-sized plants to the front occasionally and drop back some of the short ones so that all manner of pleasing shapes and contours are created. A bit of clever juggling is all part of the fun of gardening.

Bold groupings of border plants are most effective, and varying the shape and height of the individual groups increase the visual impact

Plants like the kniphofias, which take quite a long time to settle down after re-planting, are best moved in the autumn rather than the spring

Soil Improvement. No soil can possibly be ideal for every purpose for, in addition to the many plants which are happy in any reasonable soil, there are those which need an acid rooting medium and others which grow best under alkaline conditions. What most of us would like—but we usually have not got—is a nice fibrous medium loam with plenty of body in it. What we are most likely to have to cope with is a soil which is either too light and free-draining or too heavy and water-retentive for the needs of most plants. Fortunately there is a fairly easy way of overcoming both tendencies. This is by digging in moist granulated peat or other humus-forming material to improve the soil texture. Heavy clay can be vastly improved by such treatment and in time can be turned into a first-class growing medium. Light, sandy soils can be so treated but the effect is more short-termed.

I like to work clay soils in the autumn so that they can be left rough for the rain and frost to break down. Never attempt to dig such soils when they are so wet that the particles pick up on the boots. Some plants do best in, and in fact prefer, acid soil; others will only grow well in limy soil, so it is a good idea before sowing seed or planting to find out which kind you have. This is easily done with a simple soil testing kit, obtainable from garden sundriesmen for a few shillings. If it is shown to be very acid, or the plants you are putting in prefer limy soil, then a dressing of lime should be made about a couple of months after digging. The best type to use is hydrated lime and the amount to apply will depend on the degree of acidity. Three ounces to the square yard would be an average dressing. On light soils, though, I would recommend the use of ground limestone for this purpose as this is less easily leached through the soil. On average, this needs to be applied at about twice the rate for hydrated lime.

Planting

In general, March and early April is the most favourable time to plant herbaceous perennial plants (although when weather conditions are favourable I like to start this job in February), but there are some exceptions to this recommendation. For example, many gardeners like to move peonies, kniphofias, hellebores and other plants which take quite a time to settle down in the autumn so that they get away quicker the following spring and are not so badly affected by dry spells at that time.

The trouble which one experiences with autumn planting in any soil which tends to be on the heavy side is that heavy rainfall and a tendency to water-logging can delay establishment and so result in damage or loss. Intense cold such as we sometimes have also leads to losses among autumn planted perennials.

I like to prepare the planting holes for the smaller plants with a trowel but for the rest, of course, I use a spade. The important thing is to work the soil well in among the roots. Be especially careful, too, to firm the soil well around the roots for if they do not come in close contact with the soil they wither and die.

Annuals. Where annuals are to be grown the soil texture may need improving but rich soil is certainly not wanted for annuals which will produce a profusion of leaves and fewer flowers when given such conditions. At most, on hungry soils, I would put down a light dressing of bonemeal some weeks before sowing. (This fertiliser is given up slowly, and in modest quantities will do nothing but good.)

There is no doubt at all that the key to success with annuals is the preparation of the seed bed, and I have outlined the complete sowing operation later in this chapter when discussing methods of increase in general (see p. 108).

Aftercare

Perennials. Staking, thinning, weeding, forking over the soil when it has been panned down by rain, removing dead flowers to encourage further flowering, feeding with slow-acting fertilisers, and, of course, lifting and dividing (except in the case of those plants which resent being moved) every three or four years—these are the jobs which we must attend to whenever we grow perennial flowers.

For staking perennials, I use only short pea sticks—last year's pea sticks as far as I can. Those gardeners who do not have a vegetable garden should still buy pea sticks for this purpose, for I am sure that this is the cheapest way to provide support for the plants, in the long run. Pea sticks for use with herbaceous plants will usually last for a couple of years. To provide an example of what I do with these sticks, let us suppose that the plant we want to support has an ultimate height of

Pea sticks make admirable supports for perennial plants. If sticks are chosen which are shorter than the plants when fully grown they will soon be completely hidden

about 2½ft. The sticks I put around the clump would be 1½ to 2ft. tall, and if these make the border look like a timber yard when they are first put in I know that by the middle of June they will be completely hidden from view. The plants grow up through the sticks and are given all the support they could possibly want.

The only exceptions I make to this kind of supporting are for delphiniums and lupins; these I give separate stakes for individual spikes. When a flower spike is pressed by the wind, the support should allow it to move a little and then bring it back gradually and without damage when the pressure is no longer there.

For securing perennials to stakes some soft material must be used, like raffia, fillis or twine. Also, with plants like these which make rapid growth, it is important to allow plenty of room for expansion when the ties have been made. One

way of doing this and yet having a firm tie at the same time is to take the tying material round the stake, knot it and then take it loosely round the stem, knotting it again. The stem can then expand and still be held firm enough for all practical purposes.

The most important time to work over the soil in between the plants, without of course disturbing the plants' roots, is just before growth gets into its stride in spring, and this is the time, too, to work in a light dressing of a slow-acting fertiliser like bonemeal or hoof and horn. Never feed plants with quick-acting fertilisers, or unwanted growth will be produced at the expense of the flowers.

Dead-heading is a good old gardener's term for removing the flowers of ornamentals as soon as they have gone over. This is a good practice for the very worth while reason that it often encourages plants to produce another crop of flowers, but it also keeps the border tidy and that is important, too, I always feel.

If tall bearded irises are grown, cut down the foliage to within about 6in. of the ground in the autumn to lessen the chances of damage from disease, and with delphiniums cut back the stems almost to ground level at the end of the season and cover the crowns of the plants with well-weathered ashes as a protection against winter damp. The crowns of kniphofias get useful protection from tying the leaves together at this same time.

Thinning, if one is a newcomer to gardening, is a task which one does with some misgivings but as one soon finds out it can pay handsome dividends. To take delphiniums as an example again, if the growths on each plant are reduced to about six in number it will be found that the spikes are far superior to those on plants left unthinned.

In three to four years after planting a border of perennials it will be noticeable that the quality of the flowers is in many cases falling off. This is due to the exhaustion of the plants and the gradual depletion of food reserves in the soil. All this can be put right by lifting the plants. How to divide the plants and improve the soil is explained on p. 105.

Annuals. I have explained on p. 106 the necessity for thin sowing and how the plants can be finally thinned about two to three weeks after the seedlings have been first thinned.

I think a lot of gardeners go wrong over spacing. Annuals want plenty of room to develop; the minimum for the smallest should be 9in. and most will fill out to 15 to 18in. It is worth remembering, too, that in showery weather in May or early June it is quite possible to lift some plants, if this is done carefully, and replant them successfully where there are gaps. If the weather is not favourable after sowing there are quite likely to be gaps which will have to be filled.

After this stage, one's main preoccupation will be to keep the weeds under control—either by hand or by using a hoe delicately in among the plants. In dry weather it will be necessary to water the plants for, with their root systems still only partially developed, any serious shortage of moisture will have dire results.

Annuals which need staking I like to stake early. I stake all my annuals—except those like larkspur, which get a double stake—with twiggy pieces of birch. These I insert when the plants are about 8 to 9in. tall and they then grow up through them and obscure them from view. It is very much like the way I treat perennials (see p. 104).

Do not allow finished flowers to remain on the plants: dead-head them as quickly as possible so that the plants will use all their energy to produce more flowers and not divert it to the production of seed.

Methods of Increase

To become really efficient as a propagator takes time and patience. One must expect to learn from one's mistakes. Many of these can be avoided, though, if one keeps certain basic principles firmly in mind.

Briefly, I would say these are as follows: (1) Propagate vegetatively only from completely healthy stock. (2) Observe strict hygiene at all times, and when pots, boxes or other containers are used make sure that these are absolutely clean. (3) Use sterilised loam in composts used as growing mediums. Propagate only at the right time of the year for the particular plant it is desired to increase, bearing in mind the method of propagation to be adopted. (4) Keep as close as possible to the correct temperature and humidity when propagation is being carried out under glass.

Perennials

Division. The most usual method of increasing herbaceous perennial plants is by division of the roots. This is an easy task to carry out and should be done in spring or autumn. There is no doubt that autumn is the best time to divide plants which do not establish themselves reasonably quickly, as they then have the advantage of a long growing season ahead. Kniphofias, Oriental poppies, peonies and alstroemerias—which resent being moved in any case and should be left undisturbed as long as possible—are a case in point.

Division, with some exceptions, is only necessary once every three or four years. By that time there will be a marked deterioration in the quality of the growths and flowers of the plants, and the soil will need improving by the addition of compost or other humus-forming material after a long spell without attentions of this kind. One should always take the opportunity to improve the soil when, on rare occasions like this, it is cleared of plants. (See p. 104 for advice on aftercare.)

Many herbaceous plants can be divided merely by pulling them apart, the portions which are to be replanted having healthy looking roots and at least one shoot. These retained portions should always be from the outside of the plants, the older, woody centres being discarded altogether. Plants which form a mass of roots and which can be dealt with in this way include such favourites as the varieties of *Chrysanthemum maximum* (the Shasta Daisy), solidago and Michaelmas daisies.

Some plants, though, are made of sterner stuff and different tactics are necessary to cope with these. The large clumps consist of really rough roots and the best way here is to dig around the clump with a spade before lifting it out bodily with a garden fork. Then drive two forks back to back into the centre of the clump and split it by

Tough clumps of perennial plants can be divided by placing two forks back to back and pressing the handles outwards. Roots still difficult to sever can be cut with a knife

pressing the fork handles outwards. From then on all will be plain sailing, with any roots which are reluctant to part being cut with a knife.

The parts which are to be replanted should be out of the ground as short a time as possible and they should be covered with sacking or some other material to avoid drying out by sun or wind. For this reason many gardeners like to renovate the border in sections, lifting and replanting, say, one third before starting on the rest. I have never found this necessary if the lifted plants are protected as suggested, but that is not to say that I do not consider it a good scheme to adopt.

In addition to the humus-forming material which I add at this time I apply a dressing of bonemeal, a good all-purpose fertiliser, or both. Bonemeal is an excellent feed but it is slow acting; it will still be there to assist the plants when the general fertiliser has been absorbed.

Seeds. Provided one has the facilities and does not mind waiting for results it is possible to raise a collection of herbaceous perennials entirely from seed. A large number of popular species can be raised in this way, including, for example, delphiniums, kniphofias, *Anemone hupehensis (japonica), Rudbeckia speciosa,* lupins, *Monarda didyma* and *Scabiosa caucasica.* Varieties, though (with exceptions like *Geum* Mrs Bradshaw), are a different proposition and some of the characteristics which make them so desirable as garden plants are almost always lost when attempts are made to increase them from seed. Colour variations occur, flower size and formation is liable to change and the habit of growth may well be much inferior to that of the parent. With such plants vegetative propagation is essential.

But if most varieties are 'out' so far as increase from seed is concerned, strains give us other opportunities for this form of increase. A strain is a stock of a particular variety in which desirable characteristics like colour, habit or freedom of flowering are transmitted, through careful selection, to the progeny. These characteristics only vary from plant to plant within fairly narrow limits. A case in point is the Pacific Giant strain of delphinium.

Although the majority of herbaceous perennial plants can be raised in specially prepared beds in the open garden, it is a great advantage to have a garden frame or greenhouse available for seed raising. With such controlled conditions one's scope is immediately widened.

Outdoor Sowing. Choose a sheltered site for the seed bed for perennials, dig the plot and work the soil down to a fine tilth. With the variable weather conditions we experience in Britain it is most unwise to be too wedded to dates for gardening operations, and with seed sowing in particular it is worth waiting several weeks, if necessary, until conditions are just right, but some time in May or June will be found suitable. If the soil is loose firming will be an important part of the preparation and this can be done with the feet or a roller if the size of the plot warrants its use. Some gardeners like to use a wooden hay rake (a very useful tool) for the initial breaking down process, finishing the job off with an ordinary rake. If the seeds are small—as so many are—it is doubly important that the tilth should be really fine.

Seeds can be sown broadcast but I much prefer to sow in drills as this makes thinning, weeding and other jobs which may be necessary so much easier to carry out. A garden line is needed now to mark out the rows and a hoe to draw out the drills.

The depth of the drill depends on the seeds being sown but about half an inch is right for the general run of seeds with the appropriate adjust-

After thorough firming by treading, the drills for seeds of herbaceous perennial plants are drawn out with a hoe, as above

ments being made for larger or smaller seeds. As with all seed sowing, sow thinly so that the resulting seedlings do not have excessive competition for light and air. It is also less wasteful, of course. If the soil is very dry, as it can become sometimes in spring, water along the drills before sowing the seeds. After sowing fill in the drills with the feet or with the back of a rake.

When the seedlings are large enough to handle transplant them into a reserve or nusery bed. By the autumn or following spring the young plants will be ready for planting in their flowering quarters.

Sowing Indoors. If a cold frame or unheated greenhouse is available it is, of course, possible to start sowing a full two months sooner than out of doors—in late February or early March. Sow the seed in boxes or pans which have been carefully cleaned. The seed mixture used could be the John Innes Seed Compost and the boxes standard seed boxes (see p. 229), because these can be so much more easily arranged on benches and shelves and can be much more neatly stored. If the boxes are treated with copper naphthanate before use this will greatly prolong their active life.

(The 1½ in. Jiffystrips which are made to fit into a standard seed box are very convenient. Each box then accommodates 60 plants. Another advantage is that there is less disturbance of the roots when planting. These strips can be used for

Thin sowing, in addition to being less wasteful, will make subsequent thinning out of the seedlings and weeding easier to carry out

When seeds of herbaceous perennials are sown indoors in boxes or pots, a home-made presser should be used to make the compost suitably firm

pricking out seedlings or, in some cases, for direct sowing.)

Cover the drainage holes in the pots (or drainage gap in the boxes) with broken crocks and place a layer of roughage over this before adding the compost, which should be pressed firm, particularly along the edges and in the corners. A home-made presser which consists of a shaped piece of wood fitted with a handle is very useful for this job, giving the compost just the right consistency —not too hard—after initial firming with the finger-tips. The final level of the soil should be such that it reaches to within $\frac{1}{4}$ to $\frac{1}{2}$in. of the top of the box.

Sow the seeds thinly and sieve over them very fine soil to twice the depth of the seeds, but before doing this make the compost moist. To do this I fill a container with water and hold the box with its base in the water for just a few minutes. Alternatively, you can water from overhead with a watering can fitted with a fine rose, but I consider the first method most satisfactory. The containers are now ready to move to the frame or greenhouse and they must be covered with a sheet of glass and shaded with newspaper until germination takes place. The condensation which forms on the underside of the glass must be wiped away each day.

As soon as it is possible to handle the seedlings prick them out in boxes filled with John Innes

No. 1 potting compost, using a dibber to make the holes and taking special care not to damage the fine roots. Aim to plant the seedlings so that their seed leaves are just above the surface of the compost. This potting compost has considerably more food value than the mixture in which the seeds were started into growth and can be bought ready mixed or be made up at home as described on pp. 231 and 232.

Now follows the hardening off process with the plants gradually being acclimatised to life in the open air. This should be done progressively with the plants getting more time in the open air each day until, after several weeks—the timing depending on the weather conditions prevailing—they will be ready to move to nursery beds (reserve beds) or direct into the border where they will flower when mature.

Softwood Cuttings. These, as the term suggests, are cuttings made from young shoots. They provide a very useful means of increasing numerous hardy and half-hardy herbaceous perennial plants, and suitable growths from which these can be made are normally available in spring and early summer or indeed at almost any time of the year under glass. As I have already explained (p. 106) this or some other means of vegetative reproduction is almost always essential with named, man-made varieties if we want the offspring to have identical characteristics to the parent plant.

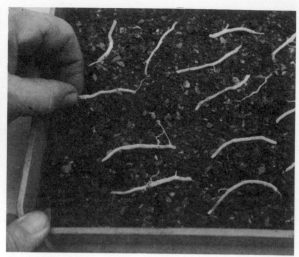

Root-cuttings of anchusa being inserted in the rooting mixture. To make sure that the cuttings are planted right way up, a slanting cut should be made at the bottom end

Plants, like phlox, with very thin roots are laid on the surface of the compost and covered with a thin layer of the same mixture

To prepare such cuttings the procedure is as follows. Look for healthy, strong looking shoots between 2 to 3in. long and remove these so that a clean cut can be made just below a node or joint. The lower leaves should also be removed. Pure sand or vermiculite can be used as the rooting medium (and indeed this is often desirable if the plant concerned is known to be what is called 'shy-rooting') or a mixture of sand and loam. If pure sand or vermiculite is used as the rooting medium, repotting will be necessary as soon as the cuttings form roots. Plants which are known to be difficult to root can be treated with a growth-promoting, root hormone preparation—available from garden stores—which will encourage the formation of roots, when applied as directed (it is essential that the manufacturer's instructions should be followed closely). These preparations are obtainable with captan added as a protection against the diseases to which cuttings are prone.

The cuttings may be inserted in the selected rooting medium in either boxes or pots and are best stood in a propagating frame where the atmosphere can be kept suitably moist and the heavy initial moisture losses by the as-yet unrooted cuttings can be replaced. Alternatively, insert them directly into the soil in a frame. Shade must also be provided against strong sunshine as in the early stages of growth this also can cause an excessive amount of moisture to be lost through the leaves of the cuttings.

As soon as the cuttings have formed a good root system they will need potting on and hardening off so that they will later be ready for planting in the border during the same season or for planting in the nursery bed for growing on.

Root Cuttings. Propagation by root cuttings is the best way of increasing many thick-rooted perennials like verbascums, Oriental poppies, phlox, anchusa and *Limonium latifolium (Statice*

latifolia). Some fibrous-rooted plants, like the gaillardias, may also be readily increased in this way. There is great latitude in the time when cuttings may be rooted by this method but for home gardeners spring and early summer, with months of good growing weather ahead, is probably the best time. Some gardeners find it an advantage, though, to put the cuttings in very early in the year, as by late spring hardened off young plants are available for planting outdoors.

To obtain the cuttings lift the plants from which they are to be taken and wash the roots. Then cut them into lengths of 1 to 3in., depending on the thickness of the particular roots. To ensure that the cuttings will be planted the right way up make the cut at the base end slanting (which provides the maximum rooting surface) or mark it with a distinctive nick. Then insert them in boxes of free-draining compost so that the tops are just covered, water them in and place in a garden frame or in a sheltered part of the garden. With plants like phlox which have thin roots it is usual just to lay them on the surface of the compost and cover with a thin layer of the same mixture. Once the cuttings have made growth they are potted on —or re-boxed—and grown on for subsequent planting out.

Annuals and Biennials

Hardy Annuals. The term 'hardy annual' is used to describe any annual flower which can be raised from seed sown in the open ground in spring and needs no protection whatever during its cycle of growth. Quite a few (the hardiest of the hardy annuals) can be sown in the open ground in August or September to come into flower in late spring or early summer, but one takes a gamble with the weather when sowing at this time and one must be prepared for losses if the winter is severe. Calendulas, larkspurs,

If annual seeds are sown in prepared drills this makes subsequent thinning and weeding much easier. The seeds here are being raked into the surface soil

Remove half-hardy annuals from their boxes with care when planting them out. Cut between the roots with a sharp kitchen knife or other suitable sharp surface

nigella, godetias and sweet peas are some of the annuals often grown in this way.

The time to sow in spring depends very much on the locality and the weather prevailing (late and early seasons occur as frequently as 'normal' seasons in Britain). In favoured parts it is possible to sow in the latter part of March while in cold districts it may be advisable to wait until late April or early May. A fine day must be chosen for carrying out this task.

Gardeners have their own ideas about how best to sow annuals. With small seeds I like to sow with my finger and thumb, holding the seeds in the palm of my left hand and just lightly sprinkling them over the ground with my right. I find that this allows me to sow thinly and reasonably accurately. Afterwards, I rake the seeds into the surface soil. Sowing broadcast, though, has its disadvantages, for weeding can be very difficult. A better way if the ground is known to be weedy is to sow in prepared drills which can be cleaned and thinned much more easily. If this method is used the depth of the drills will have to be adjusted to the size of the seed being sown, the smallest only having just a light covering of soil and the largest as much as 1½ to 2in.

Make the first thinning as soon as the seedlings can be conveniently handled. It should then, after a further two to three weeks, be possible to make the final thinning.

Half-hardy Annuals. These are annuals which cannot be planted out in the garden until all danger of frost has passed. If one has a heated greenhouse or frame in which a day temperature of about 13°C. (55°F.) can be maintained, raising these plants is quite straightforward. If such facilities are not available an alternative is to make spring sowings in the open ground and cover these with cloches until the danger of frost has passed. Also, many half-hardy annuals may be sown in the open in early May but they will be much later coming into flower.

There is no point in starting seed sowing under glass too early in the year or the plants may be ready too soon for planting out, which cannot be safely done until late May or early June when they have been hardened off. This being so I consider late February to be about right for sowing seeds of antirrhinums, lobelias, salvias, ageratums, *Begonia semperflorens*, nicotianas, petunias and verbenas, and late March for stocks, asters, French and African marigolds, nemesias, alyssum and cosmeas which germinate and grow more rapidly.

When the time comes for setting out the plants in the border (after hardening off), remove them from the boxes with care so that the delicate roots are not damaged. Plant them firmly with the aid of a trowel and water them in afterwards.

Biennials. A great deal of what I have had to say about increasing annuals also applies to biennials. Any experienced gardener will stress the importance of early sowing for biennials. This is done outdoors in a prepared seed bed during late spring or early summer, with the seedlings being transplanted into a nursery bed in July and into their flowering quarters by the autumn – or the following spring if they are not ready for the move soon enough for establishment before the onset of winter.

Perennial Plants

So far as I am concerned, it would be unthinkable to have a garden which did not include some perennial plants, including both herbaceous perennials and half-hardy perennials like the invaluable pelargoniums (geraniums), marguerites, dahlias and chrysanthemums, the last two so important that I propose dealing with them in chapters of their own (pp. 169 to 182 and pp. 183 to 202 respectively). These, with the exception of the geranium, are grown in a conventional herbaceous border or in one or other of the alternative ways which I have outlined on p. 101. After all, they have so much to offer between April and November and their wide diversity of form and colour make them perfect flowers for the gardener with a taste for experimenting. When you have planted trees and shrubs you *have* planted them in the sense that, if it becomes obvious later that they are in the wrong place, moving them to a new position is an operation fraught with some hazard. Herbaceous plants, on the other hand, mostly need to be lifted and divided once every three or four years, and a general re-shuffle to give new plant associations presents no problems, provides new interests and can give the garden a distinctive new look.

With the remarkable upsurge of interest in flower arrangement in recent years I am sure that we are all more conscious nowadays of the value of flowers for brightening up the home, and once one becomes keen on herbaceous plants it is natural to pay particular regard to those—and there are plenty of them—which play a dual role as good garden plants and 'geese laying golden eggs' in the shape of flowers for cutting. These can be grown either in a border or in a special small plot hidden away in a corner of the garden. A lot of gardeners find it too painful to see their wives raiding the border and a special cutting border then eliminates one source of matrimonial strife!

What I want to do now is to draw attention to the perennials which have much to offer the home gardener in their own individual ways. Some are old friends, some are more recent introductions which are superior to other varieties which have been popular in the past. Omissions there must be if this list is to be kept within reasonable bounds, and all I would claim for my recommendations is that they form the basis on which a splendid collection of plants could be made.

List of Plants

Acanthus (Bear's Breeches). An acanthus I have considerable regard for is *A. mollis latifolius*. This is what I would call an 'architectural' plant with large, distinctive leaves, rather like those of the thistle, and 4- to 5-ft. spikes of white, purple-marked flowers in late summer. This is a good plant for a natural garden or for planting by water. It responds to good soil conditions and is increased by division or root cuttings.

Achillea (Yarrow). I would imagine that the 4-ft. *A. filipendulina* Gold Plate is one of the best-known and most popular of all herbaceous plants. It is certainly distinctive with its large, flat, yellow flower heads borne proudly above grey, ferny

foliage between June and September. Half its height and equally garden worthy is the white-flowered *A. ptarmica* The Pearl. This has small, double flowers and their colour I find especially useful in the border for separating other plants with flowers of discordant colours. The achilleas like full sun and are good plants for dry borders. Increase by division or cuttings.

Aconitum (Monkshood). Growing well in shade and in sun the aconitums are useful plants but remember that they are poisonous and so could be a danger to children. Several reach 5ft. in height but others are considerably shorter and my choice would be the 3-ft. Bressingham Spire, which usually needs no supporting. Its flowers are violet-blue and it is July-August flowering. Increase by division.

Alkanet, see Anchusa

Alstroemeria (Peruvian Lily). The 3-ft.-tall *ligtu* hybrids have become very popular in recent years and these are rather lovely with their lily-like flowers in shades from pink to orange. The crowns of the plants must be protected in winter and they need a sheltered position in full sun with a light, well-drained soil. I would not personally grow them in an herbaceous border because I find that they spread too freely, but I consider them excellent for a bed under a warm south-facing wall. Increase by division in October, the best planting time, but leave undisturbed as long as possible.

Althaea (Hollyhock). These never look better than when planted against stone walls, and with their height equalling or exceeding that of a tall man they can look very fine indeed. With this height staking is necessary, of course. They like a rather rich soil but if they do not get it they will still make a good show. Increase by cuttings if you want to perpetuate choice varieties, otherwise raise plants from seed sown outdoors in May. These are often treated as biennial plants, although true perennials. They are not grown as much nowadays, though, as they used to be because they get badly affected by rust.

Alum Root, see Heuchera

Anchusa (Alkanet). I have my reservations about anchusas but I find that they are so attractive with their pretty blue flowers and so useful that I would be loath to do without them. The summer-flowering *azurea* (*italica*) varieties vary in height from 3 to 5ft., and an outstanding introduction of recent years is the gentian-blue Loddon Royalist. Unfortunately, all these varieties deteriorate rather quickly. *A. caespitosa* makes a plant of only a little over 1ft. in height and flowers in late spring and summer. The flowers have a compelling richness of colouring. Plant in a sunny position. Increase the species by seeds, which may be sown outdoors in May, and named varieties by root cuttings taken in winter.

Anemone. For late-season display I am especially fond of the Japanese anemones (varieties of the 2½-ft. *A. hupehensis*). I find the grace of the pink, red and white flowers most appealing. They flower well in partial shade and even near large trees, provided the cover is not too dense. They will grow well in almost any soil, but they do not like to be moved once established. The plants do tend to run, though, and unless cut back or restricted they will grow into other plants. Increase by division, when this is necessary.

Anthemis. The 2½-ft.-tall, rich yellow *A. tinctoria* Beauty of Grallagh and *A. t.* Grallagh Gold are admirable border plants for dry conditions, and they flower over a long period in summer. They are also excellent for cutting. Increase by cuttings taken in summer and rooted in a frame.

Aquilegia (Columbine). The natural, effortless grace of the aquilegias is so well known to all that I need not spend time extolling it. These are lovely flowers, and the long-spurred hybrids in particular are favourites for adding interest to the early summer border. The McKana Hybrids are a good strain with flowers of good size in a pleasing range of colours. These are 2½- to 3-ft.-tall. Also much grown are the Mrs Scott-Elliot Hybrids, also with a splendid colour range and a height of about 3ft. Increase by seed sown in a cold frame in spring or outdoors in April or May.

Artemisia. If it is possible to find room for it, the 5-ft.-tall *A. lactiflora* is a useful plant for the late summer-autumn border. Its flowers are white to creamy-white and are carried in bold panicles up to 2ft. long. It needs a sunny position and well drained soil. Increase by division.

Aster. Autumn is, inevitably, a melancholy time for gardeners, beautiful though it is when the leaves of so many deciduous trees and shrubs take on their brilliant colours. But before the greyness of winter there is always the joy of the Michaelmas daisies, which from late summer to November provide us with a glorious range of what I would call rich yet 'gentle' colours. My choice for sheer usefulness would always be the dwarf varieties, around 9in. to 1½ft. tall, as opposed to the 2½ to 4ft. of the admittedly lovely standard *novae-angliae* and *novi-belgii* varieties. One *novi-belgii* variety I do find very useful is Blandie, whose creamy-white, double flowers are a perfect foil for Michaelmas daisies of other colours. Of the dwarfs I like especially the pink, semi-double Countess of Dudley, the white Snow Sprite, the rose-pink Margaret Rose, and Lady in Blue, a rich blue variety. The intermediate-sized (2½ft.) *A. amellus* varieties start to come into flower in July, the main flush comes in August and September, and numerous varieties continue into October. All Michaelmas daisies, with the exception of the *amellus* varieties, need increasing by division each year, the latter at three-yearly intervals.

Alstroemeria ligtu hybrids, with flowers in shades from pink to orange, are handsome plants to grow in a warm, south-facing bed

Japanese anemones, graceful and easily pleased plants for August to October display, will grow well in partial shade and in almost any soil

For early-summer colour and general interest the long-spurred aquilegia hybrids are a natural choice

Astilbe. Most gardeners now know the astilbes as astilbes, but to older generations of gardeners they are still spiraeas and I suppose will remain so. Still, we need not worry too much about the name switches of the botanists; more important is the garden value of the plants in question. I delight in the showy plumes of the *arendsii* hybrids which are a natural choice for any position which is well supplied with soil moisture. Popular varieties are the dark red Fanal, the rose-pink Rhineland, the peach-pink Peach Blossom and the white Avalanche. All these are between 2½ to 3ft. tall. Increase by division.

Balloon Flowers, see Platycodon
Bear's Breeches, see Acanthus
Bee Balm, see Monarda
Bellflower, see Campanula
Bergamot, see Monarda

Bergenia. Perhaps many gardeners would not agree with me, but I would not plant the useful bergenias (these also have suffered at the hands of the botanists, for they have in the past been called megaseas and saxifragas) in an herbaceous border. Their large, leathery leaves make splendid ground cover but their right place so far as I am concerned is beside a pool, on top of a rock wall or even in a planting of their own beside a path which they overlap somewhat. The pink or white flowers are carried in bold spikes in March and April and the plants are as happy in the sun as in the shade. The 1-ft.-tall *B. cordifolia* is perhaps most widely grown (it has rose-magenta coloured flowers) and the method of increase is by division. It is tolerant of widely differing soil conditions.

Bleeding Heart, see Dicentra

Campanula (Bellflower). Two of my favourite flowers for July-August flowering are the 5-ft.-tall light blue *C. lactiflora* and its rather shorter mauve-pink variety Loddon Anna. Shorter again (mostly 2 to 2½ft.) are the varieties of *C. persicifolia* in shades from violet-blue and dark blue to lavender and white. These provide fine flowers for cutting. Then for early summer flowering there is the 1-ft. *C. glomerata dahurica* which carries heads of violet-purple flowers which are very showy. All these can be grown best in sun although the varieties of *lactiflora* will also do well in light shade. Increase by division.

Catananche. A good front-of-the-border plant which can also be dried for winter decoration is *C. caerulea major*. This variety has shapely lavender-blue flowers and makes its display during the latter half of the summer. Give it a sunny position and increase by division.

Catmint, see Nepeta

Centaurea. The pink-flowered, 2-ft.-tall *C. dealbata*, handsome in its own right, has a better variety in *steenbergii*, of richer colouring. These cornflowers are late summer-early autumn flowering and the colours of these fine border plants are

The mauve-pink *Campanula lactiflora* Loddon Anna, which flowers in July and August. It grows to about 4½ft. tall and has become a favourite of many gardeners

enhanced by the greyish foliage. Increase by division.

Christmas Rose, see Helleborus

Chrysanthemum. Among the perennial chrysanthemums one usually thinks first of the varieties of *C. maximum*, the Shasta Daisy. Of these splendid white-flowered varieties I would give pride of place to the double Wirral Supreme. Esther Read, also double, and Mount Everest, single, are others to remember. All are 2½ to 3ft. tall and July-August flowering. The flowers are excellent for cutting. The autumn-flowering *rubellum* varieties are also good border plants. Increase by division.

Another chrysanthemum which few will know by its botanical name of *C. frutescens* but many will know by its common name of Marguerite is an invaluable plant for summer bedding. The yellow and white flowers on 3-ft. stems make a splendid show when the plants are grown *en masse*. The Marguerite is tender and is raised under glass from cuttings, the plants being bedded out in late May or early June. They can be lifted before frost threatens and overwintered under glass, if so desired.

Coneflower, see Rudbeckia

Coreopsis. By far the best known coreopsis is *C. grandiflora* Badengold, a superb variety for June-September flowering. It gives of its best when subjected to poor soil conditions. In rich soil flower production is likely to be poor. The bright yellow flowers are carried on 3-ft. stems and are good for cutting. For the front of the border the shorter, rich yellow *C. verticillata* is recommended. Increase by seeds or from cuttings rooted in a frame in April.

Dahlia. These showy border flowers are dealt with in detail in Chapter Seven and I am confining my remarks now to the dwarf bedding type—splendid plants for a quick display of colour.

The foxtail Lilies (eremuri) are the aristocrats of the flower border, their lofty and stately flower spikes appearing in May and June

These grow about $1\frac{1}{2}$ to 2ft. tall and carry their colourful flowers—including purple, pink, red, yellow and white—from July to October, or November if frost holds off until then. All dahlias are tender and bedding out cannot be safely done until late May or early June. Dwarf bedding dahlias are usually raised from seeds. For details of this method of increase see Chapter Seven (p. 176).

Day Lily, see Hemerocallis

Delphinium. The large-flowered *elatum* varieties, which vary in height from 3 to 7ft., offer the gardener a wonderful range of soft colours—from purple and mauve to blue, lavender and white. The dainty Belladonna varieties, which have a freely branching habit and carry their flowers in loose sprays, are 3 to 4ft. tall and also have delightful soft colourings. All are June-July flowering and need a well-drained, rich soil and shelter from strong winds. Increase from cuttings rooted in a cold frame in spring or from seeds sown under glass in August or spring.

Dianthus (Pinks). The garden pinks are among our most-loved border flowers, the best-known variety without doubt being the white Mrs Sinkins. This, although it splits its calyces, is still highly valued by thousands of gardeners. Another type of pink is a cross between the typical pink and the perpetual-flowering carnation. This type is called *Dianthus allwoodii* and there are many very attractive named varieties in an excellent range of colours.

Pinks are easy plants to please, needing only a well-drained soil of reasonable quality which is not of an acid character. If lime is deficient in the soil add old mortar rubble. The best planting months are March and October. Increase is by means of seed sown in light, sandy soil in spring or by cuttings rooted in a cold frame in summer.

Dicentra. *Dicentra spectabilis*, the Dutchman's Breeches or Bleeding Heart, is one of the most beautiful of all herbaceous plants but, unfortunately, it cannot be relied upon from year to year. The rosy-red blooms are carried on arching stems during May and June, and very handsome these are when seen in conjunction with the deeply cut, fern-like foliage. This species grows to $1\frac{1}{2}$ to 2ft. tall. The smaller *D. eximia*, 1 to $1\frac{1}{2}$ft. tall, and with reddish-purple flowers, is a hardier and more reliable plant to grow. I have often grown *spectabilis* as a greenhouse plant for forcing and it is very well suited for this treatment. In the border, though, I have found that it is liable to die back very suddenly leaving a space to be filled at short notice. What it needs is plenty of moisture at the roots and shade from strong sunshine. A peaty soil mixture is also appreciated. Increase by division.

Doronicum (Leopard's Bane). The doronicums are especially useful plants because of their early flowering, and their qualities as garden plants and as cut flowers. The best known variety is Harpur Crewe (*D. plantagineum excelsum*) with golden-yellow flowers borne on $2\frac{1}{2}$-ft. stems. This flowers in April and May, sometimes continuing into June. Another good variety is the rich yellow Miss Mason. Increase by division.

Dutchman's Breeches, see Dicentra

Echinops (Globe Thistle). The common name comes, of course, from its globular, thistle-like flower heads which stand out prominently in any collection of hardy plants. These flowers are excellent for cutting. The best-known species is *E. ritro*, 4ft. tall, which carries its deep blue flowers over a long period in summer. Rather taller and even more imposing is the variety Taplow Blue. This has flowers of an intense dark blue. Deep digging is advisable for this plant so that the tap root it develops can delve down easily. Leave it undisturbed as long as possible. Increase by division.

Eremurus (Foxtail Lily). One has only to see a planting of eremuri to know that here are natural aristocrats of the flower border. Their tall flower spikes are a splendid sight in May and June. These plants have fleshy roots which makes them very susceptible to damage when the drainage is rather poor—hence the need for efficient drainage if one hopes to succeed with eremuri. The lovely pink *elwesii* is fairly well known and this makes a plant some 5ft. tall—often considerably more. Mulch the plants with straw or manure after the growth has died down in autumn and leave them undisturbed for many years. Increase by seeds sown under grass in spring.

Erigeron (Fleabane). The erigerons are attractive for garden decoration and cutting, and their daisy flowers are very much part of the summer scene, providing colour from June to September and often to the end of October. Varieties like the rich pink Gaiety, the lilac-blue Festivity, the pale pink Charity and the mauvy coloured Sincerity are well worth including in the planting scheme. The first two are 2ft. tall and the last about 2½ft. Increase by division.

Eryngium (Sea Holly). The highly decorative eryngiums have teasel-like flower heads encircled with spiny bracts. The foliage, too, is attractive and the flower stems are cut as ever-lasting flowers for the winter. I do think, though, that these plants need to be given quite a prominent position in the border for they are liable to be rather lost among plants of similar height. The violet-blue *E. oliverianum* and *E. tripartitum*, with steely blue flowers and both around 2 to 2½ft. tall, are grown for July-August flowering; the variety Violetta, violet-blue, will continue the display into September. Eryngiums make deep-delving roots, so, once planted, leave them alone for as long as they retain their vigour. Increase the species by seed sown in a cold frame in March or outdoors in early June; alternatively, take root cuttings in February and start these into growth in a cool greenhouse.

Evening Primrose, see Oenothera
Fleabane, see Erigeron
Foxtail Lily, see Eremurus
Fuchsia. The fuchsias are shrubs—in most cases tender, although there are some hardy varieties—but I am including them here because of their great value as bedding plants for the border. When used for bedding they are planted out in late May and taken back into the greenhouse at the end of September. Good bedding varieties include the red and purple Mrs Popple, the pink and white Mrs W. P. Wood and the cerise and mauve Tom Thumb. Increase by cuttings in spring or summer in a cool greenhouse.

Gaillardia. These are especially free-flowering plants and their season lasts from late June to late August but in some parts of the country, as with me, they are liable to be lost during the winter. Spartan soil conditions suit them best and the lighter the soil and the freer the drainage the better. Plant in full sun. Excellent varieties are the reddish-bronze Wirral Flame, the deep yellow and red Ipswich Beauty and Golden Queen. All are about 2½ft. tall and all are splendid for cutting. Increase by division.

Geranium, see Pelargonium
Geum. The numerous colourful varieties of *G. coccineum* are easily grown border plants for flowering in late summer and autumn. Like the gaillardias they grow best in poorish soil which is really free-draining, especially in winter. The scarlet Mrs Bradshaw and the yellow Lady Stratheden, both 2ft. tall, are popular varieties. Named varieties come fairly true from seed, but with this form of increase variations must be expected. For completely true reproduction division in spring must be resorted to.

Globe Thistle, see Echinops
Golden Rod, see Solidago
Gypsophila. Perhaps the best known of the elegant gypsophilas is the 3-ft.-tall Bristol Fairy. With its feathery growth and masses of small, double, white flowers it acts as a counterpoise to 'heavyweights' like the sunflowers and peonies. The flowers are good for cutting. Flamingo is identical to Bristol Fairy except for its pink flower colouring, and the 1-ft.-tall Rosy Veil, with double pink flowers, is another useful variety. These varieties are increased by root grafting in March or April.

Helenium. The heleniums are special favourites of mine, both because I like their appearance and because they are exceptionally free-flowering and undemanding. The flowers are borne from June until the end of summer in shades of yellow, bronze and crimson. Good varieties are the 3-ft.-tall, bronze-red Moerheim Beauty, and the late-

The bright yellow *Helenium* Butterpat, some 3ft. tall, flowers during August and September. Heleniums are especially free-flowering plants

flowering Butterpat of similar height. Give them an open, sunny position if possible, although they will not mind light shade. Any ordinary soil will do. Increase by division.

Helianthus (Sunflower). The late summer-early autumn flowering Loddon Gold with golden-yellow, double flowers on 5-ft. stems is a splendid variety for general planting. Increase by division.

Helleborus. The Christmas Rose, *H. niger*, 1ft. tall, is always a joy with its mid-winter flowering. The large white flowers are easily weather-damaged though, and it pays to protect them with cloches. Give them a sheltered position in partial shade and a good loamy soil, if possible. The same goes for the Lenten Roses (forms of *H. orientalis* and a little taller than the Christmas Rose) which flower a little later. Increase by division.

Hemerocallis (Day Lily). The lily-like flowers

last only one day, hence the common name, but as the succession is maintained through several weeks in summer this is not of moment. The colour range has been greatly improved in recent years and there are now varieties in shades of buff, pale yellow, pink, reds and orange. One can hope that the range will be extended still further. They are easy to grow, liking moist soil but doing well under drier conditions. They are suitable for full sun or partial shade. Most come within the height range of 2 to 3ft. I do not think they are really plants for the herbaceous border. They look better in a natural setting—say around a pool in a wild garden or in a separate bed with a background of shrubs informally planted. Increase by division.

Heuchera (Alum Root). These first-rate border plants for a front-row position carry delicate sprays of pink or red flowers above very handsome foliage. The Bressingham Hybrids, up to 2ft. tall, give a delightful selection of shades of

with Hemerocallis, there are reservations again: they are ideal for the mixed border planted amongst and in front of shrubs but I do not think they are so good for the purely herbaceous border. Some readers may disagree. The large, broad leaves of the hostas are glaucous or variegated, and the lilac or lavender flower spikes are an additional adornment, although secondary to the foliage in decorative value. These plants are particularly useful for north-facing aspects or for planting under trees which cast light shade. Soil enrichment with well-rotted manure and deep digging before planting certainly pays. Species include the blue-grey-leaved, 2-ft.-tall *H. fortunei*, the 2½-ft.-tall, green-leaved *H. plantaginea* (with fragrant white flowers), and *H. lancifolia*, with dark green leaves and lilac flowers. Disturb these plants as infrequently as possible. Increase by division.

Incarvillea. I would always find a place for the

The Christmas Rose, *Helleborous niger*. Their delightful white flowers, which appear in mid-winter, are best protected with cloches against the ravages of the weather

Heucherella tiarelloides, a plant of unusual grace. The rose-pink flowers, with their airy appearance, add distinction to the front of the border

these two colours. There are also splendid named varieties of *H. sanguinea*, such as the crimson Pluie de Feu and the pale pink Edge Hall. Give them an open position in sun. Increase by division.

Heucherella. The attractive 1-ft.-tall *H. tiarelloides* is a cross between *Heuchera brizoides* and *Tiarella cordifolia*. It carries its neat panicles of rose-pink flowers above the foliage and is a choice plant for the front of the border. Another splendid heucherella is the taller (almost 2ft.) Bridget Bloom, which carries its pale pink flowers in May-June and again in late summer-early autumn. Increase by division.

Hollyhock, see Althaea

Hosta (Plantain Lily). It is strange how the popularity of plants waxes and wanes with the generations. Hostas after years in the wilderness are now general favourites—and very glad I am to see this, too, for they are fine foliage plants. However, as

handsome 2-ft.-tall *I. delavayi*. With its rosy-magenta, gloxinia-like flowers it looks like an escape from the greenhouse, and it does in fact need a sunny, sheltered position. The soil should be fairly rich and retain plenty of moisture in summer. It flowers in May and June. Increase by seeds sown in the greenhouse in February or in a cold frame in March or early April. It can also be increased by division.

Iris. The May-June flowering tall bearded irises, lovely as they are, are not in my opinion good plants for the herbaceous border. The flowering period is short and the plants must be given quite an exposed position because it is essential that the sun should be able to reach the rhizomes to ripen them off between the end of June, after flowering, and late summer. Pockets of iris foliage and no flowers for weeks on end do not enhance the appearance of a border. No, these plants need beds on their own, a sunny, warm position and well-drained soil liberally dressed with old mortar rubble. The many modern varieties have exquisite colourings and range in height from 2½ft. to 4ft. Increase by division.

Kniphofia (Red Hot Poker). Kniphofias are not the hardiest of border plants but careful positioning will ensure that they come safely through more severe winters than usual. The brilliantly coloured flower spikes are invaluable in the border. The tallest species and hybrids reach up to 5ft. tall but there are others like the charming orange, September-flowering *galpinii* which only reach 1½ to 2ft. The 4-ft. *uvaria* is the most commonly grown species and splendid varieties include the yellow and scarlet Royal Standard and the deep yellow Buttercup, 3½ and 3ft. tall respectively. In spite of their height I often bring some of these near the front of the border to show off their beauty to greater effect. Give them an open, sunny position well provided with soil moisture, but remember that waterlogging in winter will not be tolerated. Spring is the best time for lifting and planting. Increase by division.

Leopard's Bane, see Doronicum

Liatris. An unusual feature of the reddish-purple *L. spicata* – a plant I especially like – is that the flower spikes open from top to bottom, a reversal of the normal practice. This is a showy 3-ft.-tall plant which needs plenty of sun and moisture in the growing season and good drainage in the winter. More than a foot taller is the purple *pycnostachya*. Plant in spring, and increase at that time by division.

Limonium. The lavender-blue Sea Lavender, *L. latifolium* (formerly *Statice latifolia*), is a delightful plant for the summer border. It carries sprays of delicate little flowers on 2-ft. stems in July and August and these can be dried for room decoration. Grow in fairly light, well-drained soil. Increase by seed or division.

Lupin. The arrival of the Russell lupins in many gay self and mixed colours transformed this genus for gardeners. Give them well-drained soil but make sure that it is not too rich or the plants will be encouraged to form fleshy roots which are liable to fall victim to wet conditions and frost. Increase at quite frequent intervals from seed or cuttings. If seed is sown you will have to be content with mixed colours. With a height of 3 to 3½ft. or rather more these are especially useful border plants.

Lychnis. The scarlet *L. chalcedonica* is a first-rate border plant for mid-season flowering. It grows 3ft. tall or perhaps a little less and has no special cultural requirements, except good drainage. Increase by seed or division.

Meadow Rue, see Thalictrum

Monarda (Bee Balm or Bergamot). For soils which do not dry out readily I would recommend the attractive *M. didyma* Cambridge Scarlet, 2½ to 3ft. tall. I find its shaggy looking scarlet flowers unusually appealing, and a pleasant aroma is given off by the leaves. The varieties of *didyma* also include Croftway Pink and the bluish-purple Burgundy but in my opinion neither these nor others unmentioned have the attractions of Cambridge Scarlet. Increase by division.

Monkshood, see Aconitum

Mullein, see Verbascum

Nepeta. The common Catmint, *N. faassenii* (often incorrectly called *N. mussinii*) is an excellent edging plant of about 1½ft. in height. It bears its lavender-blue flowers from June to September and the soft colouring of these and the grey foliage make it a suitable companion for many other plants. The variety Six Hills Giant grows considerably taller (up to 3ft.) and has a rather shorter flowering season. The nepetas love sun and light soil, so try to provide them with both, if possible. Increase from cuttings made from young growths in summer and rooted in a cold frame.

Obedient Plant, see Physostegia

Oenothera (Evening Primrose). The various forms of *O. fruticosa*, from 1½ft. to 2ft. tall, are good border plants, with yellow flowers in July and August. They need sun and a lightish soil. Increase by seed sown in March or April in a cold frame, or by division in March.

Paeonia (Peony). I am especially fond of the double-flowered peonies but I also wish the single-flowered varieties received more attention. For instance, the choice lemon-yellow-flowered, 2-ft.-tall *P. mlokosewitschii* (disregard the dreadful name) which is a joy in spring; or the lovely white June-flowering, 2-ft.-tall *P. lactiflora* and its varieties. Of the doubles I would mention the creamy-white Duchess of Nemours, the red Felix Crousse and the pink Sarah Bernhardt as representative of the many magnificent varieties available. The common peony, *P. officinalis*, of much the same height as the foregoing species, is one of the sources of our present range of herbaceous peonies – the other is *P. lactiflora* – and its variety, *albo-plena* can be recommended.

Peonies need a deeply dug, rich soil, well manured before planting and top-dressed in October or November with well-decayed manure or compost, lightly forked in. Water the plants in dry weather in the growing season. I like to give them a place near the front of the border where their flowers and the autumn colours of the foliage can be fully enjoyed. Disturb them as infrequently as possible for they take a long time to settle down after moving. Increase by division.

Papaver (Poppy). The perennial poppy, *P. orientale*, 2½ft. or a little more, has produced many fine garden varieties and these will grow well in any well-drained soil. Crimson, scarlet, orange-

The elegant *Sidalcea* Rev. Page Roberts, a 3-ft. tall perennial which flowers from June to September

The striking *Rudbeckia speciosa* Goldsturm, a splendid variety of the species
known as Black-eyed Susan. All
rudbeckias appreciate a sunny position

Above. The showy annual chrysanthemum, *C. segetum*, so excellent for providing colour in the latter half of summer

Left. An annual border is of very special interest, as it provides the gardener with an opportunity to 'ring the changes' each year both with regard to the choice of plants grown and the way in which they are associated one with another. The choice is wide and the permutations endless

There are no finer bedding plants th'an the petunias, available in many lovely colours

scarlet, cerise, salmon-pink and white are the colours one gets in these showy flowers. Their fault is that the foliage looks rather untidy after flowering has finished in June and some clever masking with other, later flowering plants is necessary. Increase by seed sown outdoors in May or early June if one is not concerned about getting true reproduction. With named varieties take root cuttings in February and start them off in a cold frame or cool greenhouse.

Pelargonium (Geranium). No plants 'earn their keep' more than the zonal and ivy-leaved pelargoniums (or geraniums as most of us call them). There are many varieties available; some of the zonals have attractively variegated leaves, and these are very showy when bedded out. The ivy-leaved varieties can either be allowed to trail or they can be trained on canes if one wants to create a rather different effect. Popular zonal pelargoniums include the vermilion Maxim Kovalevski, Gustav Emich and Paul Crampel, of much the same colouring as the first-mentioned. Two fine ivy-leaved pelargoniums are L'Elégante, white and purple, with variegated leaves, and the purple Abel Carrière. Both the zonal and ivy-leaved pelargoniums are tender and, after being raised from cuttings rooted under glass in a temperature of about 13°C. (55°F.), have to be hardened-off before planting outdoors when all danger of frost has passed. They must be brought under glass again at the end of the season before the cold weather arrives.

Peony, see Paeonia

Peruvian Lily, see Alstroemeria

Phlox. The herbaceous phlox, varieties of *P. paniculata*, mostly from $2\frac{1}{2}$ to $3\frac{1}{2}$ft. tall, are superb border plants for July and August colour. These will provide bold splashes of colour, especially if one grows varieties like the orange-red Brigadier and the salmon-orange Spitfire. There are also fine purples, violet-blues, pinks and white to choose from. They like a rich soil and shade from strong sunshine. Increase by division.

Physostegia. The Obedient Plant, *P. virginiana*, has a variety, Vivid, with bright rose-coloured, tubular flowers. These appear in September and continue into the autumn. This variety is only about $1\frac{1}{4}$ft. tall and is a good plant for the front of the border. It succeeds in either sun or light shade. Increase by division.

Plantain Lily, see Hosta

Platycodon (Balloon Flower). The $1\frac{1}{2}$-ft. *Platycodon grandiflorum* is a handsome, late-season plant which belongs to the campanula family and has deep blue flowers of the familiar campanula cup shape. The common name comes from the blown-up appearance of the buds immediately before they open. For preference it likes light shade but there is no reason to expect it to be inferior if placed in full sun. Again it will grow in any reasonable soil but a deep loam with plenty of body in it suits it best. Increase by division in spring.

Poppy, see Papaver

Primula. I could not even start to do justice to the huge genus *Primula* in the space I have at my disposal now. The colourful garden polyanthus belong here, and what a show these can put on if given a fairly rich soil which does not dry out and, preferably, a shady spot. Then for planting in moist places, again preferably in light shade, there are the Candelabra primulas like the reddish-purple *P. beesiana*, the orange *P. bulleyana*, the purplish-red *P. japonica*—all about $1\frac{1}{2}$ft. tall—and the splendid *P. pulverulenta*, rich red flowers with white meal-covered foliage, and its lovely pink Bartley Strain all about 2 to 3ft. tall. All the members of the Candelabra section are easily recognised because they carry their flowers in tiers on the stem. Another fine primula is *P. denticulata*, about 1ft. tall and with round heads of pale purple, lavender or white flowers in spring. To increase polyanthus sow seeds in a cool greenhouse or frame in March. Division of the roots in spring or after flowering is another method of increase. The same remarks apply to the other species mentioned.

Pyrethrum (Coloured Marguerite). The pyrethrums belong to the chrysanthemum family and their botanical name is in fact *Chrysanthemum coccineum*. By any name though they are splendid plants—so handsome in the garden and so useful as cut flowers. They have a reputation for being tricky to grow but the answer almost always, if there is trouble, is that they have not been given a well enough drained soil—and, of course, they must have sun. This means that they are liable to die off in heavy clay soils which lie wet in winter. One way to help overcome this question of free drainage is to grow the plants in raised beds. There is a fine range of single and double varieties of about 2 to $2\frac{1}{2}$ft. in height, the first group including the salmon-pink Eileen May Robinson and the white Avalanche as well as numerous other varieties in shades of pink and red; the second group includes the popular crimson J. N. Twerdy and White Madeleine. The best planting months are March, April and July, and increase is by division (which should be done at planting time and with especial care) or by seeds. I prefer the latter method as I do not mind having a mixture of colours. I sow the seeds in pans or boxes outdoors in May or early June and these germinate easily, giving good flowering plants for late spring and early summer the next year.

Ranunculus. If you have a moist, shady border or a suitable site by the waterside I would suggest growing *R. aconitifolius flore-pleno*, whose popular name is Fair Maids of France. This carries its double white flowers in spring and grows about

Rodgersia pinnata, a bold-leaved plant which bears pink flowers in early summer. It needs shade and much moisture at the roots

The mauve *Scabiosa caucasica* Clive Greaves, 1½ to 2ft. tall. This and the other *caucasica* varieties are among the best of all border flowers for cutting

The rosy-salmon *Sedum* Autumn Joy. Like other sedums it likes a deep, rich, loamy soil and a sunny position

2ft. tall. An essential requirement is a constant supply of moisture at the roots. Increase by division.

Red Hot Poker, see Kniphofia

Rodgersia. The rodgersias are early-summer-flowering plants also needing shade and lots of moisture at the roots. The flowers are borne in panicles on stems 3 to 4ft. tall and in autumn the leaves assume attractive tints. The leaves of the white-flowered *R. aesculifolia* resemble those of the horse chestnut. *R. pinnata* has pink flowers. Increase by division in March or early April.

Rudbeckia (Coneflower). An excellent flower for the summer border is the 2- to 3-ft.-tall *R. speciosa* (*R. newmanii*), or Black-eyed Susan as it is called. This is a spectacular plant with its deep yellow petals and purplish-black disk. Its habit is compact and the flowers are carried well clear of the foliage. Rather shorter is the splendid variety Goldsturm, with larger flowers of similar colouring. *R. nitida* Herbstsonne has yellow petals and a green disk and reaches a height of over 6ft., an attractive variety but a different proposition to the two preceding Coneflowers. Another tall variety, *R. laciniata* Golden Glow, with double flowers in autumn, is also around 6ft. tall and is first-rate for cutting. Give them a sunny position if possible and any reasonable soil. Increase by division.

Salvia. A perennial salvia which makes a splendid border plant is *S. superba,* which used to be known as *S. virgata nemorosa.* This bears many erect bluish-purple flower spikes in July and August and with a height of 3ft. is often used for the middle or forefront of the border. There is a variety of only 1½ft. named Lubeca and this has flowers similar to the type. Plant in a sunny position. Increase by division.

Scabiosa. The varieties of *S. caucasica* are noted for their excellence as cut flowers and they are, of course, also handsome border flowers. The flowers are mauve, blue or white in colour and the best known variety by far is the mauve Clive Greaves. All are 1½ to 2ft. tall. They need an open, sunny position and well drained soil. The quality of the soil should also be good—and these plants like lime, which means that they should not be planted near *Thalictrum dipterocarpum* or its variety Hewitt's Double, which are better without this ingredient. Plant only in spring, when the plants may also be increased by division.

Sea Holly, see Eryngium

Sea Lavender, see Limonium

Sedum (Stonecrop). A plant I would not leave out of the border is the August-September flowering *S. spectabile,* for I find its large, flat, pink flower heads a special delight. Even better than the type is the variety Brilliant with flowers of richer rose colouring. These are low-growing plants, about 1 to 1½ft., and so are ideal for narrow borders in front of the house or for the front of large borders. Perhaps a little taller is the recently introduced Autumn Joy, a splendid variety of rosy-salmon colouring. These stone-crops like a deep, rich and loamy soil, and a sunny position. Increase by division.

Shasta Daisy, see Chrysanthemum

Sidalcea. The pink, rose and red spikes of the sidalceas are a feature of many borders in the latter half of the summer. With me, they tend to die out but I know that in some other gardens they seed themselves freely. The pink Rev. Page Roberts, the salmon-pink William Smith and Rose Queen are good varieties. Most varieties are around 3 to 3½ft. tall. Plant in sun if possible. Increase by division.

Solidago (Golden Rod). These are accommodating yellow plants growing well in sun or shade and in any reasonably good garden soil. They vary in height from a modest 2ft. to the 5-ft.-tall Golden Wings. Shorter varieties include the autumn-flowering Goldenmosa and the rather earlier flowering, pale yellow Lemore. Increase by division.

Statice latifolia, see Limonium
Stonecrop, see Sedum
Sunflower, see Helianthus

Thalictrum (Meadow Rue). *Thalictrum diptero-carpum*, about 5ft. tall, is another plant I would not want to be without. It has character with its large, branching panicles of lavender flowers and graceful foliage. The flowers look delightful when cut but they fall rather quickly. Its double variety, Hewitt's Double, is similar in every way except for its flowers of lavender-lilac colouring. While the species may be easily raised from seed (sown in a cold frame in March or in the open garden in May), the double variety must be increased by removing and potting young shoots in spring. Leave the plants undisturbed as long as possible.

Tradescantia. The three-petalled flowers of *T. virginiana* of violet-purple colouring are often seen near the front of summer borders. The plant grows between 1½ and 2ft. tall and will thrive in partial shade or sun. Varieties of different colours include the pale blue J. C. Weguelin, the white *alba* and *rosea*, a pink variety, and all are much the same height as the type. Their real value lies in the period of flowering—from early in the season right through to autumn. There is no need to stake the plants. Increase by division.

Trollius (Globe Flower). The trollius are plants for shade or waterside and what a gay picture they make when carrying their orange or yellow flowers in May and June. Two good varieties are Gold-quelle and Orange Globe. They like a fairly rich soil and must not lack moisture at the roots. Increase by division.

Verbascum (Mullein). There are numerous highly decorative verbascums, ranging in height from 3 to 6ft. Many are biennial or are treated as

such but perennials include the yellow *V. chaixii* and *V. phoeniceum*, of variable colouring. There are good named varieties like Pink Domino and the yellow Gainsborough. Increase the species from seed and the named varieties by means of root cuttings taken in February and rooted in a frame or cool greenhouse.

Veronica. The medium-sized veronicas, like the rich blue *V. longifolia subsessilis* of 3ft. in height and the 1ft. Wendy with purple flowers and grey leaves, are easy plants to grow and their flower spikes are welcome in the late summer border. Another 2-ft.-tall herbaceous species, *V. gentianoides*, flowers in May and June. All these will do well given ordinary garden soil and a sunny position. Increase by division.

Yarrow, see Achillea

Annuals

I suppose every gardener is a dreamer of dreams on long winter evenings—and never more so than in the weeks following the arrival of the new-season's flower seed catalogues. It may be the descriptions of the novelties offered, or even more likely the coloured illustrations accompanying them, which, coupled with thoughts of many old favourites, sends most of us scurrying for pencil and paper. Annuals, when you come to think about it, are a very good way of getting almost 'instant colour' in the garden. The time interval between seed sowing and flowering makes these flowers invaluable to the garden owner who has not yet had time to make more permanent plantings, and to all of us they are a wonderful standby when gaps have to be filled at short notice. And if one has not the time or opportunity to raise one's own plants, then box-grown annuals like petunias, nemesias and matricarias can always be bought for summer bedding. Perhaps most important of all, without annuals our gardens would be much less gay than they are at present.

It might not be generally realised what tremendous efforts the seed trade in this country make to ensure that the annuals they are able to offer us not only maintain but exceed the standards of excellence of previous years. Competition is keen, of course, and individual seed houses not only have their own breeding programmes, but they bring in from abroad (from countries like Holland, Germany and Switzerland, the United States and Japan) the pick of the world's flower seed novelties.

So, for a modest outlay, we can carry out all manner of experimental plantings with annuals, in a full-scale annual border or in smaller plantings. This is a branch of gardening which can become fascinating once one really takes an interest in it.

Being a practical kind of chap I suppose it is

natural that I view with most favour those annuals which give a long period of colour. I am thinking of plants like calendulas, eschscholzias, clarkias, annual coreopsis, godetias, bartonia. *Statice suworowii*, nasturtiums, French and African marigolds, leptosyne, sweet peas and linum. I shall now give a brief guide to some of the most useful hardy and half-hardy annuals:

Hardy Annuals

Acroclinium. I have placed the 2-ft.-tall *Helipterum roseum* here because almost all seed catalogues continue to list it under its old name of *Acroclinium roseum*. This is a very useful 'everlasting' flower, rosy-red in colour, which is much used for winter decoration. It flowers from July to September.

Adonis. The Summer Adonis, *A. aestivalis*, 1-ft. tall, is a handsome annual with its crimson flowers (in June and July) and its indented, fernlike foliage.

Alyssum. Without the dwarf Sweet Alyssum, *A. maritimum* (now correctly *Lobularia maritima*, but this plant is still found in catalogues under alyssum), and its varieties in white, violet, lilac and other colours I do not know what we would do for edging summer bedding. The gay 3-in.-tall Pink Heather and the pink Rosie O'Day, about 4in. tall, introduced some years ago, are rapidly finding favour. Expect flowers from early summer to autumn. They like sun and poor soil.

Anchusa. There is one species of anchusa which is grown as an annual, *A. capensis*, and its variety Blue Bird, a lovely indigo-blue colour, is well worth growing. The flowers are borne on 18-in. stems.

Bartonia. I should really put *Bartonia aurea* under *Mentzelia lindleyi*, for that is its name now, but you will invariably find it listed under its previous name by seedsmen. This is a charmer with beautiful golden-yellow flowers on 1½-ft. stems from about June to September.

Calendula (Pot Marigold). No introduction is needed here for the Pot Marigold, as it is called, is certainly one of the best-loved and most colourful annuals and is so easily grown that it is bound to be a general favourite. There are varieties in shades from rich orange to lemon-yellow, apricot and creamy-white and with various flower shapes, and these grow to 2ft. and are splendid for cutting. Look out for Geisha Girl, with double orange blooms and incurving petals like a chrysanthemum. Also others like Radio, Orange King, Art Shades and Rays of Sunshine.

Centaurea. The annual Sweet Sultan, *C. moschata*, is a fine annual for cutting with a good range of colours. The flowers borne on 2-ft. stems, are larger than those of its much-loved relation, *C. cyanus*, the Cornflower, also renowned as a cut flower. Mauve, blue, rose, lavender and white are colours found here and in addition to the 3-ft.-tall varieties there are dwarf strains like Polka Dot of only about 1 to 1¼ft. They flower from quite early in the summer to September.

Chrysanthemum. There are various types of annual chrysanthemum, and all are very useful border flowers. They are varieties of *C. coronarium*, *C. carinatum* and *C. segetum*. All grow about 1½ to 2ft. tall. My favourites are the varieties of *carinatum*, the so-called Tricolor Chrysanthemum. The banded colours (including shades like scarlet, yellow, brown and white) are glorious and the flowers are excellent for cutting. Good mixtures are Double Monarch Mixed and Monarch Court Jesters Mixed, both 1½ft. tall. They flower throughout the latter half of summer, making a delightful show.

Clarkia. The 2-ft.-tall varieties of clarkia with their many, small, double flowers borne in delicate spikes, are popular for their fine colour range – from purple to scarlet and pink, with white thrown in for good measure – and usefulness for cutting. Flowers from July to September.

Collinsia. The collinsias are pretty flowers of about 1ft. in height. The salmon-rose *C. bicolor* Salmon Beauty is an attractive variety. Flowers July to September.

Convolvulus. Those sun-loving trailers, the annual convolvulus, just cry out for attention. These are varieties of *C. tricolor* in shades of

Dimorphothecas, whose popular name is Star of the Veldt, are sun-loving annuals from South Africa with a colour range which includes orange, yellow, salmon, apricot and white

purple and rich blue, and they include an especially good one in the wedgwood blue Royal Ensign. Flowers throughout summer.

Coreopsis. The annual coreopsis, which may be found listed under *Calliopsis* in catalogues, are useful for providing rich colours in the border—golden-yellow and deeper shades of red. The yellow varieties are strikingly zoned at the base of the petals with reddish-brown. Mostly in the 1½- to 2-ft. height range, there are also dwarf varieties some 8 or 9in. tall which are splendid for bedding purposes. Flowers July to October.

Cornflower, see Centaurea

Delphinium. The annual delphiniums (or Larkspurs) are universal favourites, superb for cutting, and various types are available in a good range of colours—from white, lavender and deep blue to crimson—and heights from 1 to 4ft. It is best to sow the seeds in autumn.

Dimorphotheca (Star of the Veldt). The sun-loving South African dimorphothecas are beautiful flowers which include orange, yellow, salmon, apricot and white in their colour range. Height 1ft.; flowers from July to October.

Eschscholzia. The 1-ft.-tall eschscholzias, with their poppy-like flowers in many lovely shades (from orange and scarlet to cerise, rose, pale yellow and white) and handsome, finely cut, grey foliage, provide welcome colour from July to October. They are, however, liable to seed themselves freely, producing progeny of inferior quality. Excellent for growing as mixtures or in groups of named varieties.

Gilia. The 1-ft.-tall *G. tricolor* with white flowers edged with blue and a yellow eye is a handsome annual well worth growing. Flowers July to September.

Godetia. What are called the azalea-flowered or Whitneyi varieties are my choice of these fine annuals. About 1½ft. tall they include pink, rose and crimson shades in their colour range. The taller *grandiflora* varieties are especially well suited for cutting. Flowers July to October.

Gypsophila. Also magnificent for cutting are the various colour forms of *G. elegans*, the many dainty little flowers on their branched, 1½-ft.-tall stems being so light and airy that one could imagine them being wafted away on the slightest breeze. The flowers are carried throughout the summer.

Flax, see Linum

Helichrysum. An attractive 'everlasting' flower of which there are many colour forms available. The double-flowered *bracteatum monstrosum* mixture of about 2½ft. in height is favoured by many gardeners. These are, of course, excellent for cutting. Flowers July to September.

Helipterum roseum, see Acroclinium

Larkspur, see Delphinium

Lathyrus (Sweet Pea). No introduction is needed for one of the loveliest and most popular of all annual flowers. Sweet peas make splendid border flowers and their soft colouring and delightful fragrance ensure them a place in countless gardens. Introductions like the long-flowering Early Dwarf Bijou, only about 1ft. tall, and the Little Sweetheart strain, even less at 8in., have given sweet peas an additional role in the garden. Flowers June to October.

Lavatera. The lavateras are handsome annuals with strong, branching stems and mallow-like flowers. Fine varieties are the rose-coloured Loveliness and the white Alba Splendens. Flowers July to September.

Layia. The 1-ft.-tall *L. elegans*, with yellow flowers and greyish foliage, is a valuable annual, for it is easy to grow and makes a brave splash of colour. Flowers July to September.

Leptosyne. The double-flowered Golden Rosette is the variety to grow. It grows about 1½ft. tall and is excellent for cutting. Flowers July to September.

Limnanthes. The easily grown *L. douglasii* is a useful little plant with its showy white and yellow flowers and spreading habit. Its height does not exceed about 6in. and it will give a display throughout the summer. It likes plenty of soil moisture but, as already indicated, it is not fussy about soil conditions.

Linum (Flax). The Scarlet Flax, *L. grandiflorum*,

The azalea-flowered or Whitneyi varieties of godetia are annuals with a good colour range. Shown above is the salmon-pink, edged white Sybil Sherwood

Nigella Miss Jekyll, a cornflower-blue variety of the well-known Love-in-a-Mist. A splendid mixture of colours is Persian Jewels, including purple, mauve, pink and rose

Ostrich Plume asters, which are excellent for garden display or cutting. These free-branching half-hardy annuals grow to a height of 1½ft.

is a plant I personally find very attractive. The open-faced red flowers on 1-ft. stems bring much colour to the border. The best variety without doubt is the scarlet *rubrum* and a useful variety is the white *album*.

Lobularia maritima, see Alyssum maritimum

Love-in-a-Mist, see Nigella

Marigold, see Calendula

Matthiola (Stock). The Night-scented Stock, *M. bicornis*, with lilac flowers on 1-ft. stems, has delightful fragrance during the evenings when the flowers open. Flowers throughout the summer and into autumn.

Mentzelia lindleyi, see Bartonia

Mignonette, see Reseda

Nasturtium, see Tropaeolum

Nigella (Love-in-a-Mist). How pretty the nigellas are with their cornflower-like flowers and airy foliage! Best to grow are the delightful cornflower-blue Miss Jekyll, Persian Jewels and Monarch Persian Rose. Persian Jewels is a splendid mixture of colours (including purple, mauve, pink and rose) and Monarch Persian Rose a lovely rose-pink variety. All grow to about 1½ft. tall.

Phacelia. The gentian-blue *P. campanularia* is a splendid edging plant or carpeter for a sunny position. Only about 9in. tall, it carries its bell-like flowers prolifically between July and September.

Reseda (Mignonette). The scented *R. odorata* has creamy-white flowers but there are varieties of this species with colours from orange-red to pale red and golden-yellow. All are about 1ft. tall. Flowers June to September.

Silene. The dwarf *S. pendula* is a well-known edging plant available in varieties with double or single flowers ranging in colour from deep red to pink and white. Six inches to 1ft. tall, these have

a long flowering season if autumn sowings are made as well as spring ones.

Star of the Veldt, see Dimorphotheca

Stock, see Matthiola

Sweet Pea, see Lathyrus

Sweet Sultan, see Centaurea

Tropaeolum (Nasturtium). The Tom Thumb varieties which do not exceed 1ft. are fine plants for the border; likewise the Jewel Mixed strain which carry their colourful flowers in abundance well above the foliage. Flowers July to autumn.

Viscaria. The gay viscarias with blue, red, pink or white flowers can be had in flower from early summer to autumn. There are both tall and short varieties, the former up to 1½ft. tall and the latter, Tom Thumb varieties, around 6 to 7in.

Half-hardy Annuals

Ageratum. I prefer the traditional blue ageratums like Blue Bedder (4in.) and the taller Blue Mink (6in.) to the other colours now available in varieties like Fairy Pink.

Antirrhinum. These are cheery flowers with their red, pink, yellow and white colourings and the display they make lasts from June into the autumn. There are various types ranging in height from 3ft. to as little as 10in. in the Tom Thumb varieties and 4in. in the Magic Carpet mixture. Rust can be a trouble and if this is likely to be experienced one should grow the rust-resistant varieties.

Arctotis. *A. grandis* is a grand flower for cutting with stems of about 2ft. in length. The flowers are handsome—white with a pale blue centre surrounded by a band of yellow. Flowers from June to October.

Aster. A variety of flower forms characterise the annual asters. There are the Ball and Powderpuff types, the Quilled, Ostrich Plume and Chrysanthemum-flowered. An exciting new development

The gazanias include a pleasing range of colours and are much used for flower arrangement. These tender perennials can be grown from seed as half-hardy annuals

Nicotianas, or Tobacco Plants, pretty in their own right, are grown primarily for their fragrance. They flourish in shade as well as in sun

is the introduction of the 2-ft. Duchess type with flowers like incurved chrysanthemums. The Ball type, Ostrich Plume and the Giant Branching Comet or Crego type are especially good for cutting. The colour range includes purple, shades of red and pink, mauve and blue, yellow and white. The height range is 8in. to 3ft. and the flowering season covers June to October.

Brachycome (Swan River Daisy). This is a splendid 9-in.-tall edging plant for sunny positions. *B. iberidifolia* has varieties in purple, blue, rose and white and the flowers have centres of contrasting colours. Flowers June to September.

Castor Oil Plant, see Ricinus

Gazania. The gazanias are firm favourites of mine. These are tender perennials but lovely hybrids can be grown from seed as half-hardy annuals, and the flowers give a long display of colour, provided the plants are given a sunny, well-drained position. The blooms, in shades such as orange, yellow, cream, red and pink, with zones of deeper colouring at the base of the petals, and carried on 1ft. stems, are also splendid for flower arrangement.

Ipomoea, see Pharbitis

Limonium. The 1½- to 2-ft.-tall varieties of *L. sinuatum* (still usually found in catalogues under its old name of *Statice sinuata*) have 'everlasting' flowers much in demand by flower arrangers. The colours include purple, blue, rose, salmon-pink and yellow. These are late-season flowers coming into bloom in August and continuing into autumn.

Lobelia. The dwarf lobelias are half-hardy perennials grown as half-hardy annuals and they are splendid plants for edging flower beds. The deep blue Crystal Palace Compacta, the deep blue, white-eyed Mrs Clibran Improved, and String of Pearls with flowers of blue, violet, rose and white, are among those which make a colourful display.

Matricaria. The matricarias are very free-flowering and useful bedding plants, some 1½ to 2ft. tall but others considerably less. They belong to the chrysanthemum family and the *eximia* varieties Silver Ball and Golden Ball (both less than a foot tall) are among the most desirable. A new double white variety, Snow Dwarf, is 9in. tall.

Marigold, see Tagetes

Mexican Sunflower, see Tithonia

Morning Glory, see Pharbitis

Nemesia. The compact (6 to 9in.) varieties of *N. strumosa* include such a selection of lovely colours that it would be difficult not to be won over by their charms. Orange Prince, Blue Gem and the scarlet Fire King are good named varieties.

Nicotiana (Tobacco Plant). The nicotianas are grown mainly for their delightful fragrance, but the flowers in white, pink and crimson are pleasing also. Varieties are available now, too, which open their flowers by day instead of only in the evenings – Sensation Mixed (3ft. tall, like most of the nicotianas) and Dwarf White Bedder, 1½ft. tall, are cases in point. A useful characteristic of the nicotianas is that they will flourish in shade as well as sun. Flowers July to September.

Petunia. Given a good summer with plenty of sun there are no finer bedding plants than the petunias, now available in so many striking colours and various single and double-flowered forms. Petunias have received close attention from the plant breeders and the Grandiflora and Multiflora types embrace a multitude of named varieties with a height range from 6 to 15in. Worthy of special mention are the Multiflora varieties Sugar Plum (lavender, veined purple) and Satellite, rose with a contrasting white band, The colour range includes reds, rose, pink, salmon and blue, purple, lavender and white. Flowers July to September.

The varieties of *Phlox drummondii* are very showy bedding plants, available in colours from red and purple to blue, yellow and white

Few plants are gayer than the marigold Naughty Marietta, with golden-yellow flowers blotched with maroon

Digitalis Excelsior Hybrids have fine decorative qualities, with flowers all round the stem, instead of only on one side as with other foxgloves

Pharbitis (Morning Glory). The handsome blue-flowered *P. tricolor*, formerly known as *Ipomoea rubro-caerulea*, is a splendid climber for a sunny, sheltered position—on a trellis, a porch support or a wall. Flowers from mid-summer to autumn.

Phlox. The annual phlox, varieties of *P. drummondii*, are splendid plants for summer bedding, mostly about 1ft. tall but including dwarf forms of only 6 to 9in. in height. The flowers are very showy in shades of red, purple, violet, blue, pale yellow and white. Flowers July to autumn.

Ricinus (Castor Oil Plant). This tropical plant is grown for its foliage, not its flowers, which are of no significance. The large leaves, though, are handsome in shades from green to reddish-purple. Most varieties are in the 5 to 6ft. height range and correctly sited can be very effective.

Rudbeckia (Coneflower). It is the introduction some years ago of the double and single-flowered tetraploid hybrids which bear the name Gloriosa Daisies which really put the annual rudbeckias 'on the map' so far as most gardeners are concerned. Between $2\frac{1}{2}$ and 3ft. in height these include numerous pleasing colours—yellow, golden, orange, bronze and mahogany-red. The flowers, with a daisy-like appearance, can only be described as huge, and the long, strong stems make them excellent as a cut flower. There are varieties also which are much shorter, like the 1-ft. Bambi, which are popular for the garden and for home decoration.

Salvia. If the need is for a bedding plant with brilliant colour I am sure that almost all of us think first of the varieties of the Scarlet Sage, *Salvia splendens*. These can be depended on to give an eye-catching display from July to the autumn and they are available in heights from 6 to 18in. or so. *S. splendens* is actually a perennial, tender in this country, which is very happy to be grown as a half-hardy annual. Other perennial salvias grown in this way are *S. farinacea*, a violet-blue-flowered species of about $2\frac{1}{2}$ft. and the rich blue *S. patens*, about 2ft. tall.

To go back to *splendens* and its varieties, I think there is a lot to be said for the 1-ft.-tall Blaze of Fire, which is one of the earliest to come into flower; the taller Tetra Red, also with strong colouring; and Scarlet Pygmy or Salmon Pygmy if you want something making only 6in. of height.

Swan River Daisy, see Brachycome

Tagetes (Marigold). I have already remarked on the fine work of the plant breeders, especially with regard to petunias. They have also done a magnificent job on the marigolds—flowers which lift our spirits with their cheerful colours from July into autumn. African varieties like the $2\frac{1}{2}$-ft. Golden Climax and Yellow Climax or the Crackerjack mixture with flowers of gold, orange and pale yellow are the choice of many gardeners; others may prefer the 1-ft. Spun Gold, a new golden-yellow variety of the same type, or little 6-in.-tall French marigolds of the Petite strain, the yellow, blotched maroon Legion of Honour and Petite Yellow.

I have left to last what is perhaps the most popular of all marigolds—the 9-in.-tall Naughty Marietta, a very gay variety with golden yellow flowers blotched with maroon. Its 'punch' is terrific when planted *en masse*.

Tithonia (Mexican Sunflower). The common name of this showy annual gives a clue to its appearance and it is a plant I would like to see grown more often than it is at present. With a height of 3 to 4ft. *T. speciosa* is useful for setting towards the back of the border, and it will give a fine display with its orange-red flowers from July into the autumn. The variety Torch, orange-scarlet, is often listed by seedsmen.

Tobacco Plant, see Nicotiana

Ursinia. I do not see how anybody could fail to admire these beautiful annuals with their pretty little daisy flowers and contrasting colourings. For instance, there is the 1-ft.-tall *U. anethoides*, with orange flowers marked with a purple zone, and *U. versicolor (U. pulchra)*, 9in. tall, orange with a black zone. These flowers need lots of sun. Flowers July to September.

Venidium. This lovely annual also has flowers with contrasting colours. *V. fastuosum*, 2 to 3ft. tall, and orange in colour with a black disk, is a popular plant, and there is a useful range of hybrids with colours varying from pale yellow and white to orange. These flowers are good for cutting. Flowers July to September.

Verbena. The verbenas, which make such splendid bedding plants and which have such a good colour range (from white and pink, through red and blue to purple) need little introduction. Especially useful is the Compact strain of about 6 to 9in., but the Mammoth, large-flowered type of twice its height is excellent for the border too. *V. rigida* (*venosa*), 1ft. tall with bluish-purple flowers, is a free-flowering species which can be recommended for the border.

Zinnia. A wide range of zinnia types are offered by seedsmen and how excellent these can be for the border and for cutting—in a sunny year. That is the one drawback with these flowers —they must have sunshine to do well. There are, for instance, the popular Dahlia-flowered varieties, the Chrysanthemum-flowered and the Mammoth-flowered varieties of about $2\frac{1}{2}$ to 3ft., the Gaillardia-flowered varieties of $1\frac{1}{2}$ft., the Mexicana and Pumila varieties of around 1ft. and $1\frac{1}{4}$ft. respectively, and, even shorter, the much-talked-of Thumbelina of 6in. in height. This last has a compact habit and bears masses of double and semi-double flowers in colours from white and yellow to scarlet and orange. Flowers from July to autumn.

Biennials

At the end of May, when most of us are pretty busy in the garden, it is only too easy to overlook the sowing of biennials, and those perennials which we habitually treat as biennials. The Canterbury Bells and Sweet Williams, the Forget-me-Nots, Wallflowers, Foxgloves and verbascum are some of these plants which have much to offer the gardener who is prepared to arrange his planting plans so that these transients—raised from seed sown outdoors this year for flowering next year—can be accommodated.

The following are excellent garden plants well worth growing:

Althaea (Hollyhock). Perennials grown as biennials (see p. 110).

Campanula (Canterbury Bell). Delightful colour forms of the $2\frac{1}{2}$-ft. Canterbury Bell (*Campanula medium*) are now available, from violet-blue and lavender shades to rose and white, and these, when mixed, can really be very lovely. The same colour range is available in the Cup and Saucer campanulas (forms of *C. m. calycanthema*). These flowers are easy to grow and splendid for early (May-June) colour.

Cheiranthus (Wallflower). The wallflower proper (*C. cheiri*) and the Siberian Wallflower (*C. allionii*) are popular plants for spring bedding. (If seeds of *C. allionii* are sown early in the season it can, in fact, be treated as a hardy annual, for late summer and autumn flowering.) The apricot-coloured Monarch Apricot Delight, 1ft. tall, is a new variety of Siberian Wallflower well worth growing, and another is the *allionii* variety Golden Bedder. The lovely colours to be had in the fragrant *cheiri* varieties include many shades of red, orange, golden-yellow and pale yellow. These range in height from just under 1ft. to about $1\frac{1}{2}$ft.; the Tom Thumb mixture, only 9in. tall and in all the typical wallflower colours, is useful for edging and other positions where low stature is an advantage.

Dianthus (Sweet William). The Sweet Williams of gardens are selections from *D. barbatus*. A popular strain is the 6-in.-tall Indian Carpet, and others include the $1\frac{1}{2}$-ft. Auricula-eyed Exhibition strain and Scarlet Beauty. Give these plants a reasonably good soil and a sunny position.

Digitalis (Foxglove). There is little I need say about these cottage garden flowers except to sing the praises of the splendid modern Excelsior Hybrids. These have the unique quality, for foxgloves, of bearing flowers all round the stems (normally they are carried on one side of the stem only). Also, the flowers are supported almost at right-angles to the stems so that the intricate markings on the petals are clearly seen by the observer. These flowers like shade and will grow well in dry conditions.

Forget-me-not, see Myosotis

Iceland poppy, see Papaver

Mullein, see Verbascum

Myosotis (Forget-me-Not). The Forget-me-Nots need no description. Seed is best sown in June or early July in the open border for if it is sown earlier the resultant plants are liable to become too far advanced before the autumn. There are good colour forms of *M. alpestris* in heights from 1ft. to 6in. and in various shades of blue, and Carmine King, 8in. tall, provides us with a variety of different colouring—and very attractive they are.

Pansy, see Viola

Papaver. The Iceland Poppies (strains of the 1-ft.-tall *P. nudicaule*) with their gay colours are popular both for garden display and cutting. The soil in which they are grown must be free-draining and not so rich that it encourages lush growth at the expense of flowering. They also need a sunny position. The best way to raise plants is from seed sown in small pots or boxes in August and germinated in a cold frame. The seedlings are subsequently reduced to one or two in each pot and are grown on until autumn, when they are moved to the open ground.

Siberian Wallflower, see Cheiranthus

Sweet William, see Dianthus

Verbascum (Mullein). The tall (6ft.) spires of *V. bombyciferum* (Broussa) with its pale yellow flowers and covering of white, woolly hairs is a plant of considerable character. Like all other verbascums it is easy to please in the most unpromising of soil conditions. A variety with rich yellow colouring is Harkness Hybrid. Even taller than these two is *V. olympicum* with handsome flowers of golden colouring.

Wallflower, see Cheiranthus

Viola (Pansy). Almost every gardener, I imagine, has a soft spot for the pansies, which form part of the genus *Viola*. These are available in splendid strains 6in. or so tall, and with light shade and any reasonably good soil they will give an excellent account of themselves over an extended period. Especially attractive, I think, are the Clear Crystals mixture with self-coloured flowers, but there are numerous others of merit, like the wavy-flowered Felix strain in which the blooms are marked with a rich yellow blotch. There are, too, the winter-flowering pansies which can be very welcome at a dull time of year—Celestial Queen, pale blue and the yellow Helios are two varieties of this type which come to mind. Raise the plants from sowings made in a cold frame in August. They can also be treated as half-hardy annuals by making a sowing under glass · in March.

Pests and Diseases

For information on pests and diseases of Perennials, Annuals and Biennials see pp. 294 and 295.

Above. Iceland Poppies (strain of *Papaver nudicaule*) make a welcome splash of colour in the garden and are also useful for cutting
Below. Pansies, with their attractive little flowers, are splendid plants for bringing colour to odd corners of the garden

Bulbs Corms and Tubers

6

I am often asked to define the terms bulb, corm and tuber for they are rather confusing to many gardeners, especially as all three are often described collectively as 'bulbous plants'. I am a gardener, not a botanist, but I will do my best to make myself understood.

A bulb is, in effect, a modified bud, usually found underground, with tightly packed scales or swollen leaf bases forming a food storage organ. It also contains the undeveloped shoot from which will arise next season's leaves and flowers. A corm, on the other hand, is what is termed a 'solid bulb'—a bulb-like fleshy stem, usually covered by a membraneous sheath, with the food storage organ developing a bud at its apex. A tuber consists either of a swollen stem or a swollen root, the first bearing buds or 'eyes' and the second making growth from a crown.

Setting the scene. Very few of us, I imagine, are so resistant to flattery that we do not take pleasure in having other people praise our plantings. A large part of being a successful gardener is having a true feeling for plants and being able therefore to do the right thing almost without thinking; but to be completely successful one must also teach oneself how to display plants attractively. I use the word 'teach' advisedly because although some of us are fortunate enough to be born with a natural aptitude for design, most of us have to acquire this facility. However this may be, I am sure that really impressive garden displays and perfect colour combinations are only achieved by putting much hard thought into the early planning stages. There is every reason for saying that forward thinking, accurate observation of the true behaviour of plants in one's own and other people's gardens, and a constant desire to increase one's knowledge in this respect are the main ingredients for success in this direction.

It is necessary to conjure up a mental picture of the result of making plantings and although this may be difficult at first it does become much easier with experience.

Naturalised Bulbs for Grass and Woodland. Naturalised bulbs, as the term implies, are those grown under as near natural conditions as possible. With the present trend towards more informal gardens—and how much more attractive these are than the gardens of the past—naturalised bulbs have become more and more a part of the garden scene.

Quite apart from the attractiveness of naturalised bulbs it has not been overlooked that these need far less attention than bulbs grown in other ways. I grow naturalised bulbs myself primarily because their beauty is so satisfying but I am keenly aware that they make practically no demands on my time. There is the initial planting, of course, but no annual round of lifting, drying off, storing and subsequent replanting, as one has

Daffodils massed informally around the base of a tree have a delightful air of informality. Numerous bulbous plants are suitable for naturalising in this way

for this purpose; and, for an early display, what better than the Hoop Petticoat Daffodil, *Narcissus bulbocodium*, in one or other of its forms, or *N. pseudo-narcissus*, the Lent Lily, another small trumpet-flowered species (for more detailed information on narcissi see pp. 148 to 150).

The large-flowered Dutch crocuses are excellent for naturalising in grass and such species as the spring-flowering *Crocus tomasinianus* and the autumn-flowering *C. speciosus*. Other plants include eranthis, Snake's-head Fritillarias (*Fritillaria meleagris*), galanthus (Snowdrop), muscari (Grape Hyacinth), ornithogalum (Star of Bethlehem) and scillas (Squills).

A surprisingly large number of gardens, I find, have at least a little light woodland, mostly left standing when the garden was first made, and such a feature provides splendid opportunities to grow many bulbs informally. Where the sun can strike through the trees—and some openings to sunlight can be made if not provided by Nature—lilies of many kinds will be at home. To name just a few—the purple-flowered, dark-spotted *Lilium martagon*, the much more showy orange-red, spotted *L. pardalinum* (the Panther Lily) and *L. tigrinum splendens* (a fine form of the Tiger Lily).

For clothing a lightly shaded bank under trees there is the delightful little *Cyclamen neapolitanum* for late summer-early autumn flowering. Its beautifully mottled leaves are also a joy in the dullest months of the year. Galanthus, *Anemone nemorosa*, *Chionodoxa luciliae*, camassias, dwarf narcissi, bluebells, *Erythronium dens-canis* and eranthis are other good plants for such conditions.

Bulbs in Borders. I am treading now on dangerous ground for I know some gardeners recoil in horror at the suggestion that one should plant daffodils, for instance, among bush roses. Well, I do it myself and I have not yet found anybody who can convince me that a rose bed looks better in April without daffodils than it does when they are painting the beds with gold. The facts are that one can use plants in many ways without offending against good taste.

A development of recent years, which I have previously referred to, is the mixed border in which herbaceous perennial plants, shrubs, annuals, biennials, bulbs, corms and tubers are grown together to provide colour and interest over many months. Born of necessity, because modern small gardens do not have space to grow so many different plants in separate borders, the mixed border is a compromise which really works.

In a border of this type, there is much scope for planting skills—arranging the plantings so that the dying foliage of early flowering bulbs will be hidden by nearby herbaceous plants as these develop to flowering size. Size, colour, the flowering time and cultural needs of the plants con-

with bulbs grown in formal beds used for other displays once the bulbs have finished flowering.

There is one aspect of growing naturalised bulbs, however, which must be considered by the owners of small gardens: the grass must not be cut until the foliage of the bulbs has died down naturally. This means waiting until June or thereabouts, something I do not mind myself, but in a small garden it is a thought to ponder, especially if the planting is near the house. The reason for this is that the bulbs cannot flourish if they are not allowed to complete their growth and so store up food resources for the next season's development. The grass when eventually cut may be an unattractive brown colour but watering and feeding with lawn fertiliser will soon bring the colour back to normal.

Quite delightful effects can be achieved from small plantings—a drift of daffodils round an old apple tree, crocuses studding a shallow bank or a blue sea of muscari lapping round the trunk of a blossom-decked flowering cherry; these and similar scenes are a lift to the heart after the subdued tones of winter.

When contemplating making plantings of naturalised bulbs in grass, one's thoughts naturally turn to the daffodils or narcissi, and in particular such handsome Trumpet varieties as King Alfred, Magnificence and Golden Harvest which look magnificent with their gay flowers nodding gracefully above the foliage and the bright green new grass. Numerous Large-cupped and Small-cupped varieties, the late-flowering Pheasant's Eye and other *poeticus* varieties are also well suited

cerned must all be taken into consideration when drawing up planting plans. Lilies, gladioli and dahlias are excellent plants for such a border, and Crown Imperials (*Fritillaria imperialis*), *Galtonia candicans*, the elegant *Dierama pulcherrimum*, tuberous ranunculus, montbretias, ixias, tigridias, camassias and irises of the English and Spanish types.

Bulbs in Formal Beds. Although, as already explained, informal gardening is much on the increase, there will always be those who prefer to grow their bulbous and other plants in formal beds. There will always, too, be those houses and gardens which, because of their style and layout, are most suited to this formal treatment. Town gardens often fall in this category, and gardens in industrial areas where permanent plants like evergreen shrubs become coated with a layer of filth. For these gardens bulbs such as daffodils and tulips are ideal.

The essentials for success with this kind of gardening are, in my experience, to give especially careful attention to the preparation of the beds, and to so select one's plants that the heights, colours and flowering times are properly in balance. This is, I know, far easier said than done, but a few quiet evenings with several good bulb catalogues, pencil and paper will put one on the road to success for a cost which is within one's means. Having mentioned cost, I am going to take this opportunity to point out that raising new varieties of bulbs, like daffodils and tulips, is a costly business and inevitably novelties and quite recently introduced varieties cost much more than others which have been in circulation for years.

Early Single, Early Double, Cottage, Broken and Darwin tulips are especially well suited for bedding purposes; so, too, are hyacinths, available in delightful shades of red, blue, pink, yellow and white. In addition there are the narcissi, galanthus, crocuses, muscari, chionodoxas and scillas which can all be successfully grown in beds.

Many bulbs make a very colourful display when planted in beds on their own but others such as the daffodils and tulips are often very effectively combined with other spring flowers. For example, pink Darwin tulips with an underplanting of blue forget-me-nots, or tulips combined with wallflowers, white arabis and yellow *Alyssum saxatile*. It is best to plant the bedding plants first and then put in the bulbs using a trowel.

Beds Against Walls. Flower beds against walls with a warm south or west aspect should certainly not be wasted on plants which would grow just as well in open, less-favoured parts of the garden. Sun-loving bulbs like the handsome pink *Nerine bowdenii*, the orange-red *Crocosmia masonorum*, modern hybrid montbretias and *Ipheion uniflorum*

The tuberous-rooted *Anemone blanda*, with flowers from blue to white in colour, is a cheerful harbinger of spring in many gardens

are obvious candidates for such a position. Others are *Acidanthera bicolor murieliae*, *Amaryllis belladonna*, *Dierama pulcherrimum*, *Ixiolirion montanum*, *Schizostylis coccinea*, *Sparaxis grandiflora*, *Tigridia pavonia* and *Zephyranthes candida*.

Stone walls and walls of mellow brick make an attractive backdrop for many plants, and this factor should not be overlooked when making one's planting dispositions. To see a bold clump of Crown Imperials against a Cotswold stone wall is to see these magnificent plants through different eyes. The massive red and yellow flower heads have their perfect foil in a wall of pale honey colour. Part of the art of garden design is to exploit to the full existing features.

Bulbs in the Rock Garden. Miniature bulbs fascinate me and where better to view them than in pockets in the rock garden. They seem to associate so perfectly with the stone outcrops and contours and to provide great interest over such a long period. Indeed it is not at all difficult to have bulbs flowering in the rock garden during every season of the year – a considerable advantage when one thinks of the rather concentrated flowering times of most rock garden plants.

In spring it is possible to have a wealth of colour from plants like the dwarf narcissi and tulips, *Anemone blanda*, *A. apennina* and varieties of *A. nemorosa*, camassias, *Bulbocodium vernum*, cyclamen, fritillarias, eranthis, leucojums, ixias and scillas – the list seems never-ending. Summer sees the alliums like *moly* and *ostrowskianum* flowering, the ornithogalums, dwarf gladioli, and zephyranthes, and in autumn one can look forward to the flowering of *Sternbergia lutea*, *Leuco-jum autumnale*, autumn-flowering crocus and the first of the snowdrops. The last-mentioned, of course, spread their flowering through the winter months, and one can also have in flower in February and March such charming dwarf irises as *Iris histrioides major* and the lovely deep blue and gold *I. reticulata* and its varieties. There need never be a dull moment with clever planting.

All these bulbs like the kind of open situation in which a rock garden is usually sited, and the good drainage demanded by the rock plants is admirably suited to the bulbs. Sunny, sheltered pockets in front of rocks are easily made for dwarf tulips, irises, crocuses and other kinds which benefit from such preferential treatment.

Bulbs in Containers. Growing bulbs in tubs and other containers has, apart from anything else, one unusual advantage—the plants can be moved at will if circumstances demand. One can have them near the front door, for example, and move them without trouble to the terrace or sitting-out area if this is desired. Paved sitting-out areas are now quite a common feature of many homes, and when the weather is suitable they provide what is virtually an outdoor living room. If the site is sheltered it is quite possible to sit out for several hours on really sunny days even as early in the year as late February and March—and tubs and boxes filled with colourful bulbous plants add to the enjoyment.

Window boxes can be made very attractive indeed with very little trouble to the householder, and even those living in flats with a balcony can grow flowers in this way. I prefer to see boxes devoted to one kind of flower rather than a mixture but there is no reason at all why, say, daffodils and muscari, or other similar combinations, should not look very well. Daffodils, tulips, crocuses and hyacinths are all popular window box plants.

Tubs give more scope than window boxes because of their greater depth, and tall daffodils and tulips look more in proportion in such containers. I have found tubs excellent for lilies, notably the popular Regal Lily, *Lilium regale*, and the Madonna Lily, *L. candidum*.

A method of growing daffodils which I have found particularly rewarding is that of planting double layers of the bulbs in deep tubs. I use a good potting compost—usually the John Innes No. 2—and it is important to make sure that there are sufficient holes in the bases of the tubs to ensure good drainage. Growing daffodils in this way gives a tremendous concentration of colour in a small area—just what is wanted for a prominent position.

The tubs must be at least 12in. deep as anything less than this would not hold sufficient compost. Place a layer of crocks over the drainage holes and

Growing daffodils in deep tubs—at least 12in. from rim to base—in a double layer provides an eye-catching concentration of colour. Here the first layer is placed in position

cover this with roughage. Add a layer of compost and set the first layer of bulbs in position. Cover the bulbs with compost so that the tips are just showing and then position the second layer of bulbs between the tips of the first layer. Work in compost among these bulbs and make this firm. With correct aftercare the results will be magnificent.

The colour of plant containers can make a considerable difference to the effectiveness of the display. Daffodils look particularly attractive in a green tub with the supporting bands painted black, and multi-coloured tulips in a white tub with black banding can look very striking indeed. Colour harmony is important and should be given as much attention as careful siting.

Planting and Aftercare

A point I would make first of all is that many gardeners are much too apt to take bulbs for granted. It is true that they are for the most part easily pleased and that if one buys good stock one can expect a first season display of some excellence. This is because bulbs, corms and tubers have reserves of food available in these storage organs. But if soil conditions are poor and growing conditions generally unsatisfactory these reserves will certainly not be made good and this will inevitably be reflected in the quality of the blooms in subsequent seasons.

Beds and Borders. We must then make sure that the soil conditions are satisfactory. Beds and borders need to be dug over deeply and a check made that the drainage is adequate. The soil texture can be greatly improved by forking in a dressing of horticultural peat at the rate recommended by the supplier. Make sure, though, that the peat is well saturated with water before it is worked into the soil. A dressing of bonemeal or hoof and

With the first layer of bulbs (see left) covered with compost so that only their tips are showing, a second layer of bulbs is positioned between the lower ones

Where largish numbers of bulbs are being planted, a special bulb planting tool which lifts a core of turf and soil is a great time-saver

horn scattered on the surface at the same time, at the rate of 4oz. per square yard, and forked in with the peat, will do much to ensure success in future. Fresh manure should never be placed in direct contact with bulbs, but it can most profitably be incorporated below the planting level.

Planting depths and distances vary considerably for different bulbs, corms and tubers, as one would expect bearing in mind their ultimate dimensions, and I have given these details in the notes on individual plants. Planting times, too, should be kept to as closely as possible, and these I have also indicated in the plant notes. It is worth mentioning that any bulbs which are particularly sensitive to moisture—and therefore rotting—should be bedded on sand when they are planted. Gladioli and lilies are two kinds which need such attention in most soils.

When bulbs used for spring bedding displays have finished flowering they have to be lifted to make way for summer bedding subjects. As I have already stressed, it is important that bulbs should be allowed to complete their growth cycle and so ripen off the bulbs completely. The answer in this case is to move the plants to a reserve bed in some secluded part of the garden where the soil is of good quality and they can then be lined out in shallow trenches, the soil being well firmed around them. The soil must be kept moist and in dry weather this may well necessitate watering.

When the foliage withers, usually by about late June, the bulbs can be lifted and dried out slowly in a cool, airy shed or room. Lifting must be done with care so that the bulbs are not pierced with the fork or spade. Before storing in boxes remove old roots and other extraneous material and place all diseased bulbs on one side for immediate destruction. Old tomato trays make very serviceable containers for bulbs and these must be stacked, until use, in a cool, airy place.

Naturally one does not lift bulbs like crocuses, daffodils, muscari and so on which are growing in beds among shrubs or other permanent plantings. These can be left for several years at least without disturbance.

Naturalised Bulbs. One faces a different situation from the foregoing when planting bulbs in turf, for unless the area involved is small it would be quite impracticable to clear the turf away, improve the soil, plant and relay the turves. There should not, however, be much to worry about if the grass in the planting area is in good shape, for this in itself is an indication that the soil conditions are reasonably good.

Again, although small quantities of bulbs or corms can be planted with a trowel it is well worth while obtaining a special bulb planting tool if large numbers are involved. This tool cuts out a core of turf and soil when it is pressed into the ground and removes it intact when it is given a twist and lift. The bulb is then placed in the hole and the core of soil and turf replaced and firmed with the feet.

When I know the soil where naturalised bulbs are planted is rather below par, I make a point of feeding them each autumn with a slow-acting fertiliser such as bonemeal or hoof and horn, at the rate of 4oz. per square yard.

Bulbs in Containers. One of the most important things when growing bulbs in window boxes or tubs is to ensure that the drainage is good. The container should have holes of about 1in. diameter at fairly frequent intervals over its base. If you are making up window boxes yourself use hardwood of not less than $\frac{1}{2}$in. thickness, and paint the inside, which will be in contact with the soil, with a non-toxic preservative. The drainage holes should be covered with crocks and these in turn covered with a layer of roughage before the soil is added. The John Innes No. 1 Potting Compost makes a first-class growing medium and one should plant the bulbs in good time so that they have the longest possible growing period. If ordinary garden soil is used—and this must have good texture—I would mix peat in with this and a dressing of bonemeal (at the rate of 4oz. to the bushel). Do not neglect watering if the compost appears to be rather dry for the wind can soon dry out compost in containers in exposed positions. Also, stake the plants against wind and storm damage as soon as it seems prudent to do so.

Methods of Increase

Many bulbs, corms and tubers can be raised easily from seed but in the case of hybrids the seedlings are liable to vary greatly from the parent plants and one may have to wait a long time for flowering-size plants. Many amateur gardeners are likely to prefer the quicker vegetative means of increase. **Increase from Seeds.** After the flowers of hardy cyclamen have faded the seed pods can be found on coiled stems, and if these are sown as soon as ripe in a cold frame the plants which result will flower in three to four years. Lilies also can be raised from seed but some take as long as six years to reach flowering size. On the other hand *L. regale*, the Regal Lily, is one of those which can be brought to this stage in two years.

Pans or boxes can be used for seed sowing and it is always advisable to sow the seed as soon as it is ripe. Use the John Innes Seed Compost for preference, just covering the seeds with the soil mixture if they are very small and then with about ¼in. of sand. For larger seeds a covering of about ¼in. each of compost and sand will be satisfactory. Drainage should be especially good as the seeds will be in the pans or boxes for a long time. With some lilies, for instance, there may be no signs of life above soil level during the first year although all is proceeding well underground. Patience is more than a virtue where this form of propagation is concerned!

When the seedlings appear the pans or boxes should be moved to a cold frame and plunged if they are outdoors and then left alone for a considerable time—perhaps another two years—until they are large enough to pot singly or plant out in nursery beds. Watering and protection from pests will be the only attention required during this time. This first big move is made when the young seedling bulbs are dormant, and afterwards the plants are grown on to flowering size in the normal way.

Vegetative Increase. Some lilies, like *Lilium tigrinum*, produce small, round bulbs, known as bulbils, in the joints of the leaves and these can be removed after the plants have flowered and grown on like seeds. Increase by this means is a relatively quick way of getting flowering plants, perhaps

two years from the time of sowing. Other lilies produce bulblets on the underground section of their stems and these, too, can be detached and grown on in a nursery bed for two or three years. Another method of propagating lilies is to lift the bulbs in early autumn and remove some of the scales. Insert these upright in a mixture of peat and sand in a seed box to about half their depth and cover with moist sphagnum moss. Place them in a cold frame and wait for bulblets to form on the scales which can then be removed and potted or planted out in a nursery bed.

One of the simplest ways of increasing daffodils and other bulbs is to lift the bulbs when the foliage has died down and simply separate the clusters of bulbs. The smaller offsets soon increase in size if they are planted in good soil in nursery rows. To induce hyacinths to produce bulblets it is common practice to cut grooves in the flat, disc-shaped base of the dormant bulb so that bulblets will form along the edge of the cuts. These bulblets are then planted 2in. deep in a sandy soil mixture and should produce flowers in their third year. Galanthus (snowdrops) are increased by offsets or division of the old clumps immediately the flowers fade. Tulips are propagated by separating the bulbs when these are lifted annually after the flowers fade.

Gladioli are reproduced normally by the cormlets (known as 'spawn') which form around the old corms or at the base of the new corms which form on top of them. The new corms are already of flowering size, and the cormlets can be grown on to produce flowering-size corms by the second year.

With tubers one's action is determined by whether these are swollen roots or swollen stems. Dahlias, which do not form buds on the tubers themselves, must have a piece of the old stem bearing a bud attached, when they are divided. Tuberous begonias, on the other hand, do bear buds on the tubers which can then be divided with a sharp knife after growth has begun. Both these plants, of course, can be raised from stem cuttings as well, taken from stock plants started into growth in heat early in the year.

Plants for the Garden

Making plant recommendations is always a pleasure, but it invariably presents the gardening writer with a problem—not what to include but what to leave out. Such is the wealth of material among bulbs, corms and tubers for outdoor planting that one must be more selective than one would wish. I have, however, done my best to cater for many different tastes and to include plants suitable for many different situations.

My advice to intending planters is to try to see as many as possible of the plants I have described

Left. Allium moly, a bulbous plant which bears its flowers on 12-in. stems at the beginning of summer. It likes to be planted in a sunny, open position

Below. Crocosmia masonorum, a delightful plant which is not fully hardy. It needs a warm, sunny home. In gardens which tend to be prone to frost, the corms should be lifted at the end of the season and over-wintered in a dry frost-proof shed or room. The flowers appear in August and September

Left. Colchicum autumnale, the Autumn Crocus or Meadow Saffron. These delightful, unassuming little flowers grow well in sun or semi-shade

Right. The spring-flowering *Cyclamen repandum*. The dainty flowers above the beautifully marbled foliage give these plants the appearance of accomplished ballet dancers. As plants for cool shade, they provide excellent ground cover

Left, below. A mixed planting of muscari (Grape Hyacinths) and erythroniums (Dog's Tooth Violets). The charm of such scenes is increased by the contrast between the line of the paving stones and the gay abandon of the bulbous plants

Above. Schizostylis coccinea, the Kaffir Lily, a handsome autumn-flowering plant for a warm, sunny border. The flowers are borne on 1 to 1½ft. stems *Left. Crinum powellii,* another. bulbous plant for a warm, sunny position, preferably a sheltered site backed by a south-facing wall. It flowers from July to September

growing in gardens under natural conditions. This serves two very useful purposes: it makes one familiar with the plants and it shows how well they associate with plants which are their neighbours. With some very fine private gardens now opened to the public for charity, and with so many public parks in all parts of the country making the fullest use of bulbs, corms and tubers in their planting schemes, it is not at all difficult to find the plants one is seeking.

A spring visit to the permanent display garden of the British bulb industry at Springfields, Spalding, in Lincolnshire, is well worth while and it is very easy, with air travel and cheap tours available, to visit the famous Keukenhof Gardens near Lisse in Holland where the Dutch bulb trade show bulbs grown in beautiful surroundings and used for every conceivable decorative and screening purpose.

A Selected List

Acidanthera. The pretty 3-ft. tall *Acidanthera bicolor murieliae* from Abyssinia, related to the gladiolus, is unfortunately only suitable for growing outdoors permanently in more favoured parts of the country, but elsewhere the corms can be planted out in spring and lifted again after flowering in autumn. The fragrant white flowers marked with a maroon blotch appear in September and October. Plant the corms in March or early April, 3in. deep and 6in. apart. Increase by offsets or seed, sown under glass in spring.

African Corn Lily, see Ixia
African Lily, see Agapanthus
Agapanthus. The blue heads of flowers and the large strap-shaped leaves give the African Lily, *Agapanthus orientalis* (syn. *A. umbellatus*), a distinctive appearance. The flowers are borne on stems 2 to 3ft. tall and this species makes an excellent tub plant, to be taken into a greenhouse in winter. Agapanthus has a creeping rootstock and the crown should be just below the surface when planting in spring. Increased by offsets, division or seed. From seed, though, it takes five to six years to flower.

Allium. The genus *Allium* includes the domestic onion but this should be no deterrent to growing some of the very ornamental species, for these only have the characteristic onion smell if the leaves are bruised. A splendid species is the June-flowering *A. ostrowskianum* with ball-like umbels of rose-coloured flowers. These flowers are carried on 6- to 9-in. stems. Another popular species is *A. moly* with yellow flowers on 12-in. stems. This flowers a little earlier. The white-flowered *A. neapolitanum* is 18in. tall. All these can be planted in autumn, 3 to 4in. deep in a sunny, open position. Increase by division of the bulb clusters in autumn or by seed sown in spring in a cool greenhouse or frame.

Amaryllis belladonna. This lovely, lily-like bulbous plant is only suitable for warm sunny positions, for preference under a south-facing wall. It bears fragrant, rosy-red trumpet flowers on 18-in. stems in late summer and early autumn. Plant 4in. deep and 12in. apart in August. Increase by offsets removed at this time or by seed sown in heat in spring. The latter is a very slow process, plants probably taking seven years to flower. It is, of course, a magnificent pot plant, with a very stately appearance.

American Wood Lily, see Trillium
Autumn Crocus, see Colchicum
Anemone (Windflower). These are lovely tuberous-rooted plants for various positions. The St Brigid and de Caen types have flowers of many colours and are, of course, well known as excellent cut flowers. The stems are up to 1ft. long. Plant the tubers 2 to 3in. deep and 6in. apart in November for spring flowering, in April or May for July flowering and in June for September flowering. There are also numerous varieties of the species *A. blanda*, with flowers from blue to white in colour, and *A. nemorosa*, with white, pink-tipped flowers, for March and April flowering. The latter will colonise borders under shrubs and other places where it gets light shade. The gay scarlet *A. fulgens* flowers in May. *A. blanda* prefers a sunny position and a sheltered site with good well-drained soil. The blue wood anemone, *A. apennina*, is a charming plant for naturalising in grass. Increase from seed, or by dividing the tubers.

Antholyza, see Curtonus
Babiana. The beautiful babianas (forms of *Babiana stricta*) can be grown outdoors in favoured areas, given a light sandy soil and a well-drained sunny border. Winter protection must be provided even then, though, if they are to come through safely – ashes, bracken or similar covering. Alternatively the bulbs can be lifted and stored. Lovely colours are available, from blue and cream to rose-pink and crimson and the flowers are borne in April and May. Plant 4in. deep and 2in. apart. Excellent pot plants. Increase is from seed or offsets.

Begonia. The tuberous begonias, from orange and red to yellow and white, are fine bedding plants for planting out after danger of frost has passed. The tubers are started into growth in February, March or April in a greenhouse with a minimum temperature of 13°C. (55°F.). Plant them in boxes of moist peat or a moist peat and sand mixture and when growth begins pot into 5-in. pots filled with John Innes No. 2 Potting Compost. Harden off before planting out. Lift and store the tubers in October. Increase from seed sown in January or February in a temperature of 18°C. (65°F.) or rather more.

Bluebell, see Scilla

Camassia (Quamash). The colour of *Camassia esculenta* may be anything from white to deep blue and I find the spikes of star-shaped flowers pleasing in the border during the early part of summer. This species can be used for naturalising in light woodland but it is more normally grown as a border plant, reaching a height of about 2ft. The flowers appear in June and July. Any ordinary soil will suit the camassias. Other species grown include the 3-ft. *C. cusickii* and *C. scilloides* (syn. *C. fraseri*) of half this height. Both have pale blue flowers. Plant 4 to 5in. deep and the same distance apart in autumn in sun or light shade. It is usual to leave them alone as long as possible, but some gardeners like to lift and replant the bulbs every four years. Increase is from seed, or rarely division.

Cardiocrinum. The spectacular 6- to 10-ft. tall *Cardiocrinum giganteum* (formerly known as *Lilium giganteum*) is a plant only for those with woodland and plenty of space. The white flowers appear in July and August and the bulbs die after flowering. Plant the bulbs in October, just covering them with soil. The offsets which are produced provide a means of increase and seeds may also be sown to produce flowering plants after about seven years.

Chionodoxa (Glory of the Snow). The lovely little chionodoxas named *luciliae* and *sardensis* are delightful, among other things, for brightening up the rock garden and window boxes in March. *C. luciliae* is blue with a white centre and *C. sardensis* a striking deep blue. They need plenty of sun and good drainage. Plant 3in. deep and 1in. apart in autumn. Propagate from offsets or seeds.

Colchicum (Autumn Crocus or Meadow Saffron). The colchicums will grow in sun or semi-shade and like well-drained, light soil for preference. The flowers appear before the leaves and the colours range from white to purple. The 9- to 12-in. tall *Colchicum speciosum* comes into flower in September and the 4-in. *C. autumnale* and its varieties flower a little earlier. Plant 3in. deep and 6in. apart in July or August and increase by division of the clumps at this time or by seeds sown in a cold frame in late summer.

Crinum. The beautiful South African crinums *C. bulbispermum* (syn. *C. longifolium*) and *C. powellii* are superb plants for sunny, sheltered positions, preferably backed by a south-facing wall. The trumpet-shaped flowers will be of better quality if the soil is rich. *C. bulbispermum*, 3ft. tall, bears rose-coloured flowers in summer; *C. powellii*, of similar height, bears its lovely rich rose-pink flowers rather later from July to September. Protect with straw or other material in winter. Plant the bulbs so that their tips are just under the soil. Increase by offsets, or by seeds sown in spring in a warm greenhouse.

Crocosmia. Like the two crinums above, *Cro-cosmia masonorum* is a plant which teeters on the edge of hardiness. This, too, needs a warm, sunny border to be a success in our gardens, and it likes a light, sandy soil best of all. The flowers, borne in arching sprays on 2½-ft. stems are an eye-catching orange-red shade, and appear in August and September. Plant the corms 6in. deep in spring and lift and store them at the end of the season in those gardens which tend to be frosty, over-wintering them in an airy frostproof shed or room. Otherwise, leave them in the ground and protect with straw, bracken, ashes or other suitable material in winter. Increase is by offsets or seed.

The montbretias are also correctly included here for the garden varieties are grouped under the name *Crocosmia crocosmiiflora*. The large-flowered hybrids which have been developed in quite recent years are a tremendous advance on the old cottage garden montbretia and in many gardens the relative tenderness of these will necessitate lifting them in autumn and over-wintering the plants in a garden frame. However, experience may prove in some gardens that it is sufficient to protect the plants with bracken, straw or similar material, and certainly this will be so in the case of the common kinds. The plants grow 2 to 3ft. tall and they need well-drained soil and a sunny position. Nothing shows off their bright yellow, orange or crimson late summer

The prettily named Glory of the Snow, *Chionodoxa luciliae*, which has blue, white-centred flowers, needs a sunny position. It is March flowering

flowers better than a light-coloured wall as a background. If the plants are left undisturbed over winter a wall or similar structure will also help give protection. Excellent named varieties are the golden-yellow Rheingold, the orange-red James Coey, and His Majesty, orange-red and crimson.

Plant the corms 3in. deep in March or April and 6in. apart. Increase by division in March and April.

Crocus. This large genus provides us with many fine garden flowers for brightening the garden in spring and autumn. All need well-drained soil and should preferably be left undisturbed for several years at least. The colour range is from white to yellow and purple. They are much used for naturalising in grass, for planting in borders, in the rock garden, and as an edging to paths or borders. Plant the winter and spring-flowering kinds like *Crocus tomasinianus* in September and October, and the autumn-flowering ones like *C. speciosus* and *C. zonatus (C. kotschyanus)* in July. Especially fine winter-flowering crocuses are *C. chrysanthus* E. A. Bowles, a deep yellow colour with bronzy-brown throat markings; *C. c.* Snow Bunting, white, with lilac-coloured markings; *C. tomasinianus* Whitewell Purple, a lovely mauvish-purple; and *C. c.* Taplow Ruby, a reddish-purple shade.

Plant all of them not more than 3in. deep and about 3in. apart. Increase is by offsets or seeds.

The dainty *Cyclamen neapolitanum* bears its rosy-pink flowers in August and September. It thrives in cool shade with a peaty root-run

Crown Imperial, see Fritillaria

Curtonus. The plant named *Curtonus paniculatus* will be better known to many gardeners under its old name of *Antholyza paniculata,* and very handsome it is in late summer when bearing its branched, arching stems of orange-red, tubular flowers. It grows up to 4ft. tall. It should be given a sunny position, preferably at the foot of a south-facing wall, and winter protection in the form of ashes, bracken, straw or other material unless the soil is particularly light and well drained. Plant the corms 4in. deep and 6 to 8in. apart in March or April. Increase by offsets or by seeds sown in a cool greenhouse.

Cyclamen. Both the autumn- and spring-flowering cyclamen are a joy if planted in the conditions they like in a cool, shady spot with rather peaty soil as a root run. *Cyclamen coum*, which flowers in February and March, is a popular species with dainty carmine blooms. *C. neapolitanum*, with rosy-pink flowers in August and September, *C. europaeum*, with carmine flowers in autumn, and *C. repandum*, with crimson flowers in spring, are species with beautifully marbled foliage. These make an extremely attractive ground cover which is with us for most of the year. Plant during August and September, 1½in. deep and 3in. apart, then leave undisturbed for a long period.

Increase by seed sown in spring in a cool greenhouse or frame, or by dividing old clumps in August or September.

Daffodil, see p. 148
Dahlia, see p. 169
Dierama. Like so many other fine bulbous plants *Dierama pulcherrimum* comes from South Africa. The foliage is grass-like and a perfect foil for the long, arched flower stems from which hang the bell-like flowers of purple or white in June and July. Unfortunately it is not very hardy and a sunny, sheltered position is needed for it with a well-drained but moist soil as a growing medium. Plant the corms 3 to 4in. deep and 3in. apart in autumn. Increase by seeds sown in spring in a greenhouse or frame or by dividing the clusters of corms in March.

Dog's-tooth Violet, see Erythronium
Endymion, see Scilla
Eranthis (Winter Aconite). The species most usually grown is the January-March-flowering *Eranthis hyemalis*, the cheery yellow blooms of which show up well against the distinctive frame of much divided leaves. The eranthis are useful plants as they like shady conditions and can be used in many parts of the garden, including near trees and in grass. Plant 2in. deep and 2in. apart in September or October and leave undisturbed as long as possible. Increase by division of the tubers in September or October.

Erythronium (Dog's-tooth Violet). These also,

Erythronium dens-canis, the elegant Dog's Tooth Violet

Fritillaria imperialis, the Crown Imperial

Iris histrioides, a handsome dwarf bulbous species. The rich blue flowers are marked with white and gold

like the eranthis, are plants for semi-shade which like to be left undisturbed. They are one of the delights of the spring garden. Various forms of the Dog's-tooth Violet, *Erythronium dens-canis,* are available in white, purple, pink and mauve and in heights from 6 to 9in. The nodding flowers appear in March and April. Slightly later flowering is the handsome *E. revolutum* with delightfully mottled leaves. This species has rose-pink flowers and there is a white variety, White Beauty, with brown markings at the base of the petals. *E. tuolumnense* is another fine species, with golden-yellow flowers and pale green leaves. They are especially well suited to the rock garden. Plant in August or September, 2 to 3in. deep and the same distance apart, in a soil which contains plenty of humus material. Increase by offsets removed in early autumn.

Flower of the West Wind, see Zephyranthes Fritillaria. There are two widely dissimilar members of this genus to which I would draw attention—the beautifully marked 1-ft. tall Snake's-head Fritillary, *Fritillaria meleagris* and the 3- to 4-ft. Crown Imperial, *F. imperialis.* The bell-shaped flowers of the elegant *meleagris* are beautifully chequered in various shades of purple and there are fine named forms available, like the white Aphrodite, the dark purple Charon, and Saturnus, a pinkish-purple shade. The Crown Imperial is an impressive plant with a close-packed tuft of leaves encircling the tops of the stems above the nodding yellow, red or orange-red bell-shaped flowers. Both flower in April, the Snake's-head Fritillary being partial to a moist but well-drained soil and the Crown Imperial liking a heavyish, fairly rich soil. The Snake's-head is a plant to grow in borders or in grass in half shade, and the Crown Imperial is best suited to a partly shaded border. Plant the bulbs in September or as soon thereafter as possible, *meleagris* 4in. deep and 6in. apart, *imperialis* 6in. deep and 12in. apart in a very well prepared soil. Leave undisturbed for as long as possible. Increase is by offsets taken at the time of planting.

Galanthus (Snowdrop). The 6-in.-tall *Galanthus nivalis* is our much-loved common snowdrop and a particularly fine variety is S. Arnott, this having beautifully formed flowers. Another outstanding variety is *atkinsii.* These snowdrops are January-February-flowering and by planting others like *byzantinus, elwesii* and *plicatus*—9 to 12in. tall—one can have several months of bloom. All have white flowers with green markings. Plant the bulbs 3in. deep and 3in. apart in sun or shade, but not in soil which is too light and free-draining. They can be grown in beds or borders, rock gardens, banks and in grass. Increase by offsets or by dividing clumps of bulbs after flowering in spring.

Galtonia. The white-flowered *Galtonia candi-*

cans, which used to be known as *Hyacinthus candicans*, is an attractive plant with its spikes of bell-shaped flowers on 4-ft. stems. These appear in August and September and it makes a useful border plant for a sunny position. Plant the bulbs 6in. deep and 1ft. apart on fairly rich soil at any time between autumn and March when conditions are suitable. Increase by offsets at the time of planting.

Gladiolus. These lovely flowers, now available in so many beautiful colours in both large-flowered and miniature types, are of course half-hardy plants which must be lifted at the end of each season. The types grown are the large-flowered Grandiflorus, the Butterfly, Miniature and Primulinus; the catalogues of gladiolus specialists should be consulted for descriptions of varieties. With gladioli, as with so many other garden flowers—roses, dahlias and chrysanthemums are obvious examples—new varieties appear each year and the present 'best' ones can be superseded at any time. Some favourites of mine, though, are the orange-yellow Acca Laurentia, the salmon-rose Alfred Nobel, the pink Dr Fleming, the yellow Flower Song, the purple Mabel Violet, the red Mansoer, scarlet New Europe, blue Ravel and white Snow Princess—all large-flowered Grandiflorus varieties. Of the Butterfly varieties Femina, pink and scarlet, and the pink Melodie are favourites of mine; also the Primulinus varieties Pegasus, creamy-white and purplish-red, and White City.

Plant all these types between March and May—planting the corms in succession extends by some weeks the length of the flowering season—putting the corms 4in. deep and 6 to 8in. apart. Good drainage and sunshine are the main requirements, and the soil should be well prepared. Propagate from seeds or the cormlets which form round the base of the corm each season.

Glory of the Snow, see Chionodoxa
Grape Hyacinth, see Muscari
Guernsey Lily, see Nerine
Harlequin flower, see Sparaxis

Hyacinthus (Hyacinth). These flowers, I always find, although much used for bedding, look more at home indoors. If they are grown outdoors, though, I consider them better for window boxes or other containers than for planting in beds. However, this is only a personal opinion and many gardeners will obviously think otherwise. Wherever they are grown there is always that wonderful scent to beguile one.

Varieties are available in red, pink, salmon, blue, yellow and white and a good bulb catalogue will give details of leading varieties. Favourites of mine are Delft Blue, a good light blue, the red Jan Bos, the salmon Lady Derby, the white L'Innocence, the deep blue Ostara and the deep pink Pink Pearl.

Hyacinths like a well-drained rich soil and it pays to feed them with bonemeal after planting. Plant the bulbs 3 to 4in. deep and 8in. apart in September, October or November. Propagate from small bulbs produced at the base of the parent bulb, or from seed, but this method means a wait of four to six years for the flowers.

Ipheion. Another plant which has suffered severely at the hands of the botanists in the past is *Ipheion uniflorum*. It has also passed under the generic names of *Brodiaea, Milla* and *Triteleia* (it is under this last that many catalogues list it) and I consider this a shame for it deserves a better fate. Only 6in. tall, *I. uniflora* has pretty white flowers shaded with blue, and it is admirably suited for the rock garden, given a warm, sunny position. The flowers appear in March and April. Plant 4in. deep and the same distance apart in September or October. Increase by offsets or seed sown very shallowly in March and raised in a cold frame.

Iris. The bulbous irises include many superb garden plants—dwarf species like *Iris histrioides* which flowers in early February and *I. reticulata* which flowers in February and March. The first is a rich blue shade with white and gold markings, the second deep blue and gold. The form of *I. histrioides* named *major* is particularly fine and popular varieties of *I. reticulata* are the blue and yellow Cantab; Royal Blue, bluish-purple and yellow; and Harmony, rich blue and yellow. Then there are the Dutch, Spanish and English irises which grow from 1½ to 2ft. tall and are splendid flowers for cutting, having a good range of colours from blue, mauve and yellow to white. The Dutch are the first to come into flower in May, followed by the Spanish and then the English, the season of the last being late June and July.

The Dutch, Spanish and English irises should be planted 3in. deep and 6in. apart in September or October in any reasonable garden soil to which a dressing of bonemeal has been added. *I. histrioides* and *I. reticulata* need a gritty mixture and if the soil is too heavy it should be lightened by adding sand and peat. A warm sunny position and good drainage are other requirements and they make fine rock garden plants. Plant the bulbs of the first-mentioned 2 to 3in. deep and 3in. apart, and of *reticulata* 4in. deep and 3in. apart, in August or September. Increase is by offsets removed when the plants are lifted in autumn or from seeds sown in pots in a frame when ripe or in March.

Ixia (African Corn Lily). The ixias are delightful, pretty, tender plants for warm gardens, with star-shaped flowers in a wide range of colours, carried in May and June on stems 1 to 1½ft. long. These colours include orange, red, violet and white, and named varieties are obtainable or mixtures. The soil for ixias must be well drained and the position sunny to achieve success. Annual lifting after the flowers die down and replanting in September and October is necessary. A border under a south-facing wall is suitable for a site. Plant the corms 4in. deep and 2in. apart in autumn and cover them with sand as a protection against damp. Increase is by offsets or seed.

Ixiolirion. A warm, sheltered position under a south-facing wall—with a light sandy soil—is also the best place for the ixiolirions, attractive bulbous plants with rather trumpet-like flowers in shades of blue, purple and white with narrow greyish-green leaves. Two which are offered are *Ixiolirion montanum* (often listed in catalogues under its old name of *I. ledebourii*), in various shades of blue, and the rosy to violet-blue *I. m. pallasii*. Both are 1 to 1½ft. tall. Plant the bulbs in March 4in. deep and 4in. apart, and lift them for storing in September in a cool, airy room until planting time comes round. Increase is by offsets.

Kaffir Lily, see Schizostylis

Leucojum (Snowflake). The leucojums, which look like outsize snowdrops, are charming flowers, both in the case of the 1½-ft. tall *L. aestivum,* the white, green-tipped Summer Snowflake, and in the much smaller (6in. to 1ft.) *L. vernum*, the Spring Snowflake, of similar colouring. The latter is a good bulb for naturalising in grass. Both species like to grow in shade and should be left undisturbed for periods of up to five years or so. Plant bulbs of both species 3 in. deep and 4in. apart in September or October.

There is also an autumn-flowering species, *L. autumnale*, which is not much grown and this needs a warm position in full sun. It is very beautiful with white flowers tinged with pink, borne on 6 to 8in. stems in September. Plant 3in. deep and 3in. apart in April. Increase by offsets removed in September or October.

Lilium, see p. 150

Meadow Saffron, see Colchicum

Montbretia, see Crocosmia crocosmiiflora

Muscari (Grape Hyacinth). The most well-known species is *Muscari armeniacum*, especially in its variety Heavenly Blue, with rich blue flowers. Another good variety is Cantab, with paler, Cambridge-blue flowers. These flowers are borne in racemes on stems some 8in. tall.

Other grape hyacinths with especial charm and garden value are *M. botryoides album,* pure white as the name suggests, only 6in. tall and very pretty in April and May when flowering freely. For March-April flowering there is the rather unusual and splendid *M. tubergenianum* which has flowers in two shades of blue on each spike—pale blue at the top, dark blue lower down. Another I must mention is the Plume or Feather Hyacinth, *M. comosum monstrosum*, or, as it is usually listed in catalogues, *M. c. plumosum*. The violet-coloured flower spikes of this variety have a feathery appearance and are altogether looser than those of the other muscari and it looks very effective on a sunny ledge in the rock garden.

Plant muscari in ordinary soil in sunny positions. They are splendid plants for borders, growing around trees or shrubs on rock gardens, and naturalising in grass. The bulbs should be set 3in. deep and 4in. apart and be planted in September or October. Increase by offsets from old bulbs at planting time.

Narcissus, see p. 148

Nerine (Guernsey Lily). This is a large genus but only one species, *Nerine bowdenii*, is suitable for garden planting, and this should be planted in a sunny, sheltered and well-drained border, preferably at the foot of a south-facing wall. This lovely plant with its distinctive heads of pink flowers borne on 1½-ft. stems in September and October is not suitable, of course, for cold gardens. When grown outdoors it should be protected from November to April with bracken or other dry

The Spring Snowflake, *Leucojum aestivum,* which has the appearance of an outsize snowdrop. This is a plant for shade; the bulbs are left undisturbed for up to five years

litter. Top-dress the plants with leafmould or well-decayed manure in August.

Plant between August and November, setting the bulbs 3in. deep and 4in. apart. Increase by offsets between July and September.

Ornithogalum. *Ornithogalum umbellatum*, the Star of Bethlehem, whose star-shaped flowers are white inside and white marked with green on the outside, is a useful plant for a mixed border, for the rock garden or for naturalising in grass or among shrubs. So, too, is *O. nutans*, an attractive species with more green in its colouring. *O. umbellatum* is up to 1ft. tall, *O. nutans* up to 1½ft., and both flower in late spring and early summer. Plant the bulbs 3 to 4in. deep and 6in. apart between August and November. Increase by offsets removed from old bulbs.

Puschkinia (Striped Squill). A distinctive dwarf plant for providing colour in early spring is the

6-in.-tall *Puschkinia scilloides*. The white, blue-striped flowers, as the specific name indicates, resemble those of a scilla, to which it is closely related. Plant the bulbs 3in. deep and 3 to 4in. apart in September or October. They are suitable for border planting on the rock garden in sunny positions. Increase by offsets from old bulbs in late autumn or by seeds raised in a cold frame and sown in August or September.

Quamash, see Camassia

Ranunculus. The tuberous-rooted ranunculuses which have been derived from *R. asiaticus* are valued for their colourful display when used at the front of borders and for their excellence as cut flowers. The Turban and French strains are 9 to 12in. tall and have double or semi-double flowers in May and June. Provide them with a sunny, sheltered home. Plant tubers in October or November, late February or March. The claws on the tubers should face downwards and the

Squill, *Scilla sibirica*. The intensely blue blooms are borne in profusion on 4-to 6-in. stems. Larger than the type is the variety *atrocoerulea* (or Spring Beauty as it is usually termed), which also has flowers of rich blue colouring. Another fine species is *S. tubergeniana*, with flowers of pale blue and white which appear in February and March. These are very effective when grown in groups in borders, on the rock garden or in grass in sunny positions.

A quite different scilla which can be grown successfully in a warm, sunny position is *S. peruviana*. This bears ball-shaped heads of flowers on 9-in. stems in May and June, and its specific name and common name of Cuban Lily are most misleading for it comes, in fact, from the Mediterranean region. This handsome plant is effective also in containers.

The Spanish Bluebell, which is often found listed under its old name of *Scilla campanulata* or

Ornithogalum umbellatum, the Star of Bethlehem—an adaptable plant which flowers in late spring and early summer. The flowers are white, marked with green

The distinctive close-petalled flowers of a *Ranunculus asiaticus* hybrid. They are excellent for border display or cutting

depth of planting recommended is 2 to 3in. with 4 to 6in. between the plants. Increase by seeds raised in a cold frame or cool greenhouse.

Schizostylis (Kaffir Lily). A handsome autumn-flowering plant for a warm, sunny border is *Schizostylis coccinea* or one of its varieties, the September-October flowering Mrs Hegarty or the November-flowering Viscountess Byng. The attractive flowers are borne on spikes 1 to 1½ft. long and are crimson in the case of *coccinea* and pink in the case of its varieties. A good loamy soil is required and protection must be provided during hard weather in the form of a covering of bracken, straw or other dry litter. Plant between October and March when conditions are suitable. Lift and divide the bulbous rhizomes every third year in March or April.

Scilla. There are few more welcome flowers in February than those of the charming little Siberian

S. hispanica is correctly now *Endymion hispanicus* but I shall describe it with the scillas. This is a fine plant for naturalising in light shade, say among shrubs or under trees, and there are fine named varieties like the pale blue Myosotis, the tall White Triumphator, with pure white flower spikes, and Queen of the Pinks, a rosy-pink variety. The English Bluebell, *E. non-scriptus* (also moved now from the genus *Scilla*), is so well known as to need no description and its grace and charm make it one of the pleasures of the springtime garden and countryside.

Plant in ordinary soil at any time between August and November, 2 to 4in. deep and 4in. apart. *S. peruviana*, though, should be planted 6in. deep and 6in. apart. Increase by offsets from old bulbs in autumn.

Siberian Squill, see Scilla
Snake's-head Fritillary, see Fritillaria

The flowers of *Tigridia pavonia*, the Tiger Flower, with bold spots at the base of their petals have an unfailing attraction. It grows 2ft. tall

Snowdrop, see Galanthus
Snowflake, see Leucojum
Sparaxis. The pretty sparaxis hybrids, with the attractive common name of Harlequin Flowers, deserve to be more widely grown. They are showy in spring when bearing their flowers in mixed colours like red, purple, black, white and yellow. They need a warm, sunny position and a fairly dry soil, and protection from bracken or other material is necessary in very cold weather. Species include the 1½-ft. *S. grandiflora* and the slightly shorter *S. tricolor*. Plant the corms in autumn, 4in. deep and 2in. apart. Increase by offsets.
Star of Bethlehem, see Ornithogalum
Sternbergia. The cheery, yellow-flowered, crocus-like *Sternbergia lutea* is a charming little plant for autumn colour. The golden-yellow flowers look attractive in a border or rock garden setting, open to the sky on their 6-in. stems and thrusting above the narrow foliage. A sunny position and good drainage is essential. Plant the bulbs 4 to 6in. deep and 6in. apart in August or September. Increase this plant from the new bulbs produced each year.
Striped Squill, see Puschkinia
Tigridia (Tiger Flower). The 2-ft.-tall *Tigridia pavonia* from Mexico is a distinctive-looking plant with its orange-red, three-petalled flowers spotted at the base with deeper colouring. There are also forms with pinkish, mauve, yellow and

white colouring. These appear in late summer and early autumn and although the glory is fleeting, for each flower lasts only a day, each stem produces several flowers and this plant is well worth the space devoted to it. A sunny border and a well-drained soil is necessary and there must be no lack of moisture in dry weather.

Plant the corms 3in. deep and 6in. apart in April and lift them for storing in October. The storage place must be airy and frostproof. Increase by offsets removed in April.

Trillium. Somehow one does not associate trilliums with plants of the kind we are considering but they are in fact tuberous perennials. The only one I am going to refer to now, though, is the Wake Robin or American Wood Lily, *Trillium grandiflorum*, a lovely plant for rather damp, shady border or woodland conditions. The white flowers, some 3in. across, are carried on 1½-ft. stems above whorls of practically stemless broad leaves. There is a pink variety, *roseum*.

These plants need a peaty soil and much appreciate topdressing with decayed leaves. They should be left undisturbed as long as possible but can be divided when necessary, in March.

Tulipa (Tulip), see p. 152
Wake Robin, see Trillium
Windflower, see Anemone
Winter Aconite, see Eranthis
Zephyranthes (Zephyr Flower, Flower of the West Wind). Only one of these intriguingly named, handsome American bulbous plants is hardy in this country, *Zephyranthes candida*, with white flowers on 6- to 12-in. stems. It is a handsome plant with the star-like flowers open above grass-like foliage in September and it is well worth trying in a warm, sunny border, preferably against a south wall. The soil is best light and sandy, and must be very well drained. Plant the bulbs between August and November 4in. deep and 4in. apart and increase by offsets during the same period.

Daffodils

I believe that most of us, even though we may never acknowledge it, have an almost emotional attachment to the narcissi, or the daffodils as they are popularly called. No other flower is associated in quite the same way with the pleasures of spring, and the variations to be found within the genus are such as to attract the interest of gardeners with many different tastes.

The eleven divisions into which the genus has been classified by international agreement, based on the formation and colour of the flowers, is an indication of the diversity to be found within the genus and I shall later deal with each of these divisions in their correct sequence. For the moment, though, I want to consider more practical aspects of daffodil growing and to place them in the garden scene.

For naturalising in grass, I think it will be generally agreed, the daffodils have no peer. They are just as at home, too, when planted informally in borders, among shrub plantings, around rose bushes—this is a use for them which may disturb some rosarians but which I find gives rose beds an interest at a time of year when they are necessarily stark and bare—or arranged in martial rows in formal beds. All the dwarf kinds are, of course, eminently suited for rock garden planting and the full range can be grown in greenhouses or in the home. This versatility and much grace and beauty gives the narcissus family a unique place in our affections.

Outdoors. Generally speaking, daffodils are extremely tolerant of widely differing soil conditions but at the same time there is no doubt that rich soil with plenty of body in it will produce the best results. If the soil is in rather poor condition I add a dressing of bonemeal or hoof and horn meal at the rate of 4oz. to the square yard at the time when the ground is being prepared. Well-rotted farmyard manure can be used, too, if it is placed in the lower spit of soil, well below the level of the bulbs; but everything considered it is probably best to rely for extra nourishment on the fertilisers mentioned above. Deep digging is a prime requirement for good results.

If there is a choice, then make the plantings in a sunny, open position, but if semi-shade it has to be then it will still be found, other things being equal, that the results are quite good. Drainage must be satisfactory to avoid rotting of the bulbs.

Plant as early in the season as possible, which means August in practical terms, but if necessary do this up to the end of November if circumstances demand it. The advantage of early planting is that the bulbs are out of the ground for the shortest possible time—which is much to their advantage—and they have the longest possible growing period in which to reach maturity. Where permanent plantings are made it is best to leave the bulbs alone until it is obvious that overcrowding is affecting quality. They can then be divided in the autumn, offsets being removed and planted in rows in a reserve bed to grow on into flowering size plants.

In grass, it doubtlessly saves time and trouble to use a special bulb planting tool if more than a small number of bulbs is being planted. Elsewhere I would always use a trowel, for with this handy tool it is easy to make sure that no air space is left beneath the bulb, which can easily hold back strong initial rooting and affect the establishment of the plants.

For the larger daffodils, 4 to 6in. is the recommended planting depth, those plantings made in heavier soils being set rather less deeply than

The attractive autumn-flowering *Sternbergia lutea*. The golden-yellow flowers are borne on 6-in. stems. A sunny position and good drainage are needed for success

others in light soils. The dwarf daffodils or narcissi are planted 2 to 3in. deep and these need a rather grittier rooting medium with sharper drainage than their more robust relatives.

When daffodils are used as bedding plants they must be lifted immediately after flowering to make way for summer bedding plants and this means that they have to be moved to a reserve plot in another part of the garden to complete their growth and ripen the bulbs. By late June, with the foliage withered, they will be ready to lift and store in a cool, airy shed until planting time comes round again after another couple of months.

The Types of Daffodil. I shall now consider the various types of daffodil, division by division, and make brief comments on them which I hope will be of interest. This is also an appropriate time to point out that daffodil breeding is an extremely long-term business as it may take eight or more years to produce stock of new varieties in commercial quantities. I mention this because it explains why the cream of recently raised varieties can be seen offered at as much as £40 each. Many first-rate older varieties will be found offered at less than 1s. each. Another thing requiring explanation also is the terms used in catalogues in connection with daffodil bulbs. A mother bulb is one with as many as four growths which will each produce flowering shoots. A double nose bulb has

The white Trumpet daffodil Broughshane—a variety of especial distinction. The large, beautifully-formed flowers are borne on strong stems and are long lasting

The soft primrose yellow Double daffodil Camellia, which can always be relied upon to provide a good show. This variety is vigorous and has strong stems

at least two 'noses' and will produce a minimum of two flowers, and a single nose will produce one or two flowers.

Division I is devoted to Trumpet daffodils and is divided into three sections, for yellow-, bicolor- and white-flowered varieties. One of the best known of all daffodils is the old but still very fine King Alfred, of rich golden-yellow colouring and splendid for the garden and for growing indoors. The same can be said of Golden Harvest, and others which can be strongly recommended are the white varieties Beersheba and Mount Hood, Unsurpassable, another golden-yellow, and the Queen of the Bicolors, white and yellow.

Division II, Large-cupped varieties, has four sections, and has one flower to each stem with the cup more than one-third but less than equal to the length of the perianth segments. Varieties I would recommend here are the yellow and red Carbineer, the large-flowered yellow Carlton, the white and orange John Evelyn, the creamy-white and orange Sempre Avanti, Orange Bride and Scarlet Elegance.

Division III, the Small-cupped narcissi, includes varieties with cups not more than one-third the length of the perianth segment, and here again one flower is carried on each stem. There are four sections in the division. An especially good variety of modest price is the red and white La Riante which is as good for indoors as it is impressive in the garden.

Division IV, the Double daffodils, is perhaps not universally popular for there are quite a few gardeners with a distinct bias against double flowers. Their main value is as pot-grown plants for greenhouse or home and they can look very impressive in this role when well grown. A splendid variety for forcing is the yellow and orange Texas, likewise the white and yellow Irene Copeland. I have considerable affection, too,

for the pleasing soft primrose yellow Camellia.

Division V, the Triandrus daffodils, is where we find the Angel's Tears daffodil, *N. triandrus albus*, the lovely pure white Thalia and the white and lemon Silver Chimes. The resemblance to *N. triandrus* can be clearly seen in all members of this division. Division VI, for *cyclamineus* daffodils, includes the dwarf species of that name, a very beautiful bright yellow narcissus, and such gems as the early-flowering February Gold, golden-yellow and Peeping Tom of similar colouring.

Division VII, the Jonquilla daffodils, includes a favourite of mine in the pale yellow Trevithian, and the Queen Anne's Double Jonquil, *N. jonquilla flore pleno*. Among the *tazetta* daffodils (Division VIII), which make such delightful plants for bowls or other containers indoors, my choice would be the white and scarlet Cragford, the white and orange-red Geranium (which also makes a good garden plant); and the creamy-white and orange-red St. Agnes.

The *poeticus* daffodils, grouped in Division IX, include a fine late-flowering variety in the white and red Actaea, and the Old Pheasant's Eye (*N. recurvus*) which is so valuable for naturalising.

The remaining two divisions, X and XI, are devoted to species and wild forms and wild hybrids in the first case, and miscellaneous narcissi which do not fall within any of the other divisions in the latter case.

Daffodils in the Cold Greenhouse and Home. For details of these forms of cultivation turn to p. 163

Lilies

One does not have to be a dedicated lily specialist to appreciate how much these plants have to offer for garden decoration, and for greenhouse culti-

The easily-grown Regal Lily, *Lilium regale*, a species with flowers suffused pinkish-purple and maroon shades on the outside and white within

Lilium Enchantment, nasturium red, is one of the Mid-Century Hybrids—all of which are fine garden plants. It grows to between 2½ and 3ft. tall

vation, too, in some instances (see p. 163). One of their main attractions for many gardeners is that the genus with its 80 odd species, hundreds of varieties and many strains has such diversity of interest. Some, such as the Regal Lily (*Lilium regale*), and Madonna Lily (*L. candidum*) and *L. brownii*, are easy by any standard while others like the pink-flowered *L. japonicum* offer a real challenge to the skilful cultivator. But quite apart from all this, the lilies are invariably plants of the greatest beauty and for this reason above all others we want them in our gardens.

In recent years our gardens have been enriched by the introduction of a great many hybrid lilies mostly introduced from the United States of America. These modern lilies are of two types—either they are propagated vegetatively from an original single bulb, by means of scales, division or bulbils, or they are what is called a seed strain, the plants being raised from seed obtained by planned hybridisation between two selected lilies.

A group of plants raised vegetatively by the first-mentioned method is called a 'clone' and each plant is identical in every way to the others in the group. As it takes some time to raise these lily clones, they tend to be quite expensive, particularly if the lily in question is a new one.

With the seed strains there are bound to be variations. The colour, for example, in one particular strain may range through shades of yellow, gold and orange and if several plants of the same strain are planted there will be differences between them, but this should not worry the home gardener. A great advantage of propagating from seed is that the plants are not affected by weakening virus diseases and thus have every chance, given the right growing conditions, to develop strongly and vigorously. Also, lower costs of production than with clones means that some of the best strains can be bought for a few shillings a bulb. For quite a small outlay one can have an impressive collection of lilies.

Before we consider these hybrids in more detail, though, let us consider the kind of growing conditions needed by garden lilies.

Cultivation. One of the first things one must understand when beginning to take an interest in lilies is that they include a proportion which are lime-haters and if this is a significant factor it is something to take into first consideration when deciding which to grow. In the recommendations which follow I shall indicate which of those I refer to is in this category. Also, some lilies are what is known as stem-rooting, i.e. they form roots on the lower part of the stem as well as basally from the bulb. This means that they have to be planted deeper than the others—8 to 9in. as opposed to 5 to 6in. for the others. *L. candidum* is a special case for this should only be covered with ½in. of soil, and another is *Cardiocrinum giganteum*, formerly *Lilium giganteum*.

All lilies like a well worked, not too heavy, soil which must be very well drained. They need, too, what is called a cool root run, in other words the lower part of the stem should be shaded by other plants to protect the roots from strong sunshine. Planting among rather open-habited, low-growing shrubs or herbaceous plants proves very successful and, carefully placed so that the sun strikes through to them, they are magnificent plants for light woodland.

If the soil in which lillies are to be grown needs to be improved one should be careful not to dig in manure where it can come into contact with the bulbs. Put it well down in the second spit where it will not cause any trouble. Bulb rotting is a worry with lilies and it is usual when planting to place each bulb on a bed of sand for this very reason.

October and November are good planting months but the best advice of all is to plant them as soon as they become available, for the bulbs deteriorate if left out of the ground for longer than is necessary. Lighten heavy soil by digging in peat, leafmould and coarse sand. Once planted, leave lilies undisturbed for as long as possible and mulch each spring with peat or leafmould if this is available. When deterioration in the planting is noticeable divide the clumps of bulbs and replant in October or November.

Planting Suggestions. With so large a group of plants as the garden lilies I can only hope to mention some of especial interest to readers in the space available. I shall start with some of the modern hybrids I mentioned earlier and then conclude with the species and hybrids.

The Mid-Century Hybrids are a good choice for lime-free soil, for these 2½- to 4-ft. lilies with an excellent colour range from pale yellow to deep red are fine garden plants for June-July flowering. One of the best named varieties is the nasturtium red 2½- to 3-ft. tall Enchantment, another is the yellow, maroon-spotted Joan Evans of similar height and with upright flowers. I am very impressed, too, with the African Queen strain of trumpet lilies in which the large flowers are apricot-yellow suffused with reddish-bronze. These are over 5ft. tall and flower in late June and early July. The Golden Clarion strain has trumpet flowers in varying shades of yellow, gold and orange, often marked with deep red and these, around 3½ to 4ft. tall come into bloom during July. Another lovely strain of trumpet lilies, named Golden Splendor, includes only golden-coloured lilies of the highest quality. Taller lilies, the Bellingham hybrids, which reach a height of 7 or more feet, are another I would mention as excellent plants for interplanting with shrubs. These have Turk's-cap flowers which are usually orange in colour and liberally marked with dark spots.

The species described below are all easy to grow and make a suitable introduction to a genus which holds out many delights for those who get to know it well.

Lilium candidum, the Madonna Lily, is a lovely sight in July when it carries its pure white, fragrant flowers freely on 6-ft. stems. Plant in August or September and set the bulbs so that they are only ½in. under the surface. It likes to be in full sun with shade for the roots, and it likes limy soil.

Lilium davidii is a handsome Chinese lily with red Turk's-cap flowers covered with black spots. Two fine varieties are Maxwill, a bright orange-red colour and about a foot taller than the type at 6ft. and *willmottiae*, deep orange, very impressive and also 6ft. tall. These are stem-rooting lilies and they like to have their heads in sun. Suitable for limy soil.

Lilium henryi is a fine lily with orange-yellow flowers marked with dark spots borne on stems some 6 to 7ft. tall. It is stem-rooting and peat must not be placed close to the roots for it objects to acid soil conditions. Flowering in August and September, it grows well in sun or shade.

Lilium martagon album is the white variety of the Turk's-cap Lily and a delightful plant for growing in the light shade cast by ornamental shrubs. It does not mind limy soil. Depending on the conditions it will reach a height of 3 to 5ft. It flowers in June and July.

Lilium pardalinum, the Panther Lily, is well named, for the orange and crimson Turk's-cap flowers are heavily marked with crimson-brown spots. These are borne in July on 5- to 6-ft. stems. It likes a damp soil and will grow equally well in sun or semi-shade. Does not object to limy soil.

Lilium regale, the Regal Lily, is very popular and its handsome trumpet flowers are suffused with pinkish-purple and maroon shades on the outside and white within, flushed yellow in the throat. This species grows up to 6ft. tall, likes plenty of sun (with shade at the roots) and is stem-rooting. It does not object to lime. As the shoots appear early in the season these can be cut back by frost and protection should be provided if frosts are forecast.

Lilium tigrinum splendens, a particularly fine variety of the Tiger Lily, bears rich salmon-orange flowers in August. Its height is 4 to 6ft. and it is stem-rooting. It does not like limy soils.

I cannot leave the lilies without mentioning the most glorious of all, although this is not so easy to establish as those described above. This is *L. auratum*, the Golden-rayed Lily of Japan. The huge, fragrant flowers 9 to 12in. across, are lined with gold on a white ground and heavily spotted with crimson; and are borne on stems 6ft. or more tall. This species likes a light soil which is well drained, and exposure of the stems to broken sunshine. It is stem-rooting and topdressing with leafmould during spring is much appreciated. It flowers in August and September. The numerous varieties are, with the exception of *platyphyllum*, much dearer than the species. These lilies do not like limy soil.

Tulips

If daffodils are the 'Number One' bulbous flower there can be no doubt about which comes second — the tulips which, with their kaleidoscope of colours, can be used to create dazzling displays in the garden. They have the happy knack of being pleased with almost any reasonable garden soil and sun or light shade makes little difference to their performance. As with the daffodils, though, the diversity of this genus can be a cause of bewilderment to the newcomer to gardening, so I shall do what I can to introduce the reader to the different types available.

More and more interest is being shown in the delightful tulip species and their hybrids, for these are in many cases excellent for planting in rock gardens or beside stone paving where they will make a colour show in spring and early summer. *Tulipa kaufmanniana,* the Water-lily Tulip, is a species which has given us a wonderful range of March-April flowering hybrids, all 6 to 8in. tall and suitable alike for rock garden or border planting. They can be left undisturbed for years, like the other tulip species, but unlike the tulips of gardens, which must be lifted and stored when the foliage has died down. Varieties of note are the coral-rose Fritz Kreisler, the white, cream and red Johann Strauss and the cream, crimson and bronze Vivaldi.

Taller (between 1 and 1½ft.) are the lovely Fosteriana tulips of which Easter Parade is an example. Another splendid variety of this type is Madame Lefeber (or Red Emperor as it is often called). The 9- to 12-in. tall *greigii* hybrids are especially notable for their lovely coloration and the handsome markings on their leaves. These are April flowering. Then for May flowering there are the Viridiflora or Green Tulips, 9in. to 2ft. tall, which have distinctive green markings on the petals.

Of the species proper one of the finest is the yellow *T. tarda,* with yellowish-green and white markings. About 6in. tall, this species flowers in late April and early May. Rather earlier to flower is the showy, scarlet 15-in. tall *T. eichleri,* whose petals are marked with a black, yellow-margined blotch. At this time, too, flowers *T. praestans* Fusilier, a spectacular sight when the orange-scarlet blooms are massed in a border planting on their 9-in. stems. A warm position with good drainage is needed by *T. clusiana,* the Lady Tulip, but this elegant species is a joy in April when its red, white and purple flowers are in the full flush of their beauty.

The other types of tulips for garden planting have now been brought to a high state of perfection with countless varieties of the greatest garden value.

The season opens with those fine bedding types, the Early Singles (12 to 15in. tall) flowering in mid-April and the 10- to 12-in. tall Early Doubles which bloom at the end of the month. These are followed by Mendel and Triumph varieties, which are described as mid-season tulips, and, in May, by the 2-ft. tall and beautifully shaped Darwin tulips, the Parrots, with their deeply cut petals in various mixed colours, the Lily-flowered varieties, and the Cottage, Breeder, Late Doubles, Multi-flowered (Branch-flowered) and Broken kinds.

All these should be planted about 4in. deep and 6in. apart in the latter part of October or early November for preference. Those used for bed-

Tulipa kaufmanniana, the Water-lily Tulip, and its numerous varieties are splendid plants for rock garden or border planting

Lily-flowered tulips, like the salmon-pink Mariette above, are graceful and make good cut flowers

Broken tulips with their lovely markings are unusually attractive. Shown above is Montgomery, white, with a rosy-red edge to the petals

ding purposes should be lifted and replanted in a reserve bed until the foliage has withered, when they can be lifted again and stored in a cool, airy room till replanting time comes round. The species and their hybrids are planted 3in. deep and are not, as already pointed out, lifted annually. Propagate from offsets or seeds.

Bulbs for the Greenhouse

A cool greenhouse, with a minimum temperature of, say, 7° to 10°C. (45° to 50°F.), is the passport to a whole world of delight with bulbs, corms and tubers. Many of those which are hardy garden plants are also splendid for growing in an entirely unheated greenhouse where, grown well and protected from the weather, they reach a perfection rarely possible to attain under garden conditions. Many of the plants mentioned in this chapter have already been described as garden plants

hanging baskets. The tubers can be started into growth at any time between February and late May for summer and autumn flowering but, of course, as they need a temperature of around 16°C. (60°F.) the earlier one starts the plants into growth the more heat has to be provided.

The John Innes No. 1 Potting Compost suits them well and the tubers are best started into growth in boxes, these being covered with about 1in. of compost. When growth is well under way move them into flowering-size pots (5in. or larger) or baskets. When using them for hanging baskets I like to plant them up the side and to have a fair number on top. Water with especial care at first but more freely when the plants are in full growth. When flowering has finished and the foliage has died down stop watering altogether and store the tubers in a frost-proof place until it is time to start them into growth again. Increase from seed, leaf cuttings, ordinary cuttings or

The fragrant hyacincths, which almost everybody likes, are excellent greenhouse plants. Specially prepared bulbs are available for Christmas flowering

Achimenes are decorative plants for pot cultivation or for use in hanging baskets. The colour range includes purple, blue, red and white

while others are only suitable for growing under greenhouse conditions. Perhaps one of the most important things to remember when growing these plants under glass is that slow, steady growing conditions will produce the sturdy, fine-flowered plants which can be grown.

Some can be forced, like the daffodils, hyacinths, freesias and early tulips, while others, like gladioli, are not suited to this treatment. To get the correct growing rhythm, therefore, one should aim to plant the bulbs, corms or tubers at the correct time. Most start off plunged in cold frames or under a covering of ashes outdoors, only being introduced to the greenhouse when growth is well under way. Brief cultural instructions are given for each genus in the notes which follow.

Achimenes. The achimenes in shades of purple, blue, red and white are becoming increasingly popular both for pot cultivation and for use in

scales rubbed off the tubers and sown like seed. The seed itself needs a high temperature for germination and the seedlings will bloom in the second year.

African Corn Lily, see Ixia

Amaryllis. The rosy-red *Amaryllis belladonna* with its striking lily-like flowers is an excellent plant for a greenhouse with a minimum temperature of 7°C. (45°F.). Pot the bulbs singly in 6-in. pots just below the surface in August using the John Innes No. 1 Potting Compost. Water sparingly until growth is well under way and then increase the amount given. Do not water at all during the resting period in early and mid-summer.

Arum Lily, see Zantedeschia

Babiana. Attractive plants with funnel-shaped flowers on 6-in. stems, the babianas are available in such colours as blue, pink, red and cream. Pot the bulbs 3in. deep in sandy compost in October or early November putting four or five in a 4-in. pot, plunge in a cold frame and move to the cool greenhouse when growth begins. Ripen off the bulbs after flowering and store till planting time comes round again. Increase by seeds.

Begonias (tuberous). It is not surprising that the tuberous double and pendulous begonias—the latter are superb for hanging baskets—are so highly regarded by gardeners. They have outstanding decorative value. March is a good month to start the tubers into growth and the plants will then begin to flower in late June, but February or April planting is quite suitable if for some reason that is preferable. Press the tubers, hollow side uppermost, into boxes of moist peat and coarse sand. Shade from strong sunshine and provide a minimum temperature of 13°C. (55°F.). Water with care and as soon as growth begins pot into 3-in. pots filled with John Innes No. 2 Potting

the compost is quite dry store the tubers in the greenhouse, under the bench or in some other frost-proof place, until the time comes round again to start them into growth.

They can be propagated from seed sown in January or February in John Innes Seed Compost and a temperature of 18°C. (65°F.) or rather more. For anyone with a small greenhouse, though, I would recommend buying say a dozen tubers and dividing these up in the second year. There must be at least one or two shoots on each piece, or division, and they should be dusted with flowers of sulphur before potting.

Blood Flower, see Haemanthus
Cape Cowslip, see Lachenalia
Clivia. An especially colourful spring- and early summer-flowering greenhouse plant is the clivia. Hybrids of *Clivia miniata* are available in various shades of yellow, orange and red. This plant is easily recognised, for the bold flower-heads on

Amaryllis belladonna, a plant which bears its rosy-red flowers on 18-in. stems, is a magnificent pot plant for early autumn colour

Tuberous begonias are colourful and superb plants, having outstanding decorative value. March is the best month to start the tubers into growth

Compost. Re-pot the plants into 5-in. or 6-in. pots when they have made a few inches of growth. Shade and moist atmospheric conditions are essential for success, for if they are lacking there will be trouble from bud dropping. Be especially careful when watering after potting as the plants are at their most vulnerable during the period when they are re-establishing themselves in the new compost. Only water the plants when the compost is becoming dry, and then water thoroughly.

When the flowers develop it will be noticed that single female flowers develop on both sides of each male double flower, and the female flowers should be removed. Provide the flower stems with support as they develop. Continue watering after flowering has ceased until the foliage starts to turn yellow and then dry the plants off by gradually reducing the amount of water given. When

strong stems are carried above deep green, strap-shaped evergreen leaves. Pot the fleshy roots in February in the John Innes No. 1 Potting Compost in 5- to 10-in. pots according to the size of the roots and place the pots in a sunny part of the greenhouse. A temperature of about 13°C. (55°F.) is suitable. Water the plants generously throughout the growing period which lasts until September. Feed with liquid manure and shade against strong sunshine. From September until spring water must be applied sparingly—just enough to avoid drying out, and the temperature during this period must be not less than 7°C. (45°F.). Increase the water supply when the plants start flowering in the spring. The important thing to remember with clivias is that they will only flower well when so potbound that they almost burst their pots.

Clivias are increased by seed sown in a heated

greenhouse or by dividing the roots at the time of repotting.

Crocosmia. *Crocosmia aurea*, which bears long spikes of orange-red flowers in summer, is a first-class plant for the unheated greenhouse. To make a good display plant five corms in a 5-in. pot in October and use a mixture of equal parts loam, leafmould and sand. Plunge in a cold frame and move to the greenhouse when growth begins. Increase by offsets or seed.

Crocus. These are very attractive as pot plants. October is the best time for potting, using a sandy compost; place four corms in a 3-in. pot and cover to a depth of $\frac{1}{2}$in. Plunge under ashes outdoors until growth begins and then move to the greenhouse. The plants can be forced by giving them a temperature of 13° to 16°C. (55° to 60°F.) in December. Do not use them as pot plants for a second year. After flowering, store these corms until planting time comes round again, when they can be planted out in the garden. Increase by offsets or seed.

Cyclamen. The greenhouse cyclamen, strains of *Cyclamen persicum*, are available in lovely shades of red, pink, salmon and white. They are among the finest greenhouse plants for autumn and winter decoration. So far as I am concerned, they are Number One.

Cyclamen must have a minimum winter temperature of 10°C. (50°F.). I sow seeds in late May or early June to produce plants that will flower in 18 months' time. I have tried sowing in January and hurrying the plants along to flower in the autumn of the same year, but it does not work—the plants tend to flop when they are brought into the house for they do not have the necessary sturdiness. Also, of course, it is not easy to provide the high temperature required when sowing at that time of year.

Sow the seed in pans or boxes filled with John Innes Seed Compost, spacing them out evenly on the surface. Cover with sifted compost and firm, and then cover the container with glass and newspaper until germination takes place, wiping the condensation from the glass daily. When the seed germinates place the container in a shaded garden frame out of the way of draughts. Water with great care, and as soon as the seedlings can be handled move them into boxes or $2\frac{1}{2}$-in. pots filled with John Innes No. 1 Potting Compost. Firm potting at all stages is essential. Keep growing steadily in a temperature of 13° to 16°C. (55° to 60°F.) with shade from direct sunshine and move into 4-in. pots when the present pots are full of roots. Always pot so that the top of the corm is at surface level. I always like to add a little dry cow manure to the compost used for cyclamen—when I can get it—and a little extra general fertiliser. I have also found that cyclamen appreciate a little riddled sphagnum moss in the potting mixture.

In autumn the plants are brought back into the greenhouse and kept growing steadily. By June of the following year they are ready for potting into the 5- or 6-in. pots in which they will flower and they should be hardened off carefully and placed in a shaded cold frame for the summer. Then, in late September, they are brought back into the greenhouse where they flower in autumn and winter. The earliest flowers to form are best removed, I find, and I then get excellent blooms from about mid-October right through to March.

The hardy cyclamen *C. repandum* can be used as a pan plant to decorate the unheated greenhouse. Pot in a mixture of equal parts loamy leafmould and sand and plunge in a cold frame until growth begins. Then bring the plants into the greenhouse.

Flower of the West Wind, see Zephyranthes

Freesia. The hybrids of *Freesia refracta* include many delightful rather soft colours and make splendid plants for the greenhouse which has a temperature of not less than 4°C. (40°F.). Their fragrance is also a great attraction. I prefer to grow my plants from seed rather than corms, and they can be obtained as separate or mixed colours. I sow the seed in March in pans containing a light, sandy compost and provide a temperature of 16°C. (60°F.). When the seedlings are about $1\frac{1}{2}$ to 2in. tall prick them out 2 to 3in. apart into really deep boxes—they have long tap roots—and keep these in the greenhouse until about mid-June when they should be stood beside the greenhouse wall outdoors in shade. Keep them well watered and bring them indoors again in September. The plants need support, as they grow nearly 2ft. tall so place twiggy sticks between them so that they can grow up through them. The plants will start flowering in October and will continue until February-March.

When growing from corms, place five or six in a 5-in. pot in John Innes No. 1 Potting Compost in August for flowering in January, and in subsequent months up to November for later flowering. Bury them 1in. deep. Put them in a cold frame and just cover with straw or leaves. When growth starts bring them into the greenhouse and place on a shelf near the glass. Introduce them to heat gradually. These plants, too, have to be supported with sticks and secured with raffia or twine.

Galanthus (Snowdrop). There is something particularly appealing about a pot or pan of snowdrops. These plants will grow well in a mixture of 2 parts loamy soil and 1 part each of leafmould and sand. Either medium-sized pots or pans are suitable and the bulbs should be set 1in. under the surface. Plunge in a cold frame until growth starts, then bring into the greenhouse.

Gladiolus. The gladioli make good greenhouse plants and the dwarf *colvillei* can be forced—but only very slightly. Do not attempt to force the

All the poetry of spring is to be found in this gay picture of naturalised daffodils

Above. The March-flowering
Crocus tomasinianus surrounded by
the debris of winter
Right. Lilium candidum, the
Madonna Lily, which carries its
fragrant flowers on 6-ft. stems, has a
gay partner here in the brilliant
Lychnis chalcedonica
Far right. Lilium pardalinum
the Panther Lily. This species
also grows up to 6ft. high

Left. Tulips in all their wide diversity add much to the gaiety of gardens. They have a happy knack of being pleased with almost any reasonable garden soil. Whether they are grown in sun or light shade makes little difference to their performance

Below. Tulipa tarda, one of the finest of the tulip species. It bears its flowers in late April and early May on 6-in. stems and is a delightful plant for the front of a mixed bed or other prominent position

large-flowered gladioli. Give all gladioli a rich, loamy, well-drained soil. Pot in January or February placing one corm only of a large-flowered gladiolus in a 6-in. pot 1in. deep. Move to a cold frame and cover the pots with straw or similar protective material during severe weather. When the flower spikes appear move plants into a cool greenhouse. Three plants of the *colvillei* type of gladiolus can be grown in a 6-in. pot at the same depth and grown on in the same way as the large-flowered type. After the flowers have gone over give decreasing amounts of water until the foliage withers. Clean the corms and store in trays or boxes in an airy, frost-proof room or shed.

Gloriosa (Glory Lily). The gloriosas are tuberous-rooted climbing plants needing warm greenhouse treatment. Those usually grown are *Gloriosa rothschildiana* with red and yellow flowers, and *G. superba* with orange and red flowers. Both bloom in summer. I start my plants into growth in early March, using a compost of equal parts loam, peat, leafmould and sand. The temperature needed is 21°C. (70°F.) and water is not given in any quantity until growth is well under way. The growths are trained up wires on the greenhouse wall. When flowering has finished in autumn the plants are dried off and the roots left in the pots until early March when they are repotted and brought back into growth. Increase from seeds, and offsets which have to be taken away from the parent bulbs with great care when they start into growth in the spring.

Glory Flower, see Gloriosa

Gloxinia. The summer-flowering gloxinias in their gay colourings of red, pink, blue and white are splendid greenhouse plants which present few difficulties in cultivation although not every greenhouse owner can raise them from seed as they require a germinating temperature of 16° to 18°C. (60° to 65°F.) to be maintained in January or February. An easier way is to grow one's plants from tubers which can be started into growth in March or April. These still need a temperature of 16°C. (60°F.) but this is more easily achieved in the spring months. Plant the tubers in boxes containing John Innes No. 1 Potting Compost or a mixture of leafmould and peat with a little compost, cow manure and sand added. The crowns of the tubers should be just above soil level. Place the box or boxes in a shaded, warm part of the greenhouse and just keep the tubers slightly moist. Also keep the atmosphere moist by overhead spraying. When the tubers have made about 2in. of growth move them into 5- or 6-in. pots filled with John Innes No. 2 Potting Compost. From now onwards plentiful supplies of water will be needed and the plants will probably want potting on later to 7- or 8-in. pots. During the later stages of growth a temperature of 10° to 13°C. (50° to 55°F.) is needed with plenty of ventilation.

If the plants are raised from seed a temperature of 16° to 18°C. (60° to 65°F.) is needed as already stated and I prefer early February to January sowing because it saves considerably on the heating costs. The seed, which is very fine, is sown on top of pots filled with John Innes Seed Compost. The pots are then stood in a propagating case until the seed germinates. Prick the seedlings out as soon as they can be handled into boxes filled with John Innes No. 1 Potting Compost. They need warmth and shade from sunshine and will soon be ready for a further move into 3½-in. pots filled with the same compost mixture. When ready, move them on again into 5- or 6-in. pots for flowering in July. They will then be in flower until September or October. Liquid feeding at intervals of about 10 days during the growing period will produce finer plants than otherwise.

Give decreasing amounts of water after flowering and allow them to ripen off their growth in a cold frame. Then, in the autumn, dry the tubers off completely and store over winter in a room with a minimum temperature of 10°C. (50°F.) and a dry atmosphere.

It is possible also to raise plants from leaf cuttings during the summer.

Grape Hyacinth, see Muscari

Guernsey Lily, see Nerine

Haemanthus (Blood Flower). The haemanthus are striking greenhouse plants which present no difficulties in cultivation. Those most often grown are *multiflorus* and *katherinae*, both of which bear red flowers in spring. These flowers are distinctive, being carried in ball-like heads on strong stems. The autumn-flowering *H. albiflos* has a quite different appearance with flat-topped heads of white flowers. All are about 1ft. tall and make big plants.

The spring-flowering species should be potted between August and November and the autumn-flowering one in March and April. Good drainage is essential and I find a mixture of loam and sandy peat very satisfactory as a growing medium. Use pots only slightly larger than the size of the bulbs and cover the bulbs over completely with compost when potting. Place in a sunny part of the greenhouse and provide a minimum temperature of 16°C. (60°F.) between April and September and 10°C. (50°F.) from then until March, preferably 13° to 16°C. (55° to 60°F.). When flowering ceases reduce the supplies of water and stop watering altogether when the foliage starts to turn yellow. Rest the bulbs until it is time to restart them into growth for the next season. Repotting is only necessary once every three years or so. Increase from offsets.

Hippeastrum (Amaryllis). The hippeastrums, often listed under the name Amaryllis, are superb for the greenhouse for spring and early summer, and with the provision of extra heat, as I shall explain, it is possible to have prepared bulbs in flower at Christmas. The showy trumpet flowers, which appear before the leaves, are available in such colours as red, pink, orange and white.

Start the bulbs into growth in February using the John Innes No. 2 Potting Compost and leaving the upper part of the bulbs uncovered. The drainage provided must be good, for these plants should not be repotted more than once every three or four years as they do not take kindly to root disturbance. Provide a temperature of 16°C. (60°F.) and start watering when growth begins, increasing the amounts as the plants develop. Remove some of the compost from the top of the pots in spring and topdress with new compost. This must be done very carefully.

After flowering and when the leaves have developed, feed the plants with liquid manure to build up the bulbs for the following year. Also, place the pots in a sunny position to ripen off the bulbs.

With the summer drawing to a close the foliage will start to turn yellow and this is the time to reduce the amount of water given, until eventually the compost is quite dry. When this stage is reached lay the pots on their side under the greenhouse staging. This is my method, but some gardeners prefer, I know, to keep the compost just slightly moist so that the bulbs are not completely rested. Increase from seeds or offsets.

The prepared hippeastrums for Christmas and winter flowering which I mentioned earlier, are a good proposition. A bright red variety offered in this way is called Winter Joy. The bulbs are started into growth in early November and are potted as described for other hippeastrums. These need, though, a temperature of 21°C. (70°F.) to flower for Christmas. Given lower temperatures they will flower from mid-January onwards.

Hyacinths. The much-loved, fragrant hyacinths need no introduction. Pot the bulbs in early September or as soon thereafter as possible in John Innes No. 1 Potting Compost. Place one bulb in a 4-in. pot or three in a 6-in. pot partly exposing the bulbs and place out of doors under ashes. After about 10 weeks move the pots into a cold frame and a week or so later bring them into the greenhouse. Start watering when growth begins and increase the quantities progressively. Plant the bulbs in the garden after flowering.

Especially prepared hyacinth bulbs for Christmas flowering are available in red, pink, blue, yellow and white. My favourite named varieties for forcing are the soft pink Princess Margaret, Pink Pearl, the blue Myosotis, the splendid red Jan Bos and the salmon-pink Lady Derby. For an early display, too, there are the Roman hyacinths with their loose spikes of fragrant white and pink flowers.

Iris. The dwarf bulbous iris like *I. reticulata* and

Lilium auratum, the Golden-rayed Lily of Japan, is one of the most spectacular of all lilies. It is suitable for greenhouse cultivation

I. histrioides may be grown in pans in a cold greenhouse. A 3-in. pot will accommodate several plants and they like a mixture of equal parts loam, leafmould and sand. They should be potted 1in. deep in September. Keep them in a cold frame until growth begins and then move them to the greenhouse. Start watering at the same time and increase the quantities later. Cease watering when the leaves wither after flowering. Dutch and English iris can be grown in this same way except that these need 5-in. pots instead of 3-in. The Dutch iris Wedgwood, pale blue with a yellow blotch, is one I find very good for growing in pots.

Ixia (African Corn Lily). I find these quite attractive as pot plants, when grown five or six to a 5-in. pot. They do not, however, like to be forced too much and should be kept in cool conditions although they like plenty of sunshine. The flowers are of many colours and are borne in spring.

Pot the corms 1in. deep in September, October or November in a mixture of equal parts loam, peat and sand. Plunge in a cold frame until growth begins and then bring them into the greenhouse. Always water with care, but especially so in the winter months.

Lachenalia (Cape Cowslip). The lachenalias are splendid cool greenhouse plants and their pretty name is descriptive, for these South African plants have bell-shaped, cowslip-like flowers. There are many named hybrids in a lovely range of colours,

Muscari armeniacum Heavenly Blue makes a charming pot plant for a cool greenhouse and is often grown in this way. The flowers are rich blue in colour

and two others which have been grown for many years are the golden-yellow *L. aloides nelsonii* and *L. bulbifera* (or *L. pendula* as it is often listed in catalogues) which includes coral-red, yellow, purple and green in its floral colours. The 'pendula' comes from the flower spikes which are rather pendulous. The leaves are also an attractive feature of the plant, often being handsomely mottled.

Pot the bulbs in August using the John Innes No. 1 Potting Compost and just covering their tips. A good show will be provided by potting six bulbs in a 5-in. pot. Place the pots in a cold frame until growth is made and then transfer the plants to a shelf in the greenhouse and provide them with a temperature of 7° to 10° C. (45° to 50° F.). Water with particular care in the early stages and give increasing amounts as the plants develop. Dry the plants off slowly after flowering and ripen the bulbs in full sun.

Lilium. Numerous lilies can be grown in the greenhouse, provided it is possible to keep the atmosphere fresh. I shall restrict my remarks to three, all of them stem-rooting: *Lilium longiflorum*, the white-flowering Easter Lily which is popular for forcing; *L. auratum*, the spectacular Golden-rayed Lily of Japan, for summer flowering; and *L. speciosum rubrum*, a lovely variety for late summer flowering. Start the bulbs into growth in autumn or winter and make quite sure

that the drainage is good, for lilies will not tolerate stagnant soil conditions. I always add a little extra peat to the John Innes No. 1 Potting Compost and a little well-decayed farmyard manure with stem-rooting lilies. I only half fill the pots with the compost and I leave just the top of the bulbs showing above the soil. Then, when the plants have made about 9in. of growth I topdress to within ½in. of the top of the pot. Lilies which are not stem-rooting are covered with about 2in. of soil and the pots are filled at once to within ½in. of the rim.

The pots are then put in a cold frame and are covered with paper until growth starts. When growth is well advanced they may be brought into the greenhouse, but always ventilate freely and be careful not to overwater. Feeding with liquid manure is helpful up to the bud stage and if possible keep the temperature at not more than 10° C. (50° F.) as higher temperatures shorten the life of the flowers. After flowering let the stems of the plants die down, trim up the plants and house them in a cold frame until the time arrives to start them into growth again.

Of the three mentioned I would only use *L. speciosum rubrum* for pot cultivation a second time. One can continue, too, with *L. longiflorum* but the plants do not seem so vigorous after the first year. It is better with these and others to plant the bulbs out in the garden after one season's use as pot plants.

Lilium longiflorum has delightful white trumpet flowers and grows to a height of about 3ft. *L. auratum* has huge flowers with white petals marked with gold and crimson, and *L. speciosum rubrum* is a rich rose and carmine variety with purple spots.

Muscari (Grape Hyacinth). The muscari make pleasing pot plants and *Muscari armeniacum* Heavenly Blue in particular is often grown in this way. Pot the bulbs in September or October and plunge in a cold frame until growth starts, then bring them into a cool greenhouse. Use a compost consisting of 2 parts loam, 1 part peat and 1 part sand, and plant about 15 bulbs in a 5-in. pot, setting them 1in. below the surface of the compost.

Narcissus. The narcissi or daffodils are, of course, magnificent bulbs for greenhouse cultivation. Pot the bulbs during August or September and plunge the pots out of doors under ashes until about 2in. of growth has been made. Then place them in a shaded frame for a short time to give the growths an opportunity to take on more colour and afterwards move them to a shelf in the greenhouse.

It is best to grow narcissi as coolly as possible so that strong growth is obtained and the flowers have more substance. The John Innes No. 1 Potting Compost suits them well. Any good bulb catalogue will state which varieties are best suited to greenhouse cultivation and to forcing. Plant the bulbs in the garden at the end of the season after the growth has died down. Many of the miniature narcissi are charming when grown in pans.

I would like to mention now a few varieties which I personally favour for greenhouse cultivation. Trumpet narcissi with yellow trumpets—Golden Harvest, King Alfred, Magnificence and Rembrandt. Bicolor Trumpets—Queen of the Bicolors and Celebrity. White Trumpets—Beersheba, Mount Hood and White Tartar. Large-cupped narcissi—Carlton, Scarlet Elegance, Golden Torch, Red Goblet, Rouge, John Evelyn, Flamenco and Orange Bride. Small-cupped narcissi—Firetail, Polar Ice and Verger. Double narcissi—Inglescombe, Mary Copeland and Texas. Triandrus narcissi—Thalia and *triandrus albus* (Angel's Tears). Cyclamineus narcissi—I do not really like these for indoor flowering but I have seen Peeping Tom and February Gold make good

die down, when the plants should be rested in cool conditions.

Scarborough Lily, see Vallota

Scilla (Squill). The scillas, and especially *Scilla sibirica*, are popular for greenhouse cultivation. Pot between August and November 1in. deep in a mixture of 2 parts loam, 1 part peat and 1 part sand. Plunge them out of doors until growth starts and then bring them into the greenhouse.

Snowdrop, see Galanthus

Sparaxis. The 1- to 2-ft. long spikes of violet-purple flowers of *Sparaxis grandiflora* are a welcome addition to the greenhouse in spring. Pot the corms in November and plunge in a cold frame until growth begins. Then move to a greenhouse with a temperature of 4° to 10°C. (40° to 50°F.) increasing to 16°C. (60°F.) in spring. Water the plants carefully at all times.

Tigridia. The colourful *Tigridia pavonia* is a fine plant for an unheated greenhouse with its showy

The ivory-white and pale shell-pink Jonquilla narcissus Chérie is an excellent pot plant and is also good for garden display. One to three flowers are carried on each stem

A wide range of attractive tulips can be grown to great effect in an unheated greenhouse and they can be planted out in the garden later

bowls. On the whole, however, they are not quite showy enough and they are better for the alpine house than the ordinary greenhouse. Jonquilla narcissi—Baby Moon and Chérie. Tazetta narcissi—Cheerfulness, Cragford, Geranium, Laurens Koster, Scarlet Gem and Yellow Cheerfulness. Poeticus narcissi—Actaea and Queen of Narcissi. To conclude, there is Paperwhite Grandiflora for its scent and its excellence for forcing.

Nerine (Guernsey Lily). The colourful nerine hybrids are excellent for autumn colour in the cool greenhouse. Pot the bulbs half exposed into a 4½-in. pot, in August or September and use a compost consisting of 2 parts loam, 1 part leaf-mould and 1 part sand. Place the pots in a cold frame and water very little until growth appears, when they should be moved to the greenhouse. The amount of water given should now be progressively increased and continued until the leaves

and distinctive orange-red, spotted flowers. Pot the corms in March or April in John Innes No. 1 Potting Compost and plunge in a cold frame until growth begins. Then move to the greenhouse. Water carefully when growth begins and more liberally later. After flowering give decreasing amounts of water and when the foliage has withered stop watering altogether until it is time to start the growth cycle again in the spring.

Tritonia. The tritonia usually offered is *T. crocata*, a pretty plant of about 1ft. in height with reddish-orange cup-shaped flowers which are borne in April and May. This is a plant for the cool greenhouse needing a minimum temperature of 4°C. (40°F.) during the winter months and 10°C. (50°F.) from March to September. I use the John Innes No. 1 Potting Compost for this plant and place eight corms 3in. deep in a 5-in. pot in September or early October. These are kept in a cold frame until growth begins and they are then placed on a shelf near the glass in the greenhouse. Water with discretion when growth starts and increase the supplies later but always exercise care when carrying out this task. Partially dry off the corms after flowering. Increase by seed or by dividing the mature plants.

Tulipa (Tulip). The tulips are superb flowers for the unheated greenhouse, both the garden types and the species like *greigii, eichleri, kaufmanniana* and *clusiana*. The John Innes No. 1 Potting Compost suits them well and they should be potted between September and November. Plunge the pots out of doors until growth begins and then move to the greenhouse. The temperature of the house should be moderate—16°C. (60°F.) or less —but it may rise after the flower buds have formed. Do not water after the foliage starts to wither. Dry off the bulbs and keep in a cool, dry place until the time arrives for them to be planted in the garden.

Vallota (Scarborough Lily). *Vallota speciosa*, known to so many gardeners under its common name, is a very easy bulbous plant to grow well, either in the greenhouse or the home. It is a showy plant, too, with its handsome red trumpet flowers. These appear in late summer.

Pot in March in John Innes No. 1 Potting Compost, one bulb to a 5-in. pot, and allow the plants to become root-bound, for in that condition they seem to flower better. Repot old plants in June or July, when this becomes necessary, say once every three or four years. They need plentiful supplies of water from spring to early summer but less from then to September. Only moderate amounts of water should be given from autumn to spring. Increase from offsets.

Zantedeschia (Arum Lily). This genus has been known in the past as Arum, Calla and Richardia, which is all a bit confusing, but this plant with its prominent spathes is well worth growing for

The white-spathed Arum Lily, *Zantedeschia aethiopica*, is grown for spring display in cool greenhouses. The fleshy roots are potted in August or September

Another white-flowered plant for the cool greenhouse is *Zephyranthes candida*, with the romantic common name of Flower of the West Wind

Tulips I find useful for greenhouse cultivation are the following, in their sections. Early Single tulips—Mon Trésor, Sunburst, Pink Beauty, Prince of Austria, Brilliant Star, Couleur Cardinal, de Wet and Vermilion Brilliant. Early Double tulips—Peach Blossom, Mr van der Hoef, Murillo, Orange Nassau, Vuurbaak, and Tea Rose. Breeder tulips—Indian Chief, Bacchus, President Hoover, Orange Delight, Southern Cross and the Rainbow Mixture. Cottage tulips— Dreaming Maid, Golden Harvest, Grenadier, Palestina and Inglescombe Yellow. Lily-flowered tulips—Painted Lily and Yellow Marvel. Multi-flowered tulips—Rose Mist, Claudette and Wall-flower. Parrot tulips—Fantasy, Double Fantasy, Black Parrot, White Parrot, Orange Parrot, Red Parrot, Texas Gold and Parrot Wonder. Fringed tulips—Fringed Beauty. Broken tulips—Black Boy and Union Jack.

spring display. The species usually grown is the white *Zantedeschia aethiopica* but there are also two yellow-flowered species, *Z. elliottiana* and *Z. angustiloba* (*Z. pentlandii*), which are rather more difficult to grow and need more heat.

Pot the fleshy roots of *Z. aethiopica* in August or September using the John Innes No. 1 Potting Compost and pots of 5-, 7- or 8-in. size depending on the size of the roots. Stand the pots outdoors or in a cold frame for about four weeks and then bring them into a cool greenhouse. They need plenty of water.

After flowering has finished decrease the supplies of water, increase the ventilation, and early in June, after danger of frost has passed, stand them in the open until it is time to start them into growth again. Increase is from suckers.

Zephyranthes (Flower of the West Wind). These dainty flowers, 6 to 8in. high, are attractive for the cool greenhouse. The most frequently grown is the white *Z. candida*; others are *Z. rosea*, pink, and *Z. ajax*, pale yellow. They like a compost of 2 parts loam, 1 part peat and 1 part sand and can be potted between August and November. Water very little until growth starts and then increase supplies considerably. Dry off the bulbs after flowering.

Bulbs in the Home

Growing bulbs for the home is rewarding in many ways. I find it an especial pleasure to have at such close proximity plants which I have nursed along carefully for several months, to provide colour during the late winter. One must, too, I feel, approve instinctively of a gardening activity in which virtually anybody, with or without a garden, young or old, well or in poor health can engage on more or less equal terms. I say 'more or less' because those of us who have gardens have an obvious advantage in having space to start the bulbs outdoors, before bringing them into the home. But countless householders manage to find a suitable cool, dark place where the bulbs can be kept for the first eight weeks or so, for example under the stairs, in an empty cupboard or an attic.

I shall now outline the various stages bulbs must go through when they are grown for the home, the treatment they should be given to get the desired results, and the types of containers and mediums which can be used and how effective these are in practical terms.

Containers and Growing Mediums. So far as containers are concerned one can say, in the modern idiom, almost anything goes—provided, of course, that the container in question is large enough to hold a reasonable number of the chosen bulbous plants, and its depth also is sufficient to allow good development of the roots.

Containers without drainage holes have the advantage that they can be placed on tables and elsewhere without anything underneath them to collect excess moisture. This is a considerable convenience but it means that bulb-fibre—a mixture of peat, coconut fibre, charcoal and oyster shell (the last two to keep the mixture sweet)—must be used, and it is generally agreed among gardeners that for absolute top results it is better to use a good soil mixture. However, provided a good bulb-fibre mixture is purchased, this is something which should not be over-emphasised. Given good cultivation—and especially watering—bulb fibre will give excellent results.

If a container with drainage holes is chosen, make sure that the soil compost used is suitable for this type of plant. It should have plenty of body in it but at the same time drainage must be good. The John Innes No. 2 Potting Compost is an excellent mixture but if one prefers to make up

Potting Paper White daffodils before forcing. It is important that the pots should be well-drained. The compost around the bulbs must be well firmed with the fingertips

Plunging pots of daffodils outdoors after potting, to ensure that a good root system is made before top growth commences —an essential ingredient of success

Hyacinths and other tall bulbous plants in containers should be supported early before growth sags

a mixture then 3 parts high quality loam, 1 part peat and 1 part sharp sand gives very good results.

Gravel is used as a rooting medium for hyacinths and the earliest daffodils as well, but as it has nothing to give the plants during their development one would not expect to get such strong-growing plants in this way. The gravel must be kept moist but too much water must not be allowed to collect in the bottom of the bowl. Vermiculite can also be used as a rooting medium but I am not altogether happy about this as it does not support the bulbs, and flowering-size plants, even when staked, tend to topple over.

Another method of cultivation which is rather fun for children is the hyacinth glass, used again for hyacinths and the earliest daffodils. Here it is possible to watch the bulbs send down their roots into the water and develop gradually until they are fully fledged flowering plants. Also available are plastic bowls with indentations in the top cover to hold the bulbs. The roots have access to the water in the container.

Before discussing these growing methods in more detail there is one point about containers which should be mentioned now. Some, like conventional clay flower pots, are made of absorbent materials. These must be immersed before first use in water and left until air bubbles cease to rise. If this is not done moisture losses through the sides will be excessive, to the detriment of the plants.

Bulbs Grown in Fibre. Make absolutely sure before using bulb fibre that it has been thoroughly soaked in water. Dry bulb fibre used for potting—and it is always sold in this state—cannot absorb anything like the amount of water needed to satisfy the moisture requirements of the bulbs.

There are several ways of soaking fibre and if only a small quantity is involved the best way is to half fill a large bowl with fibre and fill it to the brim with water. Allow this to soak overnight and then squeeze out excess moisture with the hands before using for potting the bulbs. If the amount of fibre to soak is large, spread it out on the ground and then soak thoroughly with a hose pipe or with a watering can. Allow this to drain overnight. A third method is to immerse a sack of the fibre in a water tank. This will mean taking the sack out several times usually, and moving the fibre around, for otherwise the fibre at the centre of the sack stays dry and the results are not entirely satisfactory. This saturated fibre should also be allowed to drain before being used. A simple test of the suitability of fibre for potting is to compress a handful and release it. If it falls away readily it can be considered in fit condition to use.

The potting procedure is simple but should nevertheless be carefully carried out. Place a few lumps of charcoal in the bottom of the container to keep the fibre sweet. Cover this with a layer of fibre and press this gently into place. (The reason why this must be done gently is that a hard layer of fibre under the bulbs would act as a barrier to the roots. Consequently, these would then exert pressure upwards and soon displace the bulbs.) Place the bulbs on the fibre, again gently, and fill in around them with fibre, firmed securely into place with the fingertips. Give the fibre a good soaking and after an hour or so turn the container on its side to drain. The bulbs are now ready for the first phase of their growth cycle and are plunged outdoors or stored in a cool, dark place indoors.

Bulbs Grown in Compost. I have already described two composts which are suitable for bulb cultivation (p. 135) and I want now to emphasise the importance of good drainage when bulbs are grown in this way. As in normal potting, cover the drainage holes with large crocks and these in turn with a thin layer of smaller crocks before adding a layer of fibrous material and the compost in which the bulbs will grow. Seat the bulbs at the correct depth in the compost and firm the compost around them with the fingertips.

Plunging Bulbs. With the bulbs in their containers the next move is to plunge them outdoors under ashes of about 4in. thickness or in a cold frame or garden shed. If such facilities are not available the alternative is to store them in a cool—repeat cool—dark place indoors. The objective now is to get the plants to make a good root system before any top growth is made. Low temperatures and darkness ensure that this objective is achieved.

The Next Stage. After four or five weeks of this treatment, periodic inspections should be made to see whether growth has been made. The top growth will, of course, be blanched due to the lack of light and when they are ready to take out of the plunge bed—when the growth is about an inch high for daffodils and hyacinths and rather more for tulips—they should have about a week in half light before being subjected to normal conditions.

The water requirements of the plants must now be watched very carefully for they must never be allowed to dry out. For the time being keep the bulbs in cool conditions and only take them into the house when growth is considerably more advanced. For example, daffodils and crocuses should only be brought indoors when the buds have formed. When the time comes, the introduction to heat must be gradual, only providing the plants with living room temperatures when they are about to flower. Too much heat too soon results in etiolated growths, general weakness and, inevitably, poor quality blooms.

Staking. Taller bulbous plants like hyacinths, daffodils and tulips should be staked early before the growths have a chance to sag. The staking

The attractive Small-cupped daffodil Verger, with a white perianth and deep red cup, is a variety suitable for forcing and for garden display

Vallota speciosa, the Scarborough Lily, does equally well in greenhouse or home. The red flowers appear in late summer and are borne most freely when the plants are root bound

should naturally be done as unobtrusively as possible so that the appearance of the plants is not spoilt, using light twiggy branches or thin canes as circumstances dictate.

The Plants

Without any doubt my favourite bulbs for the home are daffodils and hyacinths. I have been very successful in recent years with Cornish pre-cooled daffodils of the varieties Magnificence, Golden Harvest and Fortune. Planted in the first week of October, these will be in full flower a week before Christmas. I like growing these, too, in double layers, as described for tubs on p. 134 I use 7-in. pots and have a bottom layer of four bulbs and a top layer of five and these make a splendid show of colour. Planted out in the garden afterwards, they will flower three to four weeks earlier than other daffodils in the next season.

My general recommendations for untreated daffodils would be as follows:

Trumpet narcissi—Golden Harvest, King Alfred, Magnificence and Rembrandt. Bicolor Trumpets—Queen of the Bicolors. White Trumpet—Beersheba. Large-cupped narcissi—Carlton, Scarlet Elegance, Flamenco and John Evelyn. Small-cupped narcissi—Firetail. Double narcissi —Inglescombe, Mary Copeland and Texas. Poeticus narcissi—Queen of Narcissi and Actaea. Tazetta narcissi—Cheerfulness, Cragford, Geranium, Laurens Koster and St. Agnes.

For bowl cultivation, my favourite hyacinths are the dark red Jan Bos—a lovely variety for Christmas flowering—King of the Blues, the bright salmon-pink Lady Derby and Pink Pearl. The most important thing with hyacinths in bowls—and this applies to other bulbs too—is not to mix varieties, as these never flower at the same time and grown together the effect is bound to be uneven with some fading and others just

coming into bloom. I must mention, too, the Multiflora hyacinths which make a bold display.

Of the tulips, the early-flowering double varieties are the best for home decoration. Vuurbaak, Peach Blossom, Orange Nassau, Tea Rose, Murillo and Mr van der Hoef are all varieties I like to grow. The Rainbow Mixture are also very fine with their mixed colours and here is an exception to the rule that one should not mix bulbs in a bowl. These do all flower at the same time. After the early-flowering doubles I would choose the early-flowering single tulips in such varieties as the splendid Brilliant Star, de Wet, Mon Trésor, Couleur Cardinal, Sunburst, Vermilion Brilliant, Pink Beauty and Keizerskroon. Other types of tulip, in my opinion, are really too tall for the home—even the Parrots, lovely as they are.

Crocuses are always popular and I make quite a lot of use of the crocus bowls with planting holes round the sides. Where I think a lot of people go wrong with crocuses is that they try to bring them into flower too quickly, with unfortunate results. I always leave the plants in a frame, a cool room or outdoors until the colour of the flowers begins to show, then introducing them to a warm room.

Hippeastrums (or amaryllis as they are popularly called) are well worth growing in the home and prepared bulbs for winter flowering are available too. Some people grow hippeastrums in fibre, but if they are intended to go on from year to year then they should be given a good potting compost. Then, when the flowers go over and the plants come into leaf they can be fed, kept in a light window and the bulbs ripened off to build up reserves for next season's flowering.

Pests and Diseases

Pests and Diseases. For details of pests and diseases which attack bulbs, corms and tubers see Chapter Thirteen (pp. 295 to 297).

Dahlias 7

Dahlias have a very long flowering season, lasting from mid-summer right through until the frosts arrive in the autumn. They are splendid border flowers. Although dahlias originate from the warmer, sunnier parts of the world, it is possible to grow them in partial shade. However, there is no doubt that they flower much better in an open, sunny position in the garden. If they do not receive any sun at all, as would occur under a north-facing wall, they will produce hardly any flowers.

For Garden Display. Dahlias have many possibilities for garden display. I have sometimes planted them among hardy border plants where they blend in beautifully with the permanent occupants of the herbaceous border. Planting them between shrubs is another suitable method, especially if the border is newly planted, in which case there will be plenty of space between the young shrubs. Growing them in a border by themselves will give you a bold mass of colour, and anyone taking over a new garden could not do better than grow beds of dahlias in the first season or two, as they produce a good display of colour in a short time.

The dwarf varieties—which do not usually exceed 2½ft. in height—are especially suitable for summer bedding displays, and I think that the dahlia is one of the best plants for this purpose. I use a lot of the single yellow varieties, for yellow is a difficult colour to come by in summer bedding plants.

Good for Cutting. Dahlias should appeal to the lady of the house, especially as they are ideal flowers for cutting and bringing indoors. The plants respond to this treatment, as the more flowers that are cut off them the more they seem to produce—that is why I call them the cut-and-come-again flower. The blooms will last an exceptionally long time in water.

Various Types of Dahlia. To the novice dahlia grower the classification of dahlias may seem rather confusing at first. But basically dahlia varieties are divided into main groups according to the shape and formation of the flower. The varieties in these groups are then placed in various sections of the group according to the size of their flowers. As an example let us take the largest group, the Decorative dahlias. This contains the typical varieties with fully double flowers which have broad, flat, well-formed ray florets. This group is divided into giant-flowered, large-flowered, medium-flowered, small-flowered and miniature-flowered sections.

Another large group contains the Cactus dahlias with fully double blooms but with narrower, more spiky ray florets than the decoratives. This also has giant-flowered, large-flowered, medium-flowered, small-flowered and miniature-flowered sections. Then we have the Semi-cactus dahlias, which again have fully double blooms, but the ray florets are slightly broader than the true cactus dahlia and narrower than the decoratives. This group has the same sectional divisions as the two types just referred to. A group of really delightful varieties is that which contains the Ball dahlias which have fully double flowers, either ball-shaped or slightly flattened. This group is divided into only two sections, namely the ball and the miniature ball dahlias.

The Single-flowered group is not divided into sections, and contains those varieties which have a single outer ring of florets which may overlap, the centre forming a disc. Most varieties in this group are suitable for bedding as they are dwarf growers. Anemone-flowered dahlias have one or more outer rings of generally flattened ray florets, the centre of the flower being made up of a dense group of tubular florets, and showing no disc. Again, this group is not divided into sections. The

Collerette dahlias, which also are not split into sections, have a single outer ring of generally flat ray florets, with a ring of small florets (the collar), the centre forming a disc. Peony-flowered varieties have blooms with two or more rings of generally flattened ray florets, the centre being formed into a disc. There are no sections in this group, nor are there in the case of Pompon dahlias. These last are charming varieties with blooms similar to those of the ball dahlias, but more globular and much smaller in size.

Although the Dwarf Bedding varieties are usually listed separately in dahlia catalogues, they do have various types of flower formation, such as cactus, decorative, single and so on, and are, therefore, correctly placed in the appropriate groups which I have mentioned above. They are grouped on their own in catalogues simply for ease of selection.

The Best Types for Garden Display. The choice of types for garden display is purely a matter of personal appeal, but I would not grow the Giant-flowered and Large-flowered Decoratives, cactus and semi-cactus varieties, as the flowers are far too large for normal garden display. They are all right for those people who grow for exhibition, but not for general garden use. All the smaller-flowered varieties in these groups, and the other groups which have smaller flowers, are the ones I would use. These are also more suitable for use in flower arrangements.

If your garden is exposed and windy I would certainly advise growing the less tall varieties. Those that grow tall, for example, over 4ft. in height, will suffer badly from wind damage, and this makes extra work in staking and tying (I think that all these sorts of jobs need keeping down to a minimum).

When planting dahlias for garden display it is usual to group them in separate varieties rather than mix, say, half a dozen different sorts together. If you are putting them among herbaceous border plants then I think that one plant here and there in the odd space should be sufficient. If they are being planted in beds or borders of their own, then three to five plants of each variety is really the minimum number for any one group, in order to produce an effective display. An important point when growing plants for a display is to remove the dead flower heads regularly, as this will encourage the plants to keep up a continuous show of blooms. Bedding varieties can, of course, be mass-planted in their own beds, or mixed with other summer bedding plants such as the silver-leaved *Centaurea gymnocarpa, Cineraria maritima (Senecio cineraria)* and *Pyrethrum ptarmicaeflorum*, or salvias, ageratum and verbena. There are many colour combinations which can be planned, but this must be left to the imagination and artistic skill of the gardener.

Lifting the tuberous roots of dahlias for winter storage. The soil is removed and damaged parts cut away. Immediate labelling is advisable to avoid confusion later

I usually plant a row of dahlias across the vegetable garden specially for cutting, as this enables me to cut as many flowers as I like without interfering with the garden display.

Selection of Varieties. Recommending suitable varieties to people is really one of the most difficult things to do when it comes to dahlias. The breeders and growers of dahlias are supplying the market with literally dozens of new varieties and novelties each year. For the experienced dahlia grower, let alone the beginner, it is almost impossible to keep abreast of the new introductions. And the choice of varieties is very much a matter of personal preference, the ones that I particularly like may not appeal to you. One of the best ways of choosing varieties is to visit flower shows, dahlia nurseries and gardens open to the public when dahlias are in season. By doing this, and making notes of the names and colours, you will be able to decide on the types which most appeal to you, and then place your order with a reputable dahlia specialist. Alternatively, try to obtain a few dahlia catalogues; a number of these contain coloured plates of leading varieties and it is an easy matter to select those which catch your eye.

As with all types of plants, the older and well-tried varieties should be grown to start with, leaving the novelties until experience is gained. The beginner cannot go far wrong with these as they have been grown by countless numbers of gardeners and have proved their worth over the years for garden display.

Cultivation

Preparing the Ground. Thorough preparation of the ground for dahlias is more important than it is for many other plants. They like plenty of moisture at their roots, and it will be noticed that they always grow better in a wet season. When the

Dahlias being stored in a deep box. Dry peat is being worked among the roots. They should then be placed in a dry, frost-proof place

ground is moist it is easy for the roots to absorb plenty of the food which is so essential to the dahlia.

The ground should be dug over in the autumn, when plenty of farmyard manure, garden compost, peat, spent hops, shoddy, or any other organic material of this kind should be incorporated at the same time. This will provide humus in the soil, which helps to retain soil moisture. This humus is especially important on light or sandy soils which do not hold much moisture; and on heavy ones this organic matter will help to keep the soil open and well-drained. I do not think that double digging is necessary for dahlias; just dig to the depth of the spade.

After the autumn digging a dressing of fertiliser should be applied, and I would recommend bonemeal, applied at 2 to 3oz. to the square yard. This dressing must be put on well before planting time, either just before or just after Christmas, otherwise it will not be readily available to the plants when they are planted, as this fertiliser is slow in releasing its plant foods. Then just before the dahlias are planted apply an all-purpose fertiliser, again at 2 to 3oz. to the square yard. This should give the young plants, a really good start in life and produce excellent flowers.

When to Plant. The time of planting is very important, and this will depend on the part of the country in which one lives. It is essential to bear in mind that the young shoots should on no account be subjected to frosts, otherwise they will be severely damaged or even killed. Dormant tubers can be planted round about the middle of April in the South, late April in the Midlands, and early May in the North. If the tubers have got long, thin shoots on them when they come out of the store, as they very often have, it is better to cut them back so that no shoots are showing above the ground before mid-May in the South, the end of

May in the Midlands and early June in the North, when all chance of frost has gone.

Rooted cuttings may be planted out in mid-May in the South, the end of May or early June in the Midlands, and in the North during the first week of June. Tubers which have been started into growth and are bearing sturdy shoots, should also be planted at these times.

How to Plant. The first thing to consider is spacing, and this means that we have to have some idea of how tall the particular varieties grow, in order to know exactly what space to allow between them. The bedding ones need a minimum of 18in. from plant to plant. The taller-growing varieties, such as pompons, cactus and decoratives, need a minimum distance of 2ft. between plants, as they grow very bushy. If planting in rows it is best to allow 3ft. between the rows and 3ft. between the plants in the rows.

If dahlias are not given sufficient space they will not receive their fair share of light and air, and, consequently, unhealthy plants will result, together with poor flowering.

Insert the stakes before planting. This is especially important when planting dormant tubers, as, if stakes are inserted after planting, there is the chance that they may be pushed through the tubers. Tall varieties need a good strong stake to prevent them from being broken off during high winds. Use either oak or soft wood 1in. square stakes, and treat them with a wood preservative before use. If I use bamboo canes to support my plants, I prefer to put one in when I am planting, and when the plants are growing to put three or four more canes around each plant. Allow all the stems to remain inside the canes, and loop the twine around the entire plant, securing it to each cane. By giving three or four ties up the canes, all the stems should be held in securely. Even better support is provided, of course, if stakes are used to encircle the plants.

Now to the actual planting. Use a hand trowel for pot-grown plants or small tubers, and a spade for very large tubers. Place dormant tubers, and also those which have been started into growth, about 3 to 4in. deep. Rooted cuttings in pots should be planted so that the top of the root ball is only about ½in. below soil level. Make sure that the pots or boxes have been watered thoroughly before setting out the plants.

Tying. If you have planted started tubers or rooted cuttings, it may be necessary to tie in the shoots after planting to prevent them from breaking off during windy weather. A soft green garden twine is best for this purpose, and should be looped round each shoot, once round the stake, and tied at the back of the latter. As the shoots grow, so more ties should be given. This job should be done at regular intervals until the stems reach the tops of the stakes.

Thinning the Growth. Any large, old tubers which have not been divided up should have some of their shoots reduced before planting—provided the tubers have been started into growth previously. There may be eight or ten shoots on any one tuber, each one of which will be fighting for light and air, and if they are overcrowded they will become drawn and thin. They can be reduced to three shoots on each plant; and this applies to bedding varieties as well as tall-growing ones. It will be found that these shoots snap off easily at their bases.

If planting dormant tubers, it will be necessary to wait until their shoots have emerged through the soil, when the surplus ones may be cut off at ground level. With those grown from cuttings there will be no surplus shoots to reduce.

When dahlia plants reach about 8 to 9in. in height it is a good idea to pinch out the tips of the shoots to encourage strong side shoots to grow.

Disbudding. If dahlias are being grown for exhibition it is necessary to remove a fairly large proportion of surplus flower buds. This will result in much larger flowers than if they are allowed to grow completely unchecked. For general garden purposes I do not think much disbudding, if any, is necessary, as you want masses of flowers for a really good display.

As I have already mentioned, exhibition varieties must have their buds reduced in number, but possibly with the exception of pompons and ball dahlias, as heavy disbudding of these results in too large flowers. In this case they may not conform with the official classifications of flower size, and therefore would be disqualified at a dahlia show.

It will be seen that flower buds form at the top of each stem or shoot; one bud, called the crown bud, at the apex of a shoot, with several more just below it. As the crown bud develops more rapidly

Cuttings taken with a heel are best, where this is possible, as they seem to form excellent tubers. Otherwise make the cut just below a joint

Trim the base of the cutting neatly with a sharp knife and remove the bottom leaves. Then treat with hormone rooting powder before insertion in the rooting medium

This will give you a bushy plant and a larger number of flowers. There comes a stage, usually about August, when it is necessary to thin some of the surplus young side shoots higher up the plant, otherwise it becomes a mass of foliage which will result in poorer quality flowers.

All side shoots should be snapped or cut out for about 1½ to 2ft. from the tip of each stem. This thinning will encourage longer flower stems for cutting, and larger and better quality blooms. The shoots lower down the plant should be left to grow, as these will provide replacement stems to flower after the earlier flowers have been removed. Gardeners who grow dahlias for exhibition, or those who grow them specially for cutting, usually start this thinning much earlier than I recommend for garden display, and do this task at regular intervals. But for general garden displays thinning is not classed as an important job.

than the lower ones, this is the one to leave. The lower buds should be pinched out between the fingers and thumb, so leaving one bud only to each shoot. Buds must be removed when they are fairly small.

General Hints. After planting and throughout the growing season, dahlias should be watered abundantly. To conserve soil moisture the area of ground in which the plants are growing can be mulched. A 2 to 3in. layer of straw, peat, farmyard manure or any similar material can be used. If you are growing for exhibition then try, if possible, to mulch with manure.

As soon as the plants are established and are growing away, they can be fed occasionally with an all-purpose fertiliser, sprinkled around each plant, lightly pricked into the soil, and watered well in. If you have mulched the ground then scrape the material away from the plants first. The dahlia will respond better than any other plant to good and thorough cultivation.

Pest and Disease Control. Regular spraying against pests and diseases should also be carried out, and details of this are outlined on pp. 297 to 299.

Over-wintering

Cutting Down and Lifting. Many gardeners think that dahlias should be left in the garden until the stems and leaves become blackened by frost. Although a slight frost does no harm at all to these plants, they certainly do not benefit from being frosted. Where possible dahlias can be left to flower until the first frosts arrive, possibly in November. These last few flowers are indeed valuable for cutting. Of course, it is not always possible to leave dahlias in the ground until they are blackened; for example, if you have bedding

The tubers should be lifted very carefully with a garden fork, taking care not to damage them in any way, as they are the food stores for the plant, and provide food for the growth of the shoots the following spring. Work about 1ft. away from the stems.

Drying the Roots. Directly the tuberous roots are lifted they should be dried off in a frost-proof place. Before doing this remove any soil which may be adhering to the tubers; cut away any damaged parts of the tubers and dust the cut surfaces with a powder composed of equal parts ground limestone and flowers of sulphur. The way I dry mine off is to get a large wooden box, nail some thin strips of wood across the top, and then place the roots between these so that the stems are facing downward into the box. I leave them in this position for three or four weeks to allow the sap and any moisture to drain from the stems. If moisture collects in the stems and they

A wooden dibber is a useful aid when inserting the cuttings. These can be placed round the edge of a pot (as above) or in seed boxes

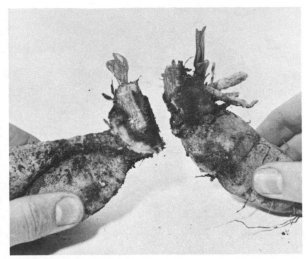

Dahlia tubers can be divided by pulling apart with the hands or cutting with a sharp knife. Each division should consist of a stem with a tuber or tubers attached

dahlias where wallflowers, polyanthus, forget-me-nots and spring-flowering bulbs are to be planted, then the dahlias must be cleared to make way for these other plants.

Before cutting down and lifting the plants I would recommend tying a label at the base of the stems of each one. On this label write down details such as the name of the variety, colour, height and type. Even if you do not know the name of the variety, you can still put down the other details, and then the following spring you will have an idea of the colours and heights of each plant when deciding on planting arrangements and colour combinations.

After fixing the labels, the stems should be cut down to within 9 to 12in. of the tuber—certainly no less than this. It is best to use either a pair of secateurs or a strong knife for this job, as the stems are usually fairly tough by this time of the year.

are stored like that, then they will be affected by fungal diseases, which could spread and result in the death of the plant.

Storing the Roots. After the tubers have been dried off thoroughly they are stored for the winter in a frost-proof place. The temperature can be allowed to drop to 1 to 2° C. (34° to 36° F.), but if it goes down to freezing point or below then the tubers will be killed. I would not recommend storing them in polythene bags because the roots give off a certain amount of moisture which collects on the inside of the bag and so encourages fungal growth on the tubers and stems. It is far better to wrap them in newspaper and place them in boxes. Alternatively, they can be placed in dry peat in boxes.

There is a tendency to keep them under greenhouse staging near heating pipes. This will cause the tubers to dry out and shrivel, and the buds

Planting a sprouted dahlia tuber. The tubers should be set 3 to 4in. deep. The stake should always be placed in position before planting

Use a hand trowel when planting pot-grown plants or small tubers. Firm the soil well

When bamboo canes are used as supports, place one in position when planting, adding three or four more around each plant later, when they are growing

will probably be killed. If this happens they will not start into growth in the spring. A frost-proof attic, cellar, garage or a cool room in the house is the ideal place for storing. While they are in store it is wise to check them over occasionally to see if any fungus is forming on the roots or stems. If this is so, dust them with flowers of sulphur.

Propagation

Cuttings

Dahlias can be increased in three ways: from cuttings, by division of the roots or from seed. If numerous new plants exactly like their parents are wanted then cuttings are the answer, but division allows those gardeners without a greenhouse to increase their plants easily. Seed is also much used as a method of increase, especially for the small-growing bedding varieties where a good range of colour is required, rather than an exact reproduction of a particular variety. Raising dahlias from seed is very easy.

Let us now consider each of these methods of increase in turn.

Dahlia tubers which are to be used for the production of cuttings can be taken out of store and started into growth in February or March in a greenhouse or frame with a temperature of at least 10°C. (50°F.). The tubers are placed on a layer of peat in seed boxes and then covered with more peat, which should be really moist. If preferred, a mixture of equal parts peat and sand can be used. Doing this encourages the dormant buds on the tubers to start sprouting and from these sprouts cuttings are taken. When boxing them up ensure that the labels are still attached to the stems. If you have a large number of dahlias it may be necessary to start them on a greenhouse bench. Place a layer of the rooting medium—peat, or peat and sand—on the bench, lay the tubers out on this, and then cover them with the rooting medium. They should be kept fairly moist until they start into growth, when they will enjoy more water still.

An important point before setting up the tubers is to check them to make sure they are still alive. The best way of doing this is to scratch the tuberous root with the thumb nail—if it is alive it will be either green or red in colour and sappy. Then scratch the base of the stem to make sure that is also sappy; if it is then the buds in this region will certainly spring into life when the tubers are set up. If the tubers are dry and hard then they are almost certainly dead, when they should be removed and destroyed.

Taking Cuttings. For the average amateur one tuber of each variety will produce an adequate number of cuttings. The remaining tubers can be left in store until planting time, or can be started into growth about one month before planting them out. The number of cuttings that can be obtained from any one root varies between five and ten.

When the shoots have grown to about 3in. in length, which takes roughly a month, each one should be cut off together with a small portion of the old stem, which is called a 'heel'. I feel that cuttings taken with a heel form better tubers, and more of them, throughout the growing season, than those that are cut above the root and just below a leaf joint. If cuttings are taken without a heel this usually allows more shoots to develop in their place, but normally far too many are produced for most amateur gardeners.

To get at the base of the shoots scrape the peat away from the tubers.

Prepare the cuttings with a very sharp knife or razor blade, in order to obtain a clean, smooth cut, rather than a torn one. Trim the bases of the cuttings neatly, to remove any ragged edges, and remove the bottom leaves, also the next pair if necessary, to give a good length of stem for insertion in the rooting mixture. When the cuttings have been prepared dip the bases in water and then into a hormone rooting powder. This will assist in the rapid growth of roots.

There is a reasonable range of suitable rooting mediums for dahlia cuttings, but the one I use is made up of equal parts loam, peat and coarse sand, the parts being by bulk. Some people use the John Innes Seed Compost, which is certainly suitable. This can either be bought ready mixed from horticultural sundriesmen, or be mixed at home. It may be made up as follows: 2 parts sterilised loam, 1 part peat, and 1 part coarse sand (parts by bulk). To each bushel of this mixture add: $\frac{3}{4}$oz. of ground chalk or limestone, and $1\frac{1}{2}$oz. of superphosphate. Mix all the ingredients thoroughly. Other rooting mediums are pure coarse sand, and vermiculite. If you are rooting them in either of these two mediums they must be potted up as soon as roots have formed as these materials contain no plant foods, and the young plants will starve.

The cuttings can be inserted in the chosen soil with a wooden dibber (a small stick with a tapered end), either around the edge of a pot, or in seed boxes. Insert the cuttings about $\frac{1}{2}$ to 1in. deep, and firm them well. Make sure that the containers are well drained beforehand by placing a layer of broken flower pots in the bottom, over which a layer of coarse peat or leaves should be placed.

After watering the pots or boxes of cuttings they should be placed on the greenhouse staging and covered with sheets of newspaper for a few days to protect them from the sun, and to prevent flagging. The temperature of the greenhouse should be maintained between 13° to 16°C. (55° to 60° F.) with a moist atmosphere.

When young plants have reached 8 to 9in. in height the growing tips of each shall be pinched out to encourage side shoots to grow

When tying in shoots use soft green garden twine for preference, looping it round the shoot and round the stake

When dahlias are grown for exhibition a fairly large proportion of surplus buds should be removed. This will result in the flowers which are left developing to a much larger size

If a closed propagating frame is available in the greenhouse the cuttings may be placed in this, where they are less likely to flag and will root much quicker. Again, pots or boxes can be used, or they can be planted directly into the bed of the frame, using any of the rooting mediums which I have recommended.

By far the quickest way of rooting dahlia cuttings is under a mist propagator. This system prevents flagging completely, and the cuttings will root in no time at all. I am not suggesting that this is an essential item in dahlia cultivation, but if you are lucky enough to possess one of these, then by all means root your dahlias under it. One important point I should mention is to keep each variety separate, and label each one.

Potting Rooted Cuttings. When the cuttings are seen to be making new growth this is a sign that they have formed roots. If they are being rooted in pots it is a simple matter to invert one of the pots, tap the edge of it on the side of a bench, and so expose the root ball. If a mass of white roots has formed then it is time to pot the young plants.

For this potting I use $3\frac{1}{2}$-in. pots, and place one plant in each. Again, it is advisable to add some drainage material before potting, i.e. a layer of broken flower pots, plus a layer of peat or leaf-mould. The soil I use is the John Innes No. 1 Potting Compost. As with the seed compost, it can either be mixed at home, or bought in from a garden sundriesman. For those who like to do their own mixing the formula is as follows: 7 parts good, sterilised medium loam, 3 parts peat, and 2 parts coarse sand (parts by bulk). To each bushel of the mixture add $\frac{3}{4}$oz. of ground chalk or limestone, and 4oz. of J. I. Base Fertiliser (obtainable from garden sundriesmen).

The young plants should be potted fairly lightly, firming the soil with the fingers, and then placed on the greenhouse bench, and watered well. If the weather is sunny they may need shading for a few days until they are established. A daily syringe with clear water will help to prevent flagging. Ventilate the house during fine spells.

If the cuttings were rooted early in the year, say February or early March, they may need to be repotted in 5-in. pots otherwise they could become short of food. Use the same compost as for the first potting. April-rooted cuttings should not need this further potting before planting-out time.

Pinching Out. When the young plants have made three or four pairs of leaves, pinch out the growing tip to encourage side shoots to grow. You will then have a good, bushy plant when the time comes to plant out, rather than a single-stemmed plant. The later-rooted cuttings will probably need to be planted out before pinching them, but after a week or so in the garden the growing tips can be removed.

Hardening Off. Before the young plants are placed out in the garden they must gradually be acclimatised to lower temperatures. If they are taken straight out of a greenhouse with a temperature around 13°C. (55°F.), into the garden, they would suffer a considerable check to growth. Therefore, the hardening-off process begins in early May, about two or three weeks before planting out, in the southern half of the country. In the Midlands and the North the time to start hardening off would be proportionately later.

The pots are placed in a cold frame, and if a frost threatens the frame lights should be covered with layers of sacking or straw. Give them plenty of ventilation, increasing the amount gradually until the lights are left off completely, both during the daytime and at night. From the middle of May they may be placed under the wall of a greenhouse or the dwelling house. They will be sufficiently protected here until planting time. All the time they are in pots do not forget to check them regularly in case they need watering.

Division

The propagation of dahlias is not restricted to those gardeners who own a greenhouse. Another method is to split up any large roots into a number of smaller ones. The dormant tubers can be divided just before planting them out; they are not boxed up in peat first to start them into growth. When doing this job it is important to note that each division should consist of a stem with a tuber or tubers attached. It is from the base of each stem that the new shoots will arise—so if there is no stem present then a new plant will not form. The tuberous roots may be either pulled apart with the hands or cut cleanly with a knife. Some of the large roots can be split into as many as four, five or six separate plants.

Those with a heated greenhouse may prefer to start the tuberous roots into growth and divide them when the shoots have developed. By this method you can be sure that each division contains one or more growths and has every chance of success. Box the tubers up during April in peat, or a mixture of equal parts peat and sand. The tubers can then be split up just before planting them in the garden, when all fear of damage from frost has passed.

Raising Dahlias from Seed

Dahlias can be grown very easily from seed and if it is not considered important to have named varieties this is an ideal method. I prefer to grow just the dwarf bedding varieties, such as Topmix, from seed as these come truer to type than other dahlias and seem ideally suited to this method of increase.

Preparations for Sowing. Dahlia seed should be sown in a heated greenhouse or frame—in

Right. Dahlias make colourful bedding plants, in particular the dwarf varieties which do not usually exceed 2½ft. in height. Although they will grow in partial shade by far the best results are obtained when they are exposed to plenty of sunshine

Below. Dahlias are ideal for cutting and taking indoors for arrangement. The blooms last well in water. What is more, the plants respond to such cutting, for the more flowers that are cut off the more they seem to produce

Right. The impressive Large Decorative dahlia Blithe Spirit, which grows 4½ft. tall. Dahlias with blooms of this size are more suitable for exhibition than garden display, but those gardeners who make this flower a special interest find them extremely rewarding to grow. Large-flowered Decorative dahlias have blooms of over 8in. but not usually more than 10in. in diameter

Right, below. The Small Decorative dahlia Amethyst, a splendid garden variety also much favoured by flower arrangers who find its colouring and form especially well suited to their needs. It grows to 4ft. in height. Small-flowered Decoratives have blooms with a diameter of more than 4in. but not usually more than 6in.

Left. The Small Decorative Chinese Lantern, a variety of especially fine quality which grows to 3ft. tall

Left, below. The Small Cactus variety Klankstad Kerkrade. This free-flowering dahlia, which grows to 3½ft. in height, is an outstanding representative of its type. Small Cactus blooms are more than 4in. but not usually over 6in. in diameter For many gardeners the cactus dahlias have a special appeal and there are sections for giant-flowered, large-flowered, medium-flowered and miniature-flowered, in addition to the small-flowered type referred to above

Left. The Miniature Ball dahlia Rothesay Superb, another variety of fine quality. It grows to a height of 4ft. Miniature Ball varieties have blooms not usually exceeding 4in. in diameter. The blooms of Ball dahlias are more than 4in. in diameter but not usually more than 6in. across

Left, below. The dwarf bedding dahlia Coltness Gem var. Scarlet. The varieties which are used for bedding purposes come from the ten classified groups and none usually grows more than 2½ft. tall. They are splendid plants for providing a massed display of colour

similar conditions to those recommended for rooting cuttings (see pp. 174 to 176). February or March is about the best time to carry out this task and I usually use John Innes Seed Compost. This must be well firmed, especially in the corners of the box and around the edges. Then make the surface fairly firm and as smooth as possible. A flat piece of wood will produce a good surface if it is pressed evenly all over the box.

Sowing the Seed. The seed must be sown very thinly over the surface of the compost and then covered with a thin layer of soil which has been passed through a fine sieve. Firm the soil lightly, give a good watering with a fine-rosed watering can, place the box on the greenhouse bench and cover it with a sheet of newspaper.

The seed should germinate in about 10 days, when the covering of newspaper should be removed. When the seedlings are large enough to handle they can be pricked out about 3in. apart in seed boxes, this time using John Innes No. 1 Potting Compost.

Planting Out. By the time the seedlings are ready to be planted out they should have been thoroughly hardened off. The same procedure can be adopted for hardening off as recommended for cuttings. They are then planted out when all danger of frost has passed. By this time, if the seed was sown early in the year, the plants will be ready to have their growing tips pinched out to encourage them to become bushy. Plant bedding dahlias 18in. apart and other types at least 2ft. apart.

Exhibiting

The vital point here is to have a healthy stock of plants—this means free from virus diseases particularly. Another important point is to start the tubers into growth earlier in a greenhouse—say from the middle to the end of January. Cuttings are then taken much earlier—as soon as they are ready, and rooted as I have described. The rooted cuttings can then be potted into 3½-in. pots, and when they are well rooted in these, they must be potted on into 5-in. pots. They are planted out from these, and you will then obtain early established plants which will produce blooms well ahead of the other dahlias—which is vital, as the flowers must be completely ready by show dates.

To obtain good plants you must have a rich soil with plenty of manure in it. It pays to apply bonemeal well before planting time. From the time the young dahlias are planted in the garden they must have an abundant supply of water—this is important. Keep them growing with no check whatsoever, and stick rigidly to the cultivation requirements as I have outlined.

Making a Start. Most people naturally tend to start exhibiting in a small, local flower show, but I do not think that this is altogether wise. I think that if you are going to exhibit then be bold and enter a larger show where there are recognised classifications—proper classes in the schedule which cater for the dahlia exhibitor. By exhibiting in a big show you will learn a lot even if you do not win a prize—and this experience will prove invaluable for future exhibiting. Watch other people when they are unpacking their flowers; see how they have brought them to the show in perfect condition without a bruise or any damage to the blooms; how they stage them on the show bench; and which varieties they are using.

While on the subject of varieties I would emphasise the importance of choosing suitable ones for exhibition. There are certain varieties that are recognised as good exhibition ones, whereas others are suitable only for garden display. If you do not start off with the right varieties then you will not obtain the right types of flowers for show work.

A good guide is to look round a dahlia show and observe which varieties have been awarded prizes; then make a list of those which appeal to you and grow these yourself for exhibiting. Try and grow varieties from as many different groups of dahlias as you can accommodate. You will then be able to enter blooms in a number of different classes in the competition.

Before you start growing for a show try to obtain the previous year's show schedule. This will be a guide as to what you will have to grow, how many plants of each variety to grow, and how many blooms of each variety you have got to enter in a class. Read the schedule carefully, and observe all the rules and regulations. For example, you will be disqualified if you place six blooms in a vase when only five are asked for.

There is a booklet published by the National Dahlia Society entitled the *Classified List of Dahlias and Rules for the Judging and Exhibiting of Dahlias in Competitive Classes* and this may be obtained from the Secretary. This is an invaluable guide to exhibitors, and should be studied thoroughly before attempting to grow dahlias for show work. As the title indicates it also contains a classified list of varieties; and it should be stressed that varieties may only be exhibited in the classes for which they have been classified.

Preparation and Transport of Blooms. The blooms to be exhibited need cutting the evening before the show, and are best placed in deep water and kept in a cool, dark place. If the show is some distance away it may be necessary to cut the blooms in the early morning of the day before the show.

There is not much in the way of dressing that you can do to dahlias—the flowers must be grown perfectly in the first place, and not titivated just before a show. However, if there should be any odd petals that are curling back into the centre of the flower, or any petals that are damaged, then

Blooms which are to be exhibited must be transported with extreme care. A large pad of cotton wool placed behind each bloom gives useful protection

Show blooms being packed in a deep wooden box. The blooms should not be allowed to touch each other, and the stems should be secured with strips of cloth tape.

these can be carefully pulled out with a pair of tweezers. This operation must be done very carefully, and not too many pulled out, otherwise it will be noticed by the judges that some of them have been removed, and the bloom will be classed as imperfect.

The first thing that a judge does is to look at the backs of the blooms, and if the back petals are not as fresh as the front ones, then these blooms do not stand a chance of a prize. Also, the centre of the bloom should not be too tight and congested, and the flower must not possess a 'daisy' centre. Exhibition flowers must be fully open, and every petal fresh.

When you are cutting flowers and packing them be careful not to touch the blooms more than you can help, as the petals are very soft and easily damaged.

Considerable attention must be given to packing the blooms, as bad packing may result in blooms being damaged during transit, and thus many months of patient work will be wasted. Large wooden or stiff cardboard boxes should be used, and must be considerably deeper than the blooms. Place pads of cotton wool behind the blooms to protect them and then lay them flat sides downwards in the boxes so that blooms are not touching each other. They may be secured by pinning strips of cloth tape across the flower stems, so ensuring that the blooms do not move while being transported.

With pompons and dahlias of similar shape, the blooms cannot be put directly into the boxes since they do not have a flat side as do the blooms of other types. Therefore, to get over this problem, the part of the stem nearest the bloom may be supported on strips of wood running across the box, a few inches from the base, and screwed to the sides. The part of the stem in contact with the support should have some cotton wool wrapped

around it; also the support may be padded with cotton wool. Here again, pin some tape across the stems to secure them.

Staging Blooms. Staging depends a lot on the particular class, and the current show schedule should be consulted in this respect. If the schedule specifies a vase of five blooms of any one variety in a certain class, then you would place two blooms at the bottom of the vase and three blooms around the top. If it specifies three large-flowered decoratives for example, not necessarily all in one vase, then I would suggest they are placed in three separate vases. It will be necessary to place some sort of packing in the vases to support the stems, for example, paper, or preferably reed packing.

If a vase of mixed varieties is required, these should preferably be of the same type. If three vases are included in one class, then it is best to place them in a row running from the back to the front of the bench. If the tables are not tiered, the back vase should be raised off the table, the centre one slightly lower, with the front one on the table. Pompons are usually shown about six or nine blooms to a vase.

When placing the blooms in the vases make sure that they are all facing the same way, and that they are facing the judges. Each bloom must be clearly visible at a quick glance—some blooms should not be hidden behind others.

Finally, there are firm rules against wiring the stems or blooms in order to hold the blooms to the required position. The dahlia flowers should be grown sufficiently well for them to be at an angle of 45 degrees to the stem.

Good cultivation, as I have outlined, will produce the most desirable exhibition blooms. Perhaps the basis of producing good blooms is to start with healthy and vigorous plants. Good results are then assured.

Chrysanthemums for Garden and Greenhouse 8

Outdoor chrysanthemums are known also as early-flowering chrysanthemums, and are grown in the open ground without any protection. Some gardeners, however, especially exhibition growers, like to give them some protection against rain by supporting frame lights or polythene sheeting over them, but I would not say this is essential by any means if growing them for normal garden decoration.

These plants have a fairly long flowering season –from about the end of July until mid-October– provided there are no frosts before then. The National Chrysanthemum Society states that 'An early-flowering chrysanthemum is a cultivar (variety), which blooms in a normal season in the open ground before 1st October without any protection whatsoever.'

Classification. Chrysanthemum varieties are grouped into various sections according to the shape and formation of the flowers. This classification is not as confusing as it may seem at first glance, and I would recommend anyone growing chrysanthemums to study the classifications in the National Chrysanthemum Society's *Classified Catalogue of Cultivars and Rules for Judging*, which is available from the Secretary. Each section is numbered and in some cases is divided into sub-sections, each of which is denoted by a letter.

Of the outdoor types, the Decoratives, with their large flowers, are probably the most popular of all. This is mainly because of the very wide range of colours available and the fact that they are ideal for cutting for use in flower arrangements. There are the Incurved Decoratives (Section 23), in which the florets turn in towards the centre of the bloom, and this section is divided into (*a*) large-flowered, and (*b*) medium-flowered varieties. Reflexed Decoratives (Section 24), have florets which grow outwards and downwards from the centre of the bloom, and these are divided into (*a*) large-flowered, and (*b*) medium-flowered varieties. Finally, the Intermediate Decoratives (Section 25), have flowers which are midway between the first two in some varieties, or they may be partially reflexed or incurved. There again we have (*a*) large-flowered, and (*b*) medium-flowered sorts.

Section 26 contains the smaller-flowered Anemones, which are appropriately named, and the blooms have the most attractive centres. These are very easy to grow and are good for cutting. Singles (Section 27) are a popular, smaller-flowered group which contains (*a*) large-flowered and (*b*) medium-flowered varieties. They are graceful plants in habit and produce masses of dainty flowers which are very attractive in floral arrangements. The Pompon varieties (Section 28), are free flowering and ideal for cutting. We have the true pompons (*a*), the blooms of which form a perfect ball of compact hard florets, and the semi-pompons (*b*), with compact florets which form a cone with a flat base. The flowers of all pompon varieties are small.

The Spray varieties (Section 29) probably rival the Decoratives in popularity as they produce an abundance of flowers in many colours. This section is divided into (*a*) anemones, (*b*) pompons, (*c*) reflexing, and (*d*) singles. All of them have fairly small flowers which make attractive floral arrangements. Finally we have Section 30 which includes Any Other Types. The best-known varieties in this section are the Koreans which form very bushy plants, the small flowers being similar in appearance to the Singles, although some varieties are double, and in a wide range of very good colours. Here again, they are favoured by flower arrangers.

Uses of Outdoor Varieties. The decorative chrysanthemums are usually grown for cut flowers rather than for garden decoration, although there is no reason why they should not be grown in a border with other hardy plants. All outdoor chrysanthemums are suitable for growing in a mixed flower border, or they may be given their own special beds. If preferred, they can be grown specially for cutting in a section of the vegetable garden. In this case it pays to disbud the plants, particularly the decorative varieties, in order to grow the blooms to perfection. The Pompons, Anemones, Singles, Sprays and Koreans are popular for cutting as I have already mentioned, but they are also excellent for garden display. They are trouble-free and need no disbudding or stopping, and very little staking. I think that these are probably the best types to create a good show in a garden.

Cultivation of Koreans and Similar Types. The Koreans and Pompons do not need such a rich soil as the large-flowered kinds, and any good garden soil will grow them successfully. Stopping

Place the supporting canes for outdoor chrysanthemums in position before planting, and plant with a trowel. Forestall bird damage to the buds by spraying with bird repellent

and disbudding is not necessary with these, as they produce masses of flowers naturally. The Anemone types do not require stopping either and the Singles, which have a very bushy habit, can be grown as natural sprays, leaving all the shoots and flower buds. The varieties in the Spray section are not disbudded or stopped, as they form bushy plants with plenty of blooms and so are grown as natural sprays.

Making a Choice. There are literally hundreds of varieties available and as with dahlias the choice is very much a personal matter. My advice would be to study the catalogues of chrysanthemum specialists; visit parks, gardens or flower shows, if possible, where chrysanthemums are known to be well displayed; and gradually build up personal experience of the types in which you are most interested and would, perhaps, particularly like to grow.

Care of Outdoor Chrysanthemums

Outdoor or early-flowering chrysanthemums must be planted in an open, sunny part of the garden if they are to give of their best. If the area is shaded for part of the day this will not matter very much, but the plants should not be in a position which receives only about one hour of sun each day, such as in a north-facing border. Under these conditions hardly any flowers will be produced, and the plants will look drawn and miserable.

The Soil and its Preparation. When it comes to soils, the early-flowering chrysanthemum is fairly tolerant. It will grow in a wide range of different types but, as with the dahlia, the better and richer the soil the better will be the growth and flowers. It is advisable to avoid giving the plants too much nitrogen, as soft, leafy growth will result. The flowers will have soft petals which will bruise easily and will not stand up to wet

Pinching out the top of a young chrysanthemum before it forms a 'break bud'. New growths will then form, each producing a 'first crown bud' at its tip

weather. So, avoid any ground which may have had a dressing of a nitrogenous fertiliser.

As with dahlias the ground should be prepared well before planting time. If the soil is heavy it needs to be dug during the autumn to allow the frost to break it down thoroughly, and if it is a light soil then digging can be carried out just after Christmas. I do not think it is necessary to double dig; just dig to the depth of a spade and at the same time incorporate well-rotted manure, garden compost, spent hops or other organic matter to improve the soil structure.

Chrysanthemums like a certain amount of lime in the soil, so I think it is a good idea to carry out a soil test to see whether liming will be necessary. If you obtain a reading of under pH 6·5, then liming will be necessary. The ideal pH should be between 6·5 and 7. Liming is best done as soon after the digging as possible, using hydrated lime which is easily obtainable.

Some gardeners like to give the ground a dressing of bonemeal soon after digging, as this gives the young plants a good start. I would not recommend hoof and horn because this is high in nitrogen and will therefore cause sappy growth. I usually give the dug ground a dressing of basic slag during the autumn or winter, as it is about the cheapest form of phosphate and contains lime also. It helps to improve the soil structure as well.

Planting. The time of planting depends to a great extent on the part of the country in which one lives. In the southern counties or the West Country early-flowering chrysanthemums can possibly be planted from late March to early April; in the Midlands not until after mid-April, or preferably the last week in April or the first week in May. In the northern counties I would not suggest planting before the first week in May, and in really exposed areas it may have to be the second or third week in May to be on the safe

side. There are no really hard and fast rules; we have to be guided by the weather.

Before planting a dressing of an all-purpose fertiliser should be raked into the soil, at 2 to 3oz. per square yard. The soil should then be made really firm.

It is best to plant with a trowel, and to insert a bamboo cane for each plant just before planting. The young plants should be placed close to the canes and the stems tied to them after planting to prevent them from being broken off. Do not plant too deeply, but firm the plants in really well. Very often, as soon as the outdoor chrysanthemums are planted out, birds will come along and start stripping the leaves and buds off and this will set the plants back. I would suggest that you spray the plants with a bird repellent, renewing it after rain.

Now to distances for planting. If I am growing chrysanthemums specially for cutting then I plant them 18in. apart in rows 2½ft. apart. These distances are also ideal for exhibition plants; if they are being planted in groups in a border then 18in. each way is enough to form a good group. I usually put three to five plants in each group. Spray varieties should not be placed less than 18in. apart, and Koreans and Pompons would probably be happier at 2ft. all round, as they are very bushy.

Staking the Plants. It is mainly the large-flowered varieties—or Decoratives—which need supporting, as they grow taller and carry more weight than the others. As I have already mentioned, I insert one cane before planting. Then, as the plants make further growth, I insert another four canes around each one, about 1ft. from the stem, tilting them slightly away from the plant. I then loop soft twine around these, so holding the stems inside the canes. It is not usually necessary to give further canes to Sprays, Koreans, Pompons and so on, although if it appears to be necessary then by all means do so.

Summer Care

Stopping. Many amateur gardeners think that stopping chrysanthemums is a complex business which they will never be able to grasp. In actual fact, it is simplicity itself once the basic facts are understood. The reason for this operation is to manipulate the time of flowering, e.g. to encourage the shoots of a plant to develop earlier than they would naturally, and therefore to produce their flowers earlier. This is vital if you are exhibiting your blooms, as they must be ready by September for the early-flowering chrysanthemum shows and also before the bad weather sets in. Stopping also encourages the production of a greater number of flowers.

Stopping simply involves pinching out the growing tip of a young plant, before it naturally forms what is known as a 'break bud'. By doing this new growths will form in the first leaf axils below the top of the stem and will develop into sturdy shoots, each of which will produce a 'first crown bud' at its tip. Usually this one stopping is sufficient for most outdoor chrysanthemums, but some varieties may need a second stopping, so the tips of these lateral growths are pinched out. This will encourage further shoots to develop from the upper leaf axils and in turn form flower buds at their tips. These are known as 'second crown buds' and there will be a greater number of these than of the first crown buds. It is not often necessary to encourage further laterals, but if these did develop they would produce the 'terminal buds'.

If no stopping is carried out, the plant will make these 'breaks' or shoots naturally, but as I have mentioned they will come into flower very much later.

It is mainly the large-flowered types, such as the Decoratives, which need to be stopped, and I think that one stopping in the early stages of growth is sufficient for normal garden use, to produce blooms from the first crown buds. Most of the small-flowered types, such as Koreans or Pompons, flower much better if left to grow naturally.

The time to carry out stopping is one of the most debatable aspects of chrysanthemum growing. There are various factors on which this depends—the part of the country in which they are being grown, the variety and the weather. Most nurserymen's catalogues give the dates of stopping each variety (this is known as a stopping 'key') and whether the particular variety needs one or two stoppings. I am sure that experience is the only way to learn the best times for stopping your own particular varieties, as much depends on your locality. I would advise you to keep records of the stopping dates of each variety that you grow as this will be invaluable information for calculating accurately the times of flowering in successive years.

As a rough guide, the first stopping can be in the first and second weeks of May, and most of it should be completed by the third or fourth week in May.

Thinning Surplus Growths. With many of the modern varieties there is not much thinning to be done, because when they are stopped they will not produce more than about five good shoots. If they produce more shoots than this then some thinning can be done if desired, cutting the surplus shoots out at their bases. Always remove the weaker ones.

For ordinary garden purposes I would allow as many as six or eight flowers on each plant of the decorative types—one to each shoot—but when they are being grown for exhibition then they must be restricted to two, three or four flowers to each plant.

Tying. This must be carried out regularly as the plants grow; one tie when they are planted out and further ties as new shoots develop after stopping. If these shoots are not tied in they will probably be broken off during high winds. Soft, green garden string is best for this job.

Disbudding. This simply means the removal of all surplus flower buds and is particularly important if growing plants for exhibition, or if large blooms on long stems are required for cutting. As I have already mentioned disbudding is not necessary for Koreans, Pompons, Anemones, Sprays and Singles when they are grown for normal garden display purposes. It is really the large-flowered kinds that respond to this operation, but if you prefer to grow these as sprays, with small flowers, then a lot of disbudding can be dispensed with.

You will notice that the tip of each shoot contains a cluster of flower buds, with one in the centre known as the crown bud. It is this centre bud which must be retained—all the surrounding ones should be rubbed out. By doing this all the energy of the plant will be diverted to these crown buds so producing good, large blooms. After disbudding, each main shoot should bear one bud only.

If sprays of small flowers are wanted, then rub out the crown bud and leave all the surrounding buds to develop. These will flower a little later than if the crown bud only is left.

Disbudding should be done as soon as the buds are large enough to snap out between the finger and thumb, without causing damage to the stem and the remaining buds. The time for doing this job is roughly towards the end of July. About this time of the year the main stems will be developing numerous side shoots which must be removed in order to divert the plant's energy into producing good blooms and long stems. These shoots will appear from the base of the stems right up to the top and they should all be rubbed out as they appear in the axils of the leaves, otherwise the flower buds will be starved. I leave about three shoots at the base of the large-flowered types as these will produce a further crop of flowers in October and November, after the first flush of blooms.

Feeding. I always start feeding my chrysanthemums about late June or early July with a balanced all-purpose fertiliser, preferably one with a fairly high potash content. This will ensure a hard plant with firm-petalled blooms which will stand up to bad weather. It is best to water the plants first if the soil is dry and then apply a diluted liquid feed, or a soluble fertiliser dissolved in water. I think that even the Koreans, Sprays and Pompons benefit from feeding at this time of the year. Usually one application is enough for these, which can be given in late July. For the large-flowered types a fortnightly feed can be given up

Removing a surplus flower bud from a chrysanthemum, of significance if blooms are required for exhibition or cutting. The centre bud in each cluster is retained

Feeding chrysanthemums with a balanced all-purpose fertiliser ensures the production of hard plants with firm-petalled blooms which will stand up to bad weather

Transparent grease-proof paper bags are used to provide show blooms with protection against rain.

to the middle of August, but after this time I do not consider feeding is necessary.

Protecting the Flowers. Bloom for exhibition and for cutting need protecting from the rain, as this makes them spotted and unsightly. In order to avoid a lot of unnecessary work I would aim at growing the older varieties which have harder petals and so withstand the weather much better than a lot of the newer ones which seem to have soft petals.

The simplest way of protecting is by placing greaseproof-paper bags—not brown ones—over the blooms before the colour of the petals begins to show. You must use bags that admit the light, otherwise up to 50 per cent. of the colour will be lost. Special greaseproof-paper bags are obtainable, with a Cellophane window in them—these are the ones I would recommend. The bags are left on until the flowers are ready for cutting. Before covering the blooms I like to dust the insides of the bags with a little DDT or BHC powder as this prevents earwigs and other insects such as aphids from getting into them and damaging the blooms.

Some growers protect their blooms by building a timber framework around the plants, and covering this with polythene sheeting or Dutch lights. In this case it is only necessary to cover the top of the structure, leaving the sides open.

Cutting Chrysanthemums. This job is best done during early morning or late evening when the flowers are fresh. If they are cut during hot weather they will not take up water when placed in vases, and they will just go limp and flop over. Some varieties are more susceptible to this than others. The blooms should be cut with at least $1\frac{1}{2}$ to 2ft. of stem. If you have disbudded correctly then this length of stem will be easily obtainable. After cutting, all the leaves, apart from three or four near the bloom, should be stripped off. The stems are then crushed at their bases, inserted in deep water, and put in a cool, dark place for about 12 hours. This will help to reduce wilting of the flowers. After this time they can be arranged in vases and put on display.

Such types as Koreans, Sprays, Pompons and so on will naturally have shorter stems. As a rule there is very little trouble with flagging and the blooms last well indoors. The large-flowered kinds will last up to three weeks when they are brought into the house.

Outdoor Chrysanthemums After Flowering

With the Koreans, Pompons, Singles and Sprays it is necessary to remove the dead flowers from about the middle of August. This will encourage new shoots to grow and so prolong the flowering period.

Lifting a chrysanthemum in October
for over-wintering. The stems are cut
down to within 9 to 12in. of the ground
and shoots growing from the base are
removed

(the name given to the plants after they have finished flowering) warm-water treatment before boxing them up. This is to control chrysanthemum eelworm (see p. 299 for description and symptoms) which may be present in the stools. If symptoms of attack have been noticed during the season then this treatment is strongly recommended.

The stools should first of all be washed clean of soil and then immersed in water at a temperature of 46°C. (115°F.) for five minutes, or, in the case of the Loveliness varieties, 43°C. (110°F.) for 20 to 30 minutes. The higher temperature is a more recent recommendation and involves less risk of damage to more sensitive varieties. The Loveliness varieties may be damaged, however, unless the plants are, apart from the eelworm damage, reasonably strong and healthy. When the stools are removed from the water they are plunged at once into cold water and then boxed up.

A few varieties are rather sensitive to this treatment, while others respond well by producing many healthy, vigorous shoots which make ideal cuttings.

I would advise the amateur gardener to invest in the proper equipment for this task, so that the water is heated to exactly the right temperature. If the water is heated above the recommended temperature, then the stools will certainly be killed, and if it is not hot enough the treatment will be ineffective.

Whether the stools have been treated or not, the next stage is to box them up into deep boxes. I use a compost consisting of equal parts loam, peat and coarse sand, but with no fertiliser. Place about $\frac{1}{2}$ to 1in. of compost in the bottom of the box, pack the stools in close together and cover the roots with more soil. It is important that the stools should be kept only just moist during the time of dormancy.

Cutting Back. Before doing this job I like to tie a label to the base of the stem of each variety. On this I write the name, colour, type and height to which it grew. By doing this one has all the details one needs when planting again the following year.

When flowering has finished, which will be about mid-October, the stems can be cut down to within 9 to 12in. of the ground. It is not important to do this job immediately after flowering – very often I leave it until the end of November and the plants do not suffer in any way.

Lifting and Boxing. After cutting down the stems the roots should be lifted carefully and all the soil shaken off them. Any shoots that are growing from the base of the stems are best cut out. There is always a temptation to leave these, thinking that they will be all right for cuttings in the spring, but they do not make good ones.

If possible, it is a good idea to give the stools

Housing the Plants for the Winter. If the weather is reasonably mild after boxing the stools, they can be placed outside under a warm wall. In severe weather they are best put in cold frames and the lights covered with straw or matting during frosty spells. Alternatively, the boxes can be placed in the frames straight away. Except during frosts, give the stools some ventilation. If you have a heated greenhouse the boxes are best transferred to this in mid-January to encourage shoots to grow for the purpose of taking cuttings (this next stage is outlined on p. 196). If a greenhouse is not available the stools can remain in the frames, if well protected during hard weather.

When the stools are brought into the greenhouse always place them on the staging where they will get the maximum amount of light to encourage sturdy shoots to develop. Spindly and 'drawn' shoots do not make good cuttings.

Greenhouse Chrysanthemums

As the outdoor or early-flowering chrysanthemums finish flowering the greenhouse varieties will carry on with blooms which can be obtained until well after Christmas. These flowers are indeed valuable because there are not a great many plants in flower at this time of the year, and a few vases of chrysanthemums in the house at Christmas are a cheerful sight.

The October-flowering varieties are the first to show their blooms and these are closely followed by the late-flowering chrysanthemums, although very often the two types overlap.

Different Types and their Uses. As with the early-flowering chrysanthemums the indoor varieties are placed in various sections. Starting with the October-flowering ones, the most impressive of all are the Decorative types with their large flowers. These are ideal for cutting and last well in water. They are divided into three sections, the first being Incurved Decoratives (Section 13) in which the florets turn in towards the centre of the bloom, and there are (a) large-flowered and (b) medium-flowered varieties. Then there are Reflexed Decoratives (Section 14) in which the florets turn outwards and downwards, and here again there are (a) large-flowered and (b) medium-flowered varieties. Finally the Intermediate Decoratives (Section 15) which have the same divisions—(a) large-flowered and (b) medium-flowered.

The Large October-flowering varieties (Section 16) are fairly new and are very similar in appearance to the late-flowering large exhibition types. Again these are useful for cutting and for floral arrangement.

The next four sections contain very few varieties and therefore are not quite so important from the average gardener's point of view. These are the Singles (Section 17), the Pompons (Section 18), the Sprays (Section 19), and Any Other Types (Section 20).

Coming now to the late-flowering chrysanthemums, we start first of all with the very popular exhibition types. These are always favoured by exhibitors as they are excellent for show work. There are three sections, namely, Large Exhibition (Section 1) of which there are reflexing and incurving varieties; Medium Exhibition (Section 2); and Exhibition Incurved (Section 3), of which there are (a) large-flowered and (b) medium-flowered varieties.

Next are the equally popular Decoratives and these include Reflexed Decoratives (Section 4), with (a) large-flowered and (b) medium-flowered varieties; and Intermediate Decoratives (Section 5), again either (a) large-flowered or (b) medium-flowered.

Section 6 contains the delightful Anemones with their attractive, anemone-like centres, but

The stools are packed close together in deep boxes, using a compost consisting of equal parts loam, peat and coarse sand. These are brought into a heated greenhouse in mid-January

I do not think they are as popular as they deserve to be. These can be grown as sprays. The Singles, however (Section 7), are widely grown and make ideal pot plants for bringing indoors. There are (a) large-flowered and (b) medium-flowered varieties of these. Pompon varieties (Section 8) are really delightful with their smallish, globular blooms, the tops of which are slightly flattened. These are ideal as pot plants, as are the Sprays (Section 9). These again have rather small flowers which are carried in light, attractive sprays, and the section is divided into (a) anemones, (b) pompons, (c) reflexing and (d) singles. There is a good range of colours in this section.

Section 10 contains the Spidery types, in other words, those varieties with very thin, spiky or thread-like petals. A good example is the old Rayonnante chrysanthemum with very shaggy blooms. The Rayonnantes, incidentally, are

Greenhouse chrysanthemums intended for exhibition need final pots of 9- or 10-in. size, but 8-in. pots are large enough for ordinary purposes

Firm potting is necessary to get the best from chrysanthemums, and a wooden potting stick makes this easy. This final potting takes place in late May or early June

coming back into fashion and are much sought after by floral-arrangement artists.

Finally we come to Any Other Types (Section 11). This section includes some of the most delightful chrysanthemums of them all. The Charm chrysanthemums make first-class pot plants and well-grown specimens may be from 2 to 3ft. across and 1½ft. in height. These bushes are densely covered with small, single flowers about the size of large Michaelmas daisy blooms, and can be obtained in a wonderful range of brilliant colours. Incidentally, the Charms are easily raised from seed, and will produce flowers within eight months of sowing.

The Cascades can also be raised from seed and they make equally beautiful pot plants. As their name suggests they are of a pendulous nature but they must be trained to a certain extent to hang down. Cascades are not really suitable for very small greenhouses as well-grown plants take up a lot of room, and also they must be placed on a shelf well above the ground. They have attractive, small flowers and are obtainable in many colours.

Soil for Indoor Chrysanthemums. The indoor chrysanthemums are grown in pots, so we must decide on a suitable potting compost in which to grow them. I always use John Innes No. 1 Potting Compost for the first and second potting, and for the final potting John Innes No. 2.

Pot Sizes and Potting. Young plants which have been raised from cuttings are first of all potted singly into 3½-in. pots. When they are established in these—about mid-April—they are transferred to 5-in. pots, and when well-rooted in these—the end of May or early June—they are potted finally into 8-in. ones. I do not think that 9-in. or 10-in. pots are really necessary—except possibly for large plants which are grown for exhibition—the 8-in. size is usually adequate for the average gardener. During the final potting I like to leave

about 2in. between the soil surface and the top of the pot for a topdressing of compost during August or early September. This supplies the plants with a little extra food which helps them along. Always pot chrysanthemums firmly to obtain the best results, using a wooden potting stick.

Starting with Young Rooted Cuttings. As soon as the cuttings have rooted early in the year (see p. 196) they should be potted off separately as I have just mentioned. If you intend to plant them in the open ground for the summer, as is sometimes practised, instead of growing them in pots, then they can be placed in seed boxes and planted out from these. Space them out about 2 to 3in. apart each way in the boxes. When removing the rooted cuttings from their containers they will receive a slight check, so cover them with sheets of newspaper after potting to help prevent flagging. Remember to insert a label with the name of the variety in each pot.

Growing On. Chrysanthemums should not be subjected to high temperatures or soft, weak plants will result. I think that a lot of people give them too much heat and they do not obtain the hard, sturdy plants which are desired. Young rooted cuttings which have been potted up should be given a temperature of around 4° to 7°C. (40° to 45°F.), and this is also an ideal temperature range to maintain for growing them on.

The young plants should be given all the light possible once they are established in their first pots. It is important to give them plenty of room on the greenhouse bench because if the pots are packed close together the plants will become drawn up or 'leggy'. By giving them adequate space the leaves will be retained right to the base of the plants.

For the first few days after the cuttings have been removed from the propagating case and

A space of about 2in. is left between the soil surface and the rim of the pot for a topdressing of compost to be added in August or early September. Label the plants clearly

When chrysanthemums are being grown for exhibition some gardeners like to provide each plant with three or four canes—one for each shoot

potted up, the greenhouse should be kept fairly close, but once they are established the ventilation should be increased gradually. After this period plenty of ventilation may be given whenever the weather is reasonable. Open the top ventilators and also the side ones when possible.

By mid-March, if some of the young plants are well hardened off, they can be placed in a cold frame and hardened off further by gradually increasing ventilation–the sooner they are in frames the better as you will then obtain good hard plants. I sometimes risk standing mine outside from early May and protect them by placing straw bales or screens of polythene sheeting around them.

Stopping. The reason for stopping is to encourage a plant to form shoots and flower buds earlier than it would naturally, which will in turn give us blooms earlier in the season and at the time we want them. Generally this task is carried out from the middle of March to the end of April or early May, giving a second stop in June or early July depending on the variety, the time when the cuttings were taken and also the area in which they are being grown. Most catalogues give stopping dates (or a stopping 'key') for each variety, and state whether it needs stopping once or twice, but this is only a rough guide and the times of stopping are best gained through experience.

All that is involved is the removal of the growing tips of the plants to encourage the side shoots to develop. Pinch out the tips between finger and thumb. It is exactly the same procedure as for the outdoor chrysanthemums which I explained earlier on (pp. 185 and 186).

Staking. This may be necessary from the time the young plants are in their first pots, in which case a thin bamboo cane will be sufficient. After they have been put into their final pots a strong cane (about 4ft. in length for tall varieties) will

be needed for each plant to prevent the stems being broken off by the wind. Some gardeners, especially those growing exhibition blooms, use three or four canes per plant, one being provided for each shoot. Use either raffia or soft green garden twine for tying; do not tie the stems too tightly to the cane, but leave room for them to develop.

Varieties. As with the early-flowering chrysanthemums there are many hundreds of varieties from which to make a choice. Again I would suggest close study of specialist catalogues; visits, if possible, to the collections of enthusiasts in your district, and to flower shows where chrysanthemums are on display.

Greenhouse Chrysanthemums in Summer

The Summer Standing Ground. During early June the chrysanthemums are removed from the cold frames—by which time they should be well-hardened off—and are placed in a specially prepared area for the summer. By this time of the year they will all be in their final pots. As I have already mentioned I sometimes place mine outside round about early May, but I stand the plants close together and protect them from frost and wind with straw bales or polythene screens.

An area—usually at the side of a greenhouse—can be prepared by covering the ground with a 2 to 3in. layer of industrial ashes. Alternatively, slates or tiles can be used for standing the pots on. These materials prevent worms from entering the pots, disturbing the root systems and blocking the drainage. If a special area is not available the pots can be placed along the side of a path.

On the standing ground the pots are placed in double rows, allowing 18in. each way between the plants, and about 3ft. between each double row. At the end of each row insert a stout wooden stake

Spraying chrysanthemums with insecticide against aphids and capsid bugs, two very troublesome pests. In hot weather the plants may need watering two or three times a day

Correct feeding is important. A proprietary chrysanthemum fertiliser can be used or a general-purpose fertiliser, in dry or liquid form

Removing side shoots when they are 1 to 1½in. long. This task is carried out from the end of August to October

about 4ft. in height. Then stretch a wire along each row securing it at the top of each stake. Canes about 4ft. in length should have been inserted in each pot after the final potting, so the tips of these are then tied securely to the wires to prevent the plants from being blown over during high winds.

I did mention that if young plants were grown in boxes they could be planted out in the open ground for the summer. For the amateur who wants just a few flowers for cutting it is the simplest way of growing them. It cuts out the need for constant watering of pots, and the fear of the pots drying out, so causing poor growth. These plants will provide really good flowers for cutting and plenty of them. The varieties Loveliness and Favourite respond particularly well to this method of cultivation.

The soil for these should be prepared as for the outdoor-flowering chrysanthemums. They may then be planted 18in. apart each way in double rows with about 3ft. between the double rows.

Watering. During the summer, in hot weather, chrysanthemums may need watering two or three times a day. The best method of finding out whether clay pots need watering is to tap them with a cotton reel on the end of a cane. If the pots 'ring' then they need watering, but if there is a dull thud then this is an indication that they contain sufficient water. With plastic pots we cannot tell by tapping them, so we must be guided by the appearance of the top of the soil. When using plastic pots it is necessary to be more careful with watering as they do not dry out so quickly as clays, and therefore there is a greater tendency to over-water them. Spraying the plants overhead with clear water each day in the early morning will help in the production of shoots.

Feeding. As the chrysanthemums are potted finally into John Innes No. 2 Potting Compost—which is fairly rich—they will not need much feeding for a number of weeks. However, some feeding may be carried out from early August, about once a fortnight, and then gradually stepped up to once a week until the colours of the petals begin to show, when it should cease. Use either a soluble or liquid general-purpose fertiliser or sprinkle a dry general-purpose fertiliser around each plant and water it well in. Choose one with a fairly high potash content. Proprietary chrysanthemum fertilisers are available.

If any of the plants appear to be making soft rather than hard growth—which can be determined by the texture of the leaves—then a watering of ½oz. of sulphate of potash dissolved in 1gal. of water will help to harden the plants and improve the quality of the flowers.

Second Stopping. If necessary this can be done from early June to the middle of July. Some varieties do not need a second stopping, so it is best to consult a chrysanthemum catalogue which should indicate the number of stops required. As I have already mentioned this task is best learned through experience – so keep accurate records of the stopping dates of each variety. At this stage the first crown buds should have formed so the second stopping merely involves pinching these out to obtain the second crown buds which will produce the flowers. Further stopping is usually unnecessary for most varieties, unless very late flowers are required.

Thinning. This involves the removal of surplus side shoots which appear from the leaf axils down the whole length of the stem. These should be nipped out when about 1 to $1\frac{1}{2}$in. long, starting from the end of August through to October. Any shoots which appear around the base of the stem should also be removed.

Disbudding. I described the technique of disbudding the earlies on p. 186, and the same principles apply to greenhouse varieties. It is usual to make a start in early August, especially with large exhibition varieties, and to carry on into early October when the Decoratives and Singles are disbudded.

Flowering Indoor Chrysanthemums

Unlike the early-flowering chrysanthemums, the indoor varieties produce their flowers in the protection of a greenhouse. The amateur can adapt a conventional greenhouse to house his taller plants by removing some of the staging and standing the pots on the floor. The shorter-growing types, such as the Charm chrysanthemums, can perhaps be accommodated on the staging. The ideal chrysanthemum house is the Dutch-light type which is fairly easy to construct at home once the lights have been purchased.

Heating. The temperature in the greenhouse does not need to be kept very high – just high enough to prevent the atmosphere from becoming damp. If this happens then mildew will become troublesome and botrytis (grey mould) attack will result in marked and spotted petals. I do not use any heat until it is absolutely necessary – when the nights start getting fairly cold – as I like to hold back the flowering so that I have blooms in December and well into January.

Electricity is, I consider, the most convenient form of heating. My house is heated by electricity thermostatically controlled, and I set the thermostat between 6° to 7°C. (43° to 45°F.). When the plants come into flower then I raise the temperature slightly to about 10° to 13°C. (50° to 55°F.), if cold foggy weather occurs. This keeps the atmosphere dry and there is no marking of the petals.

Heating with hot-water pipes can also be recommended. An oil lamp may be used but make sure the top ventilator of the house is left slightly open at all times to allow any fumes to escape, otherwise the blooms will be damaged.

Ventilation. Whenever the weather conditions permit, ventilation should be given. Free circulation of air is essential for keeping mildew in check and the plants generally healthy. Side ventilation as well as top ventilation may be given when the plants are first moved indoors.

Bringing the Plants Inside. The indoor chrysanthemums must be brought into the house before frosts begin, otherwise the flower buds and tips of the shoots will be killed. I always get mine inside before the end of September. If any varieties were planted in the open ground for the summer these should be lifted and either planted in a greenhouse border or potted in suitably sized pots. They should be prepared for lifting about a week beforehand by cutting the soil round each plant with a spade, 6in. away from the stem. This will encourage fibrous roots to form within the soil-ball. Give them a good watering before lifting. Also give them a good soaking once they are inside, spray them over with water, and shut the ventilators for at least a week to help them recover from the check of lifting. From then on give them as much ventilation as the weather will permit as this will help in preventing damping and various other troubles.

Before the chrysanthemums are taken into the greenhouse they should be sprayed to kill any insect pests and diseases which they may be harbouring. For insect control use gamma-BHC and then, some time after applying this, spray the plants with a sulphur fungicide to combat diseases. Also, the pots should be scrubbed thoroughly to remove any dirt and so on. The greenhouse must also be scrupulously clean when the plants are taken inside (see p. 231).

One important point when bringing plants inside is to carry them in pot first. This reduces the risk of damaging the buds or breaking the shoots off by catching them on the door frame. If the pots are being placed on an earth floor or on a soil border then put a tile or slate underneath each one to prevent worms and slugs entering the pot. Chrysanthemums can be placed much closer together in a greenhouse – at the risk of losing a few of the bottom leaves – than when they are outside during the summer. When staging the plants remember always to place the tallest plants at the back with the shorter ones in the front.

Cutting Back and Boxing. As soon as flowering is over the stems should be cut down to within 9in. of the pot and the stools boxed up as I explained for the early-flowering types (see p. 188). The stools can be given warm-water treatment if facilities are available. The boxes of stools are then kept in the greenhouse – not placed in cold frames as is done with the earlies.

Ring Culture of Chrysanthemums

Various plants now are grown by the ring culture method but indoor chrysanthemums respond particularly well to this treatment. The basic technique involves growing plants in bottomless pots or other containers—the 'rings'—which are placed on a bed of a free-draining material, the 'aggregate'. The pots are never watered, except when feeding the plants, but the aggregate is kept continually moist. This encourages the roots to grow down into the bed. The plants eventually form large, strong root systems in the bed and this results in larger plants and flowers. The plants, being more vigorous than when grown by the usual method, are much healthier and damage from disease will be less, if present at all. The chrysanthemums are grown on an outside bed of aggregate during the summer and are then moved into a greenhouse and grown on another bed.

Ring culture takes a lot of the worry out of growing indoor chrysanthemums, especially when it comes to watering. If the plants are grown by the usual method then the pots are liable to dry out rapidly during the summer and this results in uneven growth which is most undesirable. Once the ring culture plants have rooted into the aggregate there will be no worries about them drying out, provided the bed is watered once a day. This moisture round the base of the plants also keeps the atmosphere moist which is conductive to healthy growth.

With the ring culture method there is less trouble from soil-born pests and diseases, as the roots of the plants cannot penetrate the soil below the aggregate bed which has an impervious material, such as polythene sheeting, beneath it.
Aggregate. This must possess two main properties. Firstly, it must be capable of retaining sufficient moisture for the plants. Secondly, any surplus water must be able to drain rapidly from it, as the plants must not become waterlogged. Screened coarse industrial ashes are ideal, provided the particles are not more than $\frac{5}{8}$in. in diameter. The ashes must be screened as no dust-like particles should be present, otherwise the bed will pack down and waterlogging will occur. Ashes from a household fire are not usually suitable. The ashes should be weathered for three months before use. Screened gravel (maximum diameter $\frac{1}{4}$in.) is also very suitable.

The aggregate bed outside should be large enough to enable the plants to be spaced out as I outlined on p. 191, when I discussed the summer standing ground. The stakes and wires to support the plants are erected similarly. The bed of aggregate should be 6in. deep and placed on a sheet of polythene or similar material. Use boards or bricks to retain the aggregate. In the greenhouse a 6-in. depth of soil can be taken out, polythene sheeting put down and then filled with aggregate.

Removing shoots from chrysanthemum stools from which cuttings will be made. The best cuttings come from basal shoots and they need not be more than 2 to 3in. long

This avoids loss of height in a small greenhouse. Before setting the plants on the bed water the aggregate thoroughly so that it settles.
Containers and Compost. The containers, as I have mentioned, should be bottomless, as the compost must be in close contact with the aggregate to allow water to pass upwards. The containers or rings may be purchased in either clay—the same as ordinary flower pots—concrete or plastic. Bituminised cardboard rings are also available but these last for only one season. Bottomless wooden boxes or tins would also be suitable. The size of container is important. The usual size is 9in. deep by 9in. across—there is no need to have larger sizes than this. John Innes No. 2 Potting Compost is recommended.
Raising the Plants. Cuttings are taken at the usual times and the rooted cuttings are potted into $3\frac{1}{2}$-in. pots. These plants may then go straight into the 9-in. rings when they go outside in early June. The exceptions to this rule are large exhibition varieties and all plants intended for exhibition. These should be potted on into 5-in. pots and then into the rings, as this seems to suit them better and larger plants result.
Summer Treatment. The rings are spaced out at the distances recommended for normal culture, on the prepared aggregate bed in early June. They are then filled with compost—which should be made firm—to within $1\frac{1}{2}$in. of the top to allow for

After preparation (by cutting the base cleanly, removing the lower leaves and treating with hormone rooting powder), they are inserted round the edge of 3½in. pots

liquid feeding. Make a planting hole in the compost and insert the plant, firming it in really well. For 5-in. pots, place a suitable depth of compost in the ring first, stand the plant on this and then fill up with more compost, taking care to firm it thoroughly. Do not plant the chrysanthemums any deeper than they were in their previous pots. After planting water the rings thoroughly. From then on watering is confined to the aggregate unless the weather is exceptionally hot and dry, when the plants may start to wilt before they have rooted into the bed.

The summer treatment of ring culture chrysanthemums is the same as for normal methods of growing. The aggregate bed will not usually need watering more than once a day, but it is advisable to check it for moisture each day. On no account must it be allowed to dry out. The feeding programme is the same as for normal culture except that a liquid fertiliser must be used—dry fertilisers are not suitable because they cannot be watered in properly due to the fact that normal watering is confined to the aggregate. When feeding, ensure that the compost becomes thoroughly soaked with the fertiliser. No topdressings of compost will be necessary.

Housing the Plants. The plants are housed at the usual time. It is the practice with some gardeners to grow tomatoes in the greenhouse during the summer by the ring culture method, and then to clear these out at the end of the summer so that the same aggregate bed can be used for the chrysanthemums. Once the tomatoes have been cleared, the greenhouse should be cleaned out thoroughly as described on p. 231. Then dig out a trench about 4in. deep and 15 to 16in. wide in the greenhouse aggregate bed to accommodate the large root systems of the plants.

When bringing in the chrysanthemums from the open, it will be easier if two people can lift them as they will be rather heavy and awkward. Drive garden forks down into the aggregate on each side of the ring and about 10in. away from it. Then push the forks under the ring and gently prise it up. The mass of fibrous roots will come up easily with the ring. Then transport the plant straight into the greenhouse and place it in the trench at the same depth as it was in the open ground. Cover the root ball with aggregate so that the ring remains sitting on the surface of the bed. When the plants are in position water the aggregate in order to settle the roots. It is best to move only one plant at a time to lessen the risk of the roots drying out. This moving of plants may seem rather drastic treatment, but there is no risk whatever of a check to the plants provided the job is carried out carefully.

Propagation of All Chrysanthemums

The usual way, and indeed the easiest way, of increasing chrysanthemums is by cuttings, and the first consideration here is the time to start taking them. Unless plants are being grown for exhibition, in which case early growth and flowering is essential, there is not much to be gained from rooting cuttings too early.

Outdoor-flowering chrysanthemums are best propagated between mid-February and the end of March. For general purposes cuttings of indoor varieties can be taken between late February and early March. If they are taken any earlier then you will obtain very large plants—perhaps too large for the average small greenhouse. Possibly the exceptions to this rule are the large exhibition varieties, the cuttings of which are best taken in November or December to get large enough plants to 'stop' in time for early flowers to be obtained. Some of the incurved and decorative varieties also have to be taken in December.

Before I go on to discuss the details of taking cuttings I think this is a good opportunity to mention that I grow the Favourite and the Loveliness varieties to flower in 6-in. pots, by the following method of propagation. When I have finished taking cuttings for the early batch I put the boxes of stools in a cold frame where they remain until late May or early June. By this time they will have produced shoots about 10 or 12in.

high. I then take the tips off the shoots–about 1½ to 2in. in length–root them, then pot the rooted cuttings in 3½-in. pots, and finally transfer them to the 6-in. size. When the plants are about 5 to 6in. high I pinch out the growing tips and let them flower on the first crown buds. They make plants no more than 2 to 2½ft. high with five or six flowers per plant.

These make first-class pot plants which are ideal for bringing into the living room. They are also good for cutting. Koreans and Pompons are suited to this method of growing, and are also useful as cut flowers. These will flower in September, October and even into November. They will make bushy plants no more than 12in. high and will produce a mass of flowers.

The Kind of Cuttings. The stools of both the outdoor and greenhouse chrysanthemums which were boxed up as I described earlier (see p. 188) will produce shoots from their bases. These are the best cuttings, far better than those which grow from the stems. However, some of the modern varieties are rather shy in sending up shoots from the base, so to get the number of cuttings required it may be necessary to take a few stem cuttings of these.

The cuttings need not be more than 2 to 3in. long and should be cut off at the basal end with a sharp knife or razor blade. Cut them straight across just below the lowest leaf joint.

Rooting Compost. There are a number of rooting mediums which can be used for the successful rooting of cuttings. I prefer to use a mixture of equal parts loam, peat and sand, or even the John Innes Seed Compost. Equal parts of peat and sand may also be used or even sand on its own, but in these instances the cuttings must be potted up as soon as they have formed roots. Vermiculite is popular with many gardeners and cuttings will root rapidly in this material.

Preparing the Cuttings. There is not a great deal of preparation necessary. The main thing is to trim off the lower leaves, so leaving a clear stem for insertion in the compost. Make sure that the base of each cutting has been cut cleanly, without ragged edges. Before inserting the cuttings dip their cut ends into water and then into a suitable hormone rooting powder, to accelerate the formation of roots. I find that a rooting powder makes a difference of 10 days in the time of rooting, and it also encourages a better and stronger root system.

Inserting the Cuttings. I use 3½-in. pots for rooting the cuttings, placing four or six round the edge. Outdoor varieties are usually rooted in seed boxes in John Innes Seed Compost, about 30 cuttings per box. Make sure that the rooting medium is firm and then, with the aid of a small wooden dibber, insert the cuttings to about half their length, firming them well with the fingers. Make

sure that the base of each cutting is in close contact with the bottom of the hole, otherwise they will not root. After they have been inserted give them a good watering with a fine-rosed can. They should not need any more water until they have rooted. If they are given too much water before rooting has taken place they will probably rot off. Make sure that each pot or box has been labelled with the name of the variety; also, the date of taking the cuttings may be useful for future reference.

Where to Root Cuttings. If possible root them in a closed propagating case, or better still under a mist propagator if this is available. They do not need a high temperature–about 10° to 16°C. (50° to 60°F.) is sufficient. A simple propagating case can be made from a reasonably deep bottomless wooden box placed on a layer of moist peat on the greenhouse staging. Place the pots or boxes in this, cover the top of the box with a sheet of glass, and place some newspaper over this to protect the cuttings from the sun.

It will be necessary to turn the glass over each day because water condenses on the underside, and if this drips on to the leaves it will cause them to rot and the cuttings will almost certainly be lost.

If you have a propagating case with soil-warming cables this is ideal, as rooting will then be very rapid. The cuttings can be rooted in a layer of pure sand placed over the cables.

If a greenhouse is not available then cuttings of outdoor varieties may be rooted in a cold frame in mid-March, but ensure that they are well-protected from frost. The base of the frame should be covered with a 2- to 3-in. layer of fine ashes, followed by a layer of John Innes Seed Compost in which to insert the cuttings. The rooted cuttings can then be planted out straight from the frame.

Hardening-off Cuttings. Before the rooted cuttings are potted up they need to be acclimatised to the normal temperature of the greenhouse. If a mist propagator is being used then gradually reduce the amount of mist. The glass on a propagating box can gradually be raised over a period of a few days, so getting the cuttings used to the greenhouse atmosphere. Give the rooted cuttings as much light as possible to prevent them from becoming drawn.

When completely hardened off they should be potted individually in 3½-in. pots, or if you wish to plant the greenhouse varieties in the open ground for the summer they may be placed in seed boxes. Seed boxes can also be used for the outdoor varieties if preferred. The latter can, in fact, be planted out in cold frames once they are well hardened off. Plant them about 3 to 4in. apart each way in a mixture of equal parts loam, peat and sand placed about 2 to 3in. thick.

Right. The large-flowered Reflexed Decorative chrysanthemum Tracy Waller, an early-flowering variety
Right, below. The large-flowered Intermediate Decorative Keystone, another attractive early-flowering variety

Right. The large-flowered Exhibition Incurved chrysanthemum Mavis Shoesmith. This late-flowering variety is representative of the very popular exhibition types

Right, below. The large-flowered Reflexed Decorative chrysanthemum Yellow Symbol, a late-flowering variety with blooms of fine quality

Far right, above. The large-flowered Single chrysanthemum Crimson Crown, a late-flowering variety. Varieties of this type make ideal pot plants for bringing indoors.

Far right, below. Korean chrysanthemums, which can be grown successfully in any good garden soil. With this type, stopping and disbudding is not necessary. A wide range of good colours is available

Right. The early-flowering Spray chrysanthemum Yellow Juweeltje. The Sprays are popular chrysanthemums for cutting and excellent for garden display

Right, below. The early-flowering Spray chrysanthemum Davine

The final hardening-off of both outdoor and greenhouse chrysanthemums consists of placing the plants in a cold frame a few weeks before setting them out in the open. Whenever the weather is reasonable prop up the frame lights to allow the free circulation of air. This ventilation can be gradually increased over a few weeks and the lights can be left off completely just before the plants are removed from the frames.

Exhibiting

Growing chrysanthemums for exhibition really is an art in itself, and tests the skill of the gardener to the full. There is no greater thrill than when you gain your very first prize at a show, although it may have taken a number of years of growing chrysanthemums in order to have gained the necessary experience.

I think the best exhibition varieties are grown in the northern parts of the country, as the plants prefer the cooler climate. These conditions produce good hard plants which are able to withstand bad weather much better than plants of softer growth. Also there is not so much trouble from pests and diseases as they do not increase so rapidly in a cooler climate. The blooms seen at shows in Sheffield and in other industrial and mining areas in the North are first class.

Making a Start. Some varieties are more suitable for show work than others, so in order to get some idea of which ones are mainly used it is a good idea to visit a few chrysanthemum shows before purchasing plants for exhibition purposes. Make a list of those varieties which have been awarded prizes, and then make your own personal selection from this list. Try and get hold of a few old show schedules to see what sort of exhibits are required and to become familiar with the various classes. This will give you some idea of what types to grow and how many blooms will be required and so on.

As I stressed when writing about exhibiting dahlias, as far as possible go straight into a large show, preferably a proper chrysanthemum show that has all the correct classes. I would advise anyone who is thinking about exhibiting to study the National Chrysanthemum Society's booklet on the rules of judging. This will give you a good idea of the standard of quality required and it also mentions the common faults in exhibition blooms. It will serve as a guide to the classifications of varieties, and this is important when exhibiting. A variety must be shown in its correct class or section otherwise it will definitely be disqualified by the judges, and so many months of hard work will be wasted.

Raising Plants for Exhibition. Cuttings must be taken sooner than those for normal garden-display purposes. It is necessary to have good strong plants earlier in the season so that the stopping can be carried out correspondingly earlier, therefore producing good blooms in time for the shows. The treatment of cuttings is the same as I have already described (see p. 196).

Cultural Requirements. Stopping chrysanthemums for exhibition is rather different from that of ordinary display plants. They have to be stopped strictly according to a 'key'. The key that nurserymen supply in their catalogues is not always a good guide, as the stopping dates will vary according to the part of the country in which the plants are being grown. The nurserymen's keys are only a very general guide. I would strongly advise the novice exhibitor to make his own stopping key for each of his varieties. It will be necessary to grow a variety for a year, noting the time of stopping and how the plant behaves. Then adjust the time of stopping the next year if it blooms incorrectly for a particular show.

Extra care and feeding must be exercised when growing plants for exhibition. The plants need more room for development, because to grow a good chrysanthemum the foliage must be kept healthy right down to the base of the plant. The healthier the foliage the stronger the plant and this, of course, encourages perfect blooms. The plants will benefit from plenty of potash applied when the flower buds are seen to be developing.

Disbudding has to be carried out earlier than for the display plants, and indoor varieties have to be brought into a greenhouse early. Large and medium exhibition and exhibition incurved varieties start showing the colour of their petals sooner than the other greenhouse types. Great care must be taken to prevent damping of the blooms when the plants are in the house as this will render them useless for show work.

Preparing and Transporting Blooms. After selecting the blooms for a show they should be cut as I have explained on p. 187. The stems of large exhibition varieties should have all the leaves stripped off them as points are not awarded for foliage in this section. Additional foliage may, however, be added when arranging them. Points are awarded for foliage in all other sections, so I would suggest leaving about half the foliage on the top part of the stems. Check each bloom for any signs of insect pests and remove any present with a soft artist's brush.

Packing blooms in their boxes ready for transporting them to a show is really a work of art. Some exhibitors, especially those who travel down from the North to the London shows of the N.C.S., use huge wooden packing crates for their blooms, which need to be transported in something akin to a furniture van. The sole reason for this is that the blooms must necessarily be transported without rubbing against each other, so damaging the petals.

The 'dressing' of the untidy centre of a chrysanthemum bloom before placing on the show bench. If just a few petals are damaged they can be removed

Well-staged Single chrysanthemums. Things the judges look for are freshness, good colour, shape, form and size, good centres, petal crispness and healthy foliage

For the average amateur, who does not have to travel a great distance to a show, long, deep boxes, either of wood or thick cardboard, can be used satisfactorily. These should be equipped with a removable lid. The first thing to do is to line the box completely with soft tissue paper. At one end of the box, about 6in. away from the end, a roll or cushion of tissue paper should be placed across the bottom of the box. The top part of the chrysanthemum stems can then be laid on this support, so keeping the blooms off the bottom of the box. The same procedure can be carried out at the other end of the box. A second row of blooms can be placed at each end again. Place a roll of tissue paper over the stems already in place, just below the blooms. Then lay a further row of blooms on this support, pushing the stems carefully between the blooms at the opposite end of the box. After the blooms are in place, wedge a bamboo cane across the centre of the box to hold the stems securely. Lay a sheet of tissue paper over the blooms and then place the lid on the box.

Special boxes are constructed by some of the big exhibitors to carry the blooms upright, the stems being supported between wooden cross rails, often with the bases of the stems in small vases of water. These are fairly complicated affairs for the average amateur to tackle, but as experience and knowledge is gained so he will progress to these types of boxes, especially when he starts showing his blooms up and down the country. A good way of finding out which types of crates are most popular, and how the blooms are packed, is to watch some of the well-known exhibitors at some of the larger shows bringing in their crates and unpacking them. Very often these experts are only too willing to give a word of advice to the less experienced.

Staging Blooms. On arrival at the show the blooms should be unpacked very carefully and placed preferably in buckets of water until they can be arranged in vases. Any damaged or discoloured petals may be gently removed with a pair of tweezers, but on no account remove too many as this will be noticed by the judges who will then down-grade the blooms.

The vases should be packed with reeds or privet stems to hold the stems upright. The use of supports for blooms is a very important point and the exhibitor must become familiar with the rules laid down by the N.C.S., otherwise he will be disqualified if he uses them in a class in which supports are not allowed.

Cups or wire rings may be used to support the blooms of large exhibition and medium exhibition varieties only. These are secured underneath the bloom. The cups or rings must not exceed 3in. in diameter, including padding if used. No cups or rings may be used for exhibition incurved varieties, but supports for the stems will not disqualify. The use of supports of any kind for blooms in any other section is strictly forbidden.

It is interesting to note that a 'single' chrysanthemum is a bloom with about five rows of ray florets. Medium-sized blooms often have more than five rows. This point should be borne in mind if exhibiting singles.

Finally, I would like to point out that the main thing which judges are watching for are: freshness of blooms; good colour, typical of the variety; good shape or form, according to the variety; good size, according to the particular variety; perfect centres to the blooms; crispness of petal; clean, healthy foliage, except for large exhibition varieties; and the way in which the blooms have been staged. As I say, these are some of the main points to watch, but I would suggest that further study of the N.C.S. booklet, *Classified Catalogue of Cultivars and Rules for Judging*, will put the novice exhibitor on the road to success.

Rock, Pool and Waterside Plants

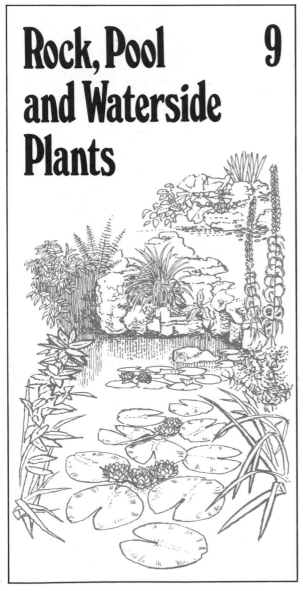

9

There are few of us who have not admired the beautiful jewel-like flowers one sees growing in rock gardens. Rock gardening, like growing chrysanthemums or dahlias, is just another specialised form of gardening, one which attracts many gardeners because there are so many delightful plants which can be grown in this way. Rock gardens can be quite small, using a few pieces of stone, covering a few square feet, in which many diminutive plants will flourish, or they may be very extensive with immense rocks and large pockets of soil, in which large plants, including shrubs, are grown. Fortunately, there is such a wide range of plants that there is no difficulty in finding suitable ones for gardens, great or small. And, although the main flowering period is in the spring, it is possible by choosing plants carefully to have something in flower for practically every week of the year, even in a quite small area.

Forms of Display

Making a Rock Garden. A rock garden is not merely a haphazard collection of pieces of rock or lumps of concrete, sticking up at all angles, with pockets of soil in between in which dwarf plants languish. A properly constructed rock garden should, as far as possible, look like a natural outcrop of rocks and give the impression that the rocks seen above the surface are far fewer than those that lie beneath—like an iceberg, which has most of its bulk below water level. To give the right impression it is not necessary to have enormous lumps of stone—quite small pieces are sufficient although if they are available one or two large pieces are helpful to provide the keystones around which the rock garden is built.

To get this naturalistic effect it is best to use stone with well-defined lines or strata. Water-worn Cumberland limestone and certain hard sandstones are the best. Granite and marble are

If a rock garden is built on a series of
rough steps this will leave pockets of
soil between each layer in which plants
can be grown in soil suited to their
needs. For example, lime-free soil for
the lime-haters, peat for peat-lovers,
gritty soil for those liking sharp
drainage and so on

Be careful when moving heavy rocks into position or you may injure yourself. When positioning really large ones it is necessary to use a block and tackle and boards. It is important when laying the rocks to make sure that they are seated firmly, packing soil behind and around each one. This will not only prevent the stones wobbling but it will ensure that there are no air pockets into which the roots may penetrate and die.

After the rock garden has been completed it is a good plan to topdress the soil. Where lime-hating plants are being grown peat makes a good topdressing. In fact peat can be used for all plants as it helps to retain moisture and is gradually absorbed into the soil to improve its consistency. If the natural soil is on the heavy side coarse sand can be worked in with a hand fork. Where the natural soil is poor it is best to remove it and replace it with a prepared compost of fibrous loam, moist peat and coarse sand.

Planting between the rocks, must be done carefully with a small trowel or hand fork. Make sure that the soil is firmed evenly. Numerous small shrubs and conifers are suitable for planting on rock gardens and these should be planted first. Do not plant too closely as many alpine plants spread rapidly.

Rock Beds. Where there is insufficient space for a rock garden or it is not desired to have such a feature but still to grow rock plants, a rock bed provides a very attractive alternative. Such a feature is ideal for a corner of the garden, open to the south or west, which needs to be given additional interest. It is surprising how many fine plants can be grown in this way in even a relatively small area. The low stone walls of the bed are easy to build, and with adequate drainage and skilful planting one can have a feature which gives pleasure out of all proportion to its size.

Dry Walls and Steps. Dry walls are useful for separating one part of the garden from another or for use as retaining walls. They need to be carefully constructed, without cement between the layers of stone. It is important, too, that they should have a slight backward slope to provide better support. Soil should be packed solidly behind the wall as it is built and construction will be easier if layers of soil are put between each row of stones.

As the stones which are used will be irregular in shape there will usually be spaces left between them in which alpines can be planted. Aubrietas, alyssums, houseleeks, saxifrages and erinus look delightful planted in the face of the wall.

Alpines can also be grown in steps which connect one part of the garden with another. Campanulas, geraniums (cranesbills), sedums, thymes and the wall plants mentioned above are suitable. Holes can be left for plants when the steps are built but if this was overlooked it is not difficult to

the most unsuitable types. The rocks are laid so that the strata lines run approximately in the same direction. Although this is easier when you build the rock garden on a bank, many successful gardens have been built on the flat. Try to ensure that the garden is built in a series of rough steps and this will leave pockets of soil behind each layer to accommodate plants. These can be filled with lime-free soil for lime-hating plants, peat for peat lovers etc, or you can put in deep beds of gritty soil for those that like sharp drainage. But ordinary loamy garden soil will do for most alpine plants. Crevices left between the rocks can be used to plant alpines which will create a natural effect. As the plants are supplied in pots it is possible to plant at almost any time of year, even when the plants are in flower. But in the latter case shade them from strong sunshine for a day or two after planting.

make holes with a hammer and cold chisel large enough for the plants.

Trough and Sink Gardens. Rock plants grown in troughs, sinks and other containers can be a tremendous source of interest. They look particularly effective on paved areas but can, of course, be successfully incorporated into the designs of many gardens.

Drainage is extremely important and there must always be a drainage hole in the container, covered with crocks, towards which excess water can gravitate freely. When deciding on the compost to use it will be necessary to consider whether lime-hating plants are to be grown and to make the necessary adjustments. The compost should have a depth of at least 4in. with ½in. space at the top for watering. If rocks are to be added position these before the sink is filled and work the soil in around them. Only a few should be used. Pack the soil in firmly and make sure that the sides and corners are firm, otherwise water may go straight down the sides instead of into the compost.

A dressing of stone chippings around the plants gives a good finish and deters weeds, but with lime-hating plants be careful not to use limestone chippings.

Some suitable plants for trough gardens are *Androsace sarmentosa*; *Armeria caespitosa*; the campanulas *arvatica* and *cochlearifolia* (syn. *C. pusilla*); *Dianthus alpinus*; *Erinus alpinus*; *Phlox douglasii* and its varieties; Kabschia saxifrages; and the small, compact, *Sempervivum arachnoideum*. *Salix arbuscula*, a very small willow which grows to about 9in. tall, and the very slow-growing and popular conifer *Juniperus communis compressa* (the Noah's Ark Juniper), are useful for such plantings.

Once planted a sink garden requires little attention other than hand-weeding occasionally and watering in dry weather. If the container is large enough, a few small bulbous plants, such as miniature narcissi and the smallest crocuses, may also be planted.

Propagation. Many alpines can be propagated quite simply either by lifting and dividing the plants or by detaching rooted offsets. Cuttings of others root readily if small pieces are detached and dibbled into pans of sandy compost. After they have been watered a pane of glass is placed over the pan and it is put either in a cold frame or in a shady place outdoors until the cuttings have rooted. The condensation must be wiped from the glass every day, and when the cuttings have rooted pot them separately into 3-in. pots using a compost containing coarse sand to ensure good drainage.

Routine Tasks. It is worth going over rock garden plants quite frequently to cut off dead flower heads, and autumn leaves which are blown by the wind into crevices should be removed

Trough and sink gardens are a source of delight and they can be successfully included in many different kinds of garden layout. Plants must, of course, be chosen which will take kindly to the confined conditions and, for obvious reasons, the drainage of such containers must be especially good

regularly for they get wet and soggy and may kill the plants they are covering.

Some delicate rock plants such as lewisias and those with hairy leaves, such as the well-known Edelweiss, survive cold damp winters better if they are protected with a pane of glass. The glass can be supported over the plants on four short sticks and be kept in place with a stone. Alternatively, a cloche can be placed over the plants.

Some Rock Garden Plants

If your soil is naturally alkaline, that is, it contains lime or is chalky, there are certain plants to avoid, unless separate pockets of lime-free soil are made on the rock garden, as I have explained. This, however, is not always satisfactory, owing to water containing lime seeping through and contaminating the soil. Plants which do not tolerate

Aethionema Warley Rose, perhaps the best-known of all the Stonecresses. The bright pink flowers are shown off well by the bluish-green leaves

Alyssum saxatile, a favourite rock garden and wall plant—not surprisingly, for its yellow flowers are gay and it is easy to please

lime include lithospermums; the bright blue autumn-flowering *Gentiana sino-ornata* and its hybrids; and lewisias and their hybrids, which in any case are not really plants for the beginner, and do best in or on top of a dry wall.

A good alpine nursery would be happy to supply a collection of, say, 50 different varieties of rock plants for the beginner. Depending on the size of the rock garden, you could have two or three plants in each variety and plant the varieties in groups. This will prove more effective than single plants dotted around. Such a collection should include armerias, named varieties of aubrieta, campanulas, dianthus, the creamy-white-flowered Mountain Avens (*Dryas octopetala*), geranium species, helianthemums (Sun Roses), *Phlox subulata* and its hybrids, potentillas, saxifrages, sedums, and sempervivums (Houseleeks).

Small bulbous plants are also excellent for the rock gardens as I have explained in Chapter Six. Detailed information on these plants will be found in that chapter.

Acaena. These are trailing evergreen plants with fern-like leaves, which make very good ground cover, and are particularly attractive draped over a dry stone wall. They may establish so well that they become invasive but can easily be kept under control and cut back if required. A good species is *A. buchananii*, which has dense mats of grey-green leaves and large yellowish fruits. *A. microphylla* has bronze leaves and large, bright red fruits, and *A. novae-zealandiae* is similar, but with purplish-red fruits. All will grow in full sun or partial shade.

Aethionema (Stonecress). These are mostly hardy evergreen perennials or small shrubs, suitable for the rock garden or dry wall. The heads of small pink flowers appear in May and June. *A. grandiflorum*, about 10in. tall, is branching, with greyish leaves covered with rosy-pink flowers in

June and July. *A. pulchellum* has a trailing habit of growth with grey-blue leaves and bright pink flowers. Warley Rose is a particularly attractive variety, up to 6in. tall, with very bright pink flowers until August, and Warley Ruber is similar but with rosy-red flowers. They can be planted in light soil in crevices or on ledges.

Ajuga (Bugle). There are some extremely attractive low-growing bugles with variously coloured leaves, and delightfully contrasting blue flowers, which make good ground cover on the rock garden, or can be restricted to grow as specimen plants. *A. reptans atropurpurea* has dark purple leaves and blue flowers in short spikes between May and July; *A. r. multicolor* (Rainbow) has bronze and white variegated leaves, with a touch of rose in the white; *A. r. variegata* has greyish-green leaves with cream variegations, and light blue flowers. They will grow in any ordinary soil, in sun or light shade.

Alyssum (Madwort). The alyssums are understandably popular spring-flowering plants for they are gay and easy to please. *A. saxatile*, with yellow flowers, is a favourite rock garden and wall plant. There is a pale yellow form of this known as *citrinum*, and also a variety with double flowers, *flore pleno*, which is even more showy. Grow the alyssums in any ordinary soil and an open sunny position. They are inclined to die off in winter if the drainage is bad, and all flower most profusely in soil that is rather poor.

Androsace (Rock Jasmine). The pink or white flowers of these hardy perennials appear in spring and summer. Some come from very high altitudes and require very careful treatment in the garden, but the popular kinds are easy to grow. *A. lanuginosa* is one of these, growing to 2in. and consisting of silvery mats of trailing rosettes with pinkish flowers from May to September. *A. sarmentosa* is also a popular kind, up to 4in. tall, with grey-green

Androsace sarmentosa, a Rock Jasmine which bears soft pink flowers in May and June. It needs a sunny position and well-drained soil

The campanulas are among the most attractive of all rock garden plants. The light bluish-violet *C. portenschlagiana* is a sheet of colour in June and July

rosettes of leaves and soft pink flowers during May and June. All androsaces are prostrate, either trailing or cushion-forming. The popular ones can be grown in a well-drained soil containing stone chippings and leafmould or peat in an open and sunny position. The more difficult androsaces, such as the lime-hating *A. alpina* (syn. *A. glacialis*) and *A. arachnoidea,* need a compost of sharp sand and stone chippings, with just a little loam, peat and leafmould in a sunny position.

Arabis (Rock Cress, Wall Cress). The rock cresses are mostly trailing, perennial plants, and the single and double forms of *A. albida* are very good for rock gardens, edgings and on dry walls. All flower in spring. *A. albida* is white, and there is a double variety of it, *flore plena,* which is rather a rampant grower. Rosabella is a variety with pretty pink flowers, which does not 'take over' the rock garden. They do best in a sunny position although *A. albida* will grow almost anywhere in reasonably drained soil. Cut back after flowering to encourage a more compact habit.

Armeria (Sea Pink, Thrift). The thrifts make useful edgings or carpetings for rock gardens which are sunny. They are herbaceous and *A. maritima* is a native plant with pink or rose flowers on 6-in. stems in spring. *A. caespitosa* is only a few inches high, and varieties are: *alba,* white; Beechwood Variety, deep pink with large flowers; and Bevan's Variety, deep rose, only 2in. Provide *A. caespitosa* with a very open compost, with plenty of stone chippings and a little leafmould or peat added.

Aubrieta (Purple Rock Cress). This is the ubiquitous purple-flowered, trailing plant seen on every rock garden. It is evergreen and, besides the mauve kinds so often seen, there are modern varieties obtainable now in pink, red, almost blue and lilac. Its sheets of colour in spring hanging over the walls or down rock garden banks provide a welcome change after the bareness of winter. Some good varieties are: Barker's Double, deep red, semi-double or single; Carnival, large, deep violet; Gloriosa, rose-pink and large-flowered; Gurgedyke, deep purple; Maurice Prichard, pale rose-pink; Mrs. Rodewald, red; and Studland, lavender. Plant in ordinary soil and a sunny position. Trim after flowering.

Avens, see Geum
Barrenwort, see Epimedium
Bitterwort, see Lewisia
Bugle, see Ajuga
Campanula. The blue, violet or white bells of the campanulas provide some of the most attractive of the rock plants and they are not difficult to grow. Flowering time is from the middle of June onwards, to September, and they prefer a sunny, well-drained position. *C. lactiflora* Pouffe produces mounds of leaves thickly scattered with blue-violet bells in July-September; it reaches 6 to 9in. *C. portenschlagiana* has light bluish-violet flowers in sprays, in June and July, and reaches a height of 3 to 6in.; it can be grown in sun or shade and has a white form. *C. porscharskyana,* with an equally unpronounceable name, is very handsome (but it spreads rapidly), with powder blue flowers from mid-summer on. These are just one or two of my particular favourites, but there are many more, to be found in good alpine plant catalogues.

Candytuft, see Iberis
Cranesbill, see Geranium
Dianthus (Pinks). The pinks make attractive specimens, with their grey-blue spiky leaves, and pink and magenta, or red and rose-pink flowers. La Bourbille has bright pink flowers and silvery-grey leaves, fringed petals and reaches 3in.; *D. alpinus* has large red flowers in May and June, and is also short, to 3in. *D. gratianopolitanus* (syn. *D. caesius*) has 6- to 9-in.-high flower stems, but the

habit of growth is very mat-like. It is very fragrant with pink flowers in May and June. Others include the late summer-flowering Little Jock, rosy-pink, 3 to 4in. and Mars, deep red and double, to 6in. and long flowering. Pinks do best where there is lime in the soil, sun, and a well-drained site.

Edelweiss, see Leontopodium

Edraianthus. These are easily grown trailing plants of the same family as the campanulas, preferring really deep—at least 1ft.—of well-drained soil and sun. They are sometimes listed under *Wahlenbergia* in catalogues. *E. pumilio* reaches 2in. and makes a cushion of grey-green leaves with blue, upturned bell-like flowers in June.

Epimedium (Barrenwort). These plants make good ground cover and do well in shady positions. Their glossy bright green leaves set off the yellow, white or rose flowers very well, and, later in the year, turn to good shades of red. *E. pinnatum* has bright yellow flowers, on stems 8 to 12in. high in summer, and *E. youngianum niveum* has white flowers above bronze leaves in spring. There is a red variety of this species called *rubrum*. A sandy soil with peat added suits them.

Erinus. A common name for this plant is the Fairy Foxglove, which aptly describes its flowers, and it does in fact belong to the same family as the foxgloves. It is a very small tufted plant, being only a few inches tall, and does best in a sunny position. *E. alpinus* has violet-purple flowers in spring, and there are some good varieties: *albus*, white; Dr. Haenaele, crimson-red; and Mrs. Charles Boyle, deep pink. They grow well on walls and in any well-drained soil, in a sunny or lightly shaded position.

Erodium (Heron's-bill). These are closely related to the hardy geraniums and produce charming, mostly pink or red flowers in summer. Among the best kinds are: *E. chamaedryoides roseum*, prostrate, pink, summer; *E. chrysanthum*, yellow, 6 to 9in., summer; and *E. corsicum*, pink, 4in., summer. They should be planted in ordinary well-drained soil and a sunny place. After planting do not move unless it is absolutely essential, as the roots of old plants are tough and do not readily form fresh fibres.

Fairy Foxglove, see Erinus

Gentiana (Gentian). The flower *par excellence* for rock gardens; with the Edelweiss this is probably the most well-known of rock garden plants, and indeed it would be difficult to find a more intense or truer range of blues than is found in the flowers of these plants. Some are easy to grow, but some are of more difficult cultivation and need exactly the right soil, site and management if they are to succeed. Some of the more easily cultivated ones are the following: *G. acaulis*, deep blue funnel-shaped flowers in May and June, 4in., prefers heavy well-drained soil in sun, with leaf-

mould and a little bonemeal; *G. farreri*, prostrate, with Cambridge blue, white-throated flowers in September and needing a sunny position, and well-drained soil with leafmould or peat added; *G. lagodechiana*, 5in. tall with deep blue flowers in autumn; *G. macaulayi*, deep blue, funnel-shaped flowers, to 4in., autumn, must have lime-free soil; *G. septemfida*, deep blue bell-shaped flowers in late summer and autumn, 6 to 12in. and one of the most easily-grown; and *G. sino-ornata*, royal blue, trumpet-like flowers in September, to 4in., must have lime-free soil. Plant gentians in a mixture of 2 parts loam, 1 part peat, and 1 part grit or coarse sand. Topdress in March with a little peat or well-rotted leafmould, and water freely where growing on dry soil in the summer.

Geranium (Cranesbill). The flowers of the geranium species are mostly rather fragile and saucer-shaped, produced mainly in early summer. They are easily grown plants, in any soil and in sunny or shady positions, and flower for a long time throughout the summer. Out of a great number which can be grown the following are particularly recommended: *G. cinereum*, grey-green leaves and pale pink flowers, June to August; *G. dalmaticum*, to 6in., shining green leaves in clumps with pink flowers in summer; *G. pylzowianum*, silvery leaves and rose flowers, June to September, 6in.; *G. sanguineum lancastriense*, a pink variety of the native plant, the Bloody Cranesbill, to 6in., in June and July; and *G. wallichianum* Buxton's Blue, to 1ft., deep blue with a white centre, from July to September, a very attractive plant. There are no special soil requirements for these plants and all will tolerate lime in the soil.

Geum (Avens). The dwarf species of these hardy perennials are suitable for the rock garden, and *G. montanum*, 6in., is one of the best, with golden flowers in May and June, followed by feathery seed heads. *G. borisii* has brilliant orange-scarlet flowers from May to August, and reaches 1ft., in height. They are easy to grow in a sunny or lightly shaded position, in ordinary soil. Cut down the flower stems to prevent the strain on the plants of developing seed, unless the seed heads are required.

Heart's-ease, see Viola

Helianthemum (Sun Rose). Low-growing evergreen shrubs with gaily-coloured flowers in early summer, and intermittently for the rest of the season. Most of the varieties available are derived from *H. nummularium* and some good ones are: Amy Baring, prostrate with orange-bronze flowers; Ben Vane, brilliant salmon; Golden Queen; Jubilee, primrose-yellow; Mrs. Earle, double, red with a yellow flush at the base of the petals; Rose of Leeswood, double, soft pink; Snowball, double, creamy-white; and Watergate Rose, very large rose-crimson flowers, grey-

A skilfully planted rock bank can be used to link the different levels of a garden sympathetically

Right. Dry stone retaining walls separating one part of the garden from another provide excellent opportunities to grow a variety of rock plants imaginatively. Indeed, few other garden features can compete as spectacles with the 'waterfall' of colour which can so easily be created

Right, below. In complete contrast to the above is the flat rock garden or rock bed which has become increasingly popular in recent years, as an alternative to the traditional rock garden. These can be extremely effective even when of very small size, and it is little short of astonishing how many fine rock plants can be grown in a restricted area

Far right, above. For those who garden on lime-free soils the September-flowering *Gentiana sino-ornata* can be a source of especial pleasure. There is something irresistibly attractive about the upturned faces of the trumpet flowers carpeting the ground with colour

Far right, below. Pulsatilla vulgaris, the Pasque Flower, is an April-May-flowering plant of great beauty. Sunshine and good drainage are necessary provisions

Left. The Kabschia saxifrage, *Saxifraga apiculata*, which is one of the delights of the rock garden in spring. Given the conditions it likes, a well-drained compost including plenty of sand, stone chippings and leafmould or peat, it will soon form a plant of good proportions

Left, below. Crevice plants of especial beauty are the lewisias. Resentful of disturbance, they are rather temperamental but are worth all the care taken to satisfy their requirements, notably good drainage and rich soil, sunshine and a warm exposure

Right. Sedum spathulifolium purpureum, a natural choice for dry, sunny positions

Left. A rock and water feature in the author's Shropshire garden. Water-lilies and associated moisture-loving plants can bring many weeks of colour and interest to gardens of all sizes. Such primulas as *P.bulleyana* and *P. japonica* are naturals for these conditions, and they are seen here with helianthemums, violas, small-growing conifers and other plants

Right. The dainty summer-flowering *Iris sibirica*, a splendid plant for the waterside

Right, below. Primula pulverulenta, another outstanding waterside plant which will naturalise itself freely in moist soil. Like other Candelabra primulas, this species and its delightful Bartley Strain grow best in partially shaded conditions

The popular water-lily, *Nymphaea* James Brydon—an excellent variety for pools containing 18in. to 2ft. of water

green leaves. Sun roses will spread over several square feet of ground and should be planted in a light, sandy soil. Prune to shape immediately after flowering. They require a sunny position.

Heron's-bill, see Erodium

Houseleek, see Sempervivum

Iberis (Candytuft). The candytufts have white, lilac, or purple flowers in spring and summer. Some of the best kinds are: *I. saxatilis*, perennial to 6in., white flowers in May and June, and *I. sempervirens* Snowflake, 1ft., a delightful variety. Plant on ledges, or in crevices in full sun, in ordinary soil.

Leontopodium (Edelweiss). Everyone knows and has an affection for the yellow, starry flowers with whitish woolly bracts and grey woolly leaves of the Edelweiss, the Swiss national flower. Although it was once thought to be very difficult to grow, and rather rare, because it was first found in inaccessible positions on mountain ledges, it is now easily obtainable from any good nurseryman. *L. alpinum* is 6in. high, and flowers in summer; *L. haplophylloides* has lemon-scented foliage. The Edelweiss is not choosy about soil, provided it is well-drained, but dislikes wet and heavy winter rains, so protect it during this sort of weather with a pane of glass.

Lewisia. These are plants of great beauty which need an especially free-draining soil and sunshine to succeed. They are most at home in the crevices of stone walls. *L. tweedyi* and *L. howellii* are the two species usually grown. The first mentioned has apricot-rose flowers, the later has flesh pink blooms.

Lithospermum (Gromwell). These are hardy trailing evergreen shrubs and perennials. They do best on ledges in the rock garden, and a good species is *L. diffusum*, better known as *L. prostratum*, prostrate, with bright blue flowers in great profusion between May and June. A particularly good form of this with larger flowers than this type is Grace Ward; remember that it requires a lime-free soil. Heavenly Blue is another good variety with smaller flowers. *L. doerfleri* reaches about 1ft. and has deep purple flowers erect in graceful clusters in May and June; this can be grown in limy soil. Plant in gritty well-drained soil and a sunny, open position.

Pansy, see Viola

Phlox. The alpine varieties of phlox make most attractive plants for the rock garden; they flower very freely over a long period and can be all shades of pink and red, and blue, purple and white. They are mostly prostrate and flower between May and July. *P. amoena* has showy pink flowers and reaches 6in., *P. douglasii* is 2 to 4in. tall and has almost stemless flowers of pale violet. Good varieties of this are: Boothman's Variety, clear mauve with a dark centre, and Rose Queen, silvery-rose. *P. stolonifera* Blue Ridge has stems up to 9in.

long, although of creeping habit, and large blue flowers. *P. subulata* is the species from which most varieties have been obtained; some good ones, all 4 to 6in. tall, are: *atropurpurea*, wine-red; *lilacina*, pale lilac; Temiscaming, deep reddish-purple; and Vivid, salmon. Plant phlox in spring in light, well-drained soil and a sunny position.

Pink, see Dianthus

Polygonum. One or two of the polygonums are native plants and the genus as a whole can be grown easily. They are often spreading plants, ideal for ground cover. *P. affine* Donald Lowndes has dense spikes of deep pink and rose in summer and autumn, and forms mats which do not spread so quickly that it becomes invasive. Sun or partially shady places are equally suitable. Darjeeling Red has bright, deep red flowers and is slightly taller than the previously mentioned variety, up to 10in. *P. vacciniifolium* is particularly good and has masses of rose-pink flowers in spikes up to 9in. tall from late summer onwards, well into autumn. Cultivation is easy, as they will grow in any soil, in sun or partial shade.

Primula. Many of these make excellent rock garden subjects; the genus is a very large one and only a few can be mentioned here. All prefer a well-drained peaty loam incorporating plenty of sand and stone chippings. A well-known primula is *P. denticulata*, up to 12in. tall with round lavender-blue flower heads in early spring. This species prefers moist soil. *P.* Garryarde Guinevere, 6in. tall, has bronzy leaves and large pink, primrose-like flowers and needs plenty of moisture and shade from strong sunshine. *P. marginata*, up to 6in. tall, has lavender flowers and white-edged leaves, with slightly fragrant flowers in March; a well-known variety is the mauve Linda Pope. *P. rosea*, rose coloured, is 6in. tall; *P. veris*, the Cowslip, deep yellow, to 9in.; *P. viallii*, 1ft., a very distinctive and handsome primula, with purple and red flowers; and finally *P. viscosa* with violet flowers in spring and rather sticky leaves, 6in. tall.

Pulsatilla. One of the loveliest flowers for the rock garden is *Pulsatilla vulgaris*, the Pasque flower. This British native has violet flowers with yellow stamen and leaves a delightful softly-hairy texture. The flowers appear in April and May and there are forms with flowers from purple, red and pink to almost white in colour.

Rock Cress, see Arabis

Rock Jasmine, see Androsace

Saxifraga (Saxifrage). This is one of the most important genera for the rock garden as it abounds in first-rate species and garden-raised varieties, many of which flower very early in the year. From the cultural viewpoint it is necessary to divide the genus into a number of subsections, as saxifrages differ considerably in their requirements.

The principal sections are: Cushion Saxifrages

Saxifrages and sempervivums (Houseleeks) in a trough. Encrusted saxifrages prefer sunshine but will grow in northern aspects. Houseleeks succeed almost anywhere

or Kabschias, which make dense hummocks or mats of growth studded in early spring with small flowers on short stalks; Encrusted saxifrages or Silver saxifrages which make handsome rosettes of silvery leaves and have sprays of flowers anything from 6in. to 2ft. in height, flowering mainly in late spring and early summer; and Mossy saxifrages, with mounds or mats of soft green leaves and flowers on 6- to 9-in. stems in mid-spring. Other species include *S. umbrosa* (London Pride), with rosettes of green leaves and 6- to 12-in. sprays of pink flowers in late spring lasting into May and June; and *S. oppositifolia* with carpets of growth and practically stemless flowers in spring. From the many varieties available, it is best to make a personal choice from the lists of those nurserymen who supply rock garden plants.

So far as cultivation is concerned, Encrusted saxifrages prefer sunny conditions in rock gardens, but will grow with northern aspects. Sand and limestone chippings should be mixed freely with the soil before planting in spring. For Mossy saxifrages, a cool, rather shady place is best, and the soil should be well supplied with humus. Their

soft green masses of growth often die off during the summer if planted in hot, dry places. Such plants should be lifted after flowering, divided, and replanted after the soil has been improved by the addition of leafmould or peat. All saxifrages of this type delight in plenty of moisture and are unhappy in sun-baked places. They can be planted in March or immediately after flowering. Cushion saxifrages delight in a very well-drained compost containing plenty of sand, stone chippings and either leafmould or peat. They should be given an open but not unduly exposed position. *Saxifraga apiculata* is quick growing, forming large mats of yellow flowers in spring.

Sea Pink, see Armeria

Sedum (Stonecrop). This is a large genus containing many useful species and varieties. Stonecrops have succulent leaves and in consequence are able to grow in dry, sunny places. Some suitable ones for the rock garden are: *S. dasyphyllum*, minute greyish leaves, bluish-pink flowers in June, 2in., rather a tufted habit; *S. lydium*, 2in., forms mats of rosettes, with reddish leaves and white flowers in June; *S. cauticola*, 3 to 4in. trailing, with crimson flowers in September; and *S. spathulifolium*, 3 to 5in., with rosettes of fleshy leaves, tinged red, and yellow flowers in May and June. *S. spurium* is 6in. high, and a good variety is Schorbusser Blut with dark red flowers in July and August and bronze tinted leaves. *S. spathulifolium purpurium* is an excellent plant, with 2in. deep purple flowers.

Sempervivum (Houseleek). The houseleeks grow with practically no attention and where nothing else will succeed. *S. arachnoideum* is the Cobweb Houseleek, with tight rosettes of pointed fleshy leaves, covered in 'cobwebs', and leafy stems with bright rosy flowers in July. The variety Commander Hay has green and plum-coloured rosettes; *S. tectorum*, the St. Patrick's Cabbage, has rosettes 2 to 6in. across, and purplish flowers on 10-in. stems in July. *S. pumilum* has rosettes only ½in. across with green hairy leaves and rosy-purple flowers on 2-in. stems in July. Gritty soil and full sun suits them best.

Shortia. Not an easy plant to grow, but so attractive that it is worth trying. *S. uniflora* is shell pink, spring flowering and 3in. tall, and its evergreen leaves become bronzy-red in autumn. Shortias do best in partial shade, and a rather moist, though well-drained soil containing a mixture of sandy peat and leaf-mould. Water freely in dry weather.

Stonecress, see Aethionema
Stonecrop, see Sedum
Sun Rose, see Helianthemum
Thrift, see Armeria
Thymus (Thyme). The fragrance and easy cultivation of the thymes make them popular plants for the rock garden, particularly as they can be

used for cooking, to add that certain something. *T. vulgaris aureus* has bright yellow leaves, *T. v.* Silver Queen, white variegation. Both reach a height of about 6 to 8in. *T. serpyllum* is the creeping thyme with small leaves which forms mats close to the soil. *T.s. albus* has white flowers; Annie Hall is pink, and *coccineus* red. Ordinary soil, preferably in the sun, suits these plants.

Viola (Pansy, Heart's-ease, Violet). The genus that includes the violets has some attractive species for the rock garden among which are *V. cornuta*, violet-blue flowers, 4 to 9in., May to October, tall and summer flowering, and *V. c. alba*, white; *V. gracilis* is deep violet, 4 to 6in.; and *V. labrodorica purpurea*, with deep violet-blue flowers and purple-tinged leaves, is 4 to 6in. tall and provides its display in spring. *V. saxatilis aetolica*, a prostrate, deep yellow species, flowers in May and June and is 2in. tall. Supply a soil enriched with peat or leafmould and plenty of light, but not full sun.

Wall Cress, see Arabis

Garden Pools

A garden pool, however small, can be one of the most attractive features in the garden. Such pools are not difficult to construct in concrete, and once the edges have been disguised with plants or paving slabs, it does not look nearly such a harsh substance.

Nowadays it is possible to buy pools, formal or informal, already made in glass fibre and plastic. These are available in various sizes and shapes and it is only necessary to dig a hole to receive them. Pools can also be made with the aid of heavy-grade polythene sheeting. The method of construction is simple. A hole of the right shape and size is dug and the bottom and sides beaten flat, taking out any sharp stones in the process. The polythene sheeting is placed in position and a little water run in to hold it in place. Then the edges of the sheeting, which should overlap the pond edges by a foot, are folded under and held in place with flat paving stones. The pool can now be filled, although planting is done most satisfactorily when pools are empty.

Many plants, such as water-lilies and other aquatics, are best if planted either in beds on the bottom of the pool or in baskets which are weighted with stones. In any case it is best to plant in a fibrous, turfy loam. Water-lilies are available for pond depths of less than a foot to $2\frac{1}{2}$ to 3ft. and when ordering make sure you get the kinds suitable to your pond. Excellent varieties for pools of medium depth, 18in. to 2ft. or so, include the red James Brydon, the yellow *N. marliacea chromatella*, the deep yellow *N. odorata sulphurea* and the rosy-red *N. odorata turicensis*, and the rose *N. laydekeri lilacea*. The very attractive yellow *N.*

A pool is always an asset to a garden. In addition to the water-lilies and other plants which can be grown, reflections are a source of interest

pygmaea helvola is one of those which can be grown in only a few inches of water (4 to 12in.).

But in addition to the water-lilies there are other suitable plants including the Water Hawthorn (*Aponogeton distachyus*), the Water Violet (*Hottonia palustris*), the Golden Club (*Orontium aquaticum*), and floating plants such as Frog Bit (*Hydrocharis morsus-ranae*) and Water Soldier (*Stratiotes aloides*). If you are going to have fish in the pond you should also have a few oxygenating plants, like the Water Violet mentioned above, Blanket Weed and elodea. These help to provide oxygen for the fish and, although they are often referred to as water-weeds, they are a necessary part of the pond planting scheme. It may occasionally be necessary to remove surplus oxygenating weeds such as Blanket Weed, with a rake or fork. Pieces of broken off and floating weed can be easily removed by using a small fishing net.

Fish and water snails help to keep the water clean and will provide a good deal of interest in themselves. Breeding pairs of goldfish and shubunkins are not expensive. In some districts it may be necessary to take precautions against herons by stretching nets below the surface of the water.

Planting Water-lilies and other Aquatics.
One way of planting a water-lily is to make a special bed on the bottom of the pool. Pieces of stone can be used to make the bed which is filled with good fibrous loam in which the water-lily is planted. A few stones placed over the top will hold the soil and the plant in place. Less elaborate planting is needed for such oxygenating plants as elodea. A rubber band placed round the root ball will be sufficient to hold it together. It is then placed at the bottom of the pool and water is run in gently from a hose.

Some Waterside Plants

Many waterside or bog plants are most colourful, and if thought is given to their siting, remarkably beautiful effects may be achieved by reflections in

stems and the leaves are dark green and glossy. There are various forms including a fine double variety with long-lasting flowers, *C.p. flore pleno.*
Day Lily
(see Hemerocallis, Chapter Five, p. 114).
Giant Cow Parsnip (see Heracleum).
Globe Flower
(see Trollius, Chapter Five, p. 123).
Gunnera. Where space permits its use, *Gunnera manicata* is a spectacular foliage plant for the waterside. Its rhubarb-like leaves can be as much as 6ft. across. It needs a rich loamy soil and in autumn the crowns should be covered with dry straw or bracken for protection, or they can be covered with their own leaves if these are cut off and inverted. The leaves used in this way must be fixed in position with netting, branches or in some other way.
Hemerocallis
(Day Lily, see Chapter Five, p. 114).

Caltha palustris flore-pleno, the decorative double-flowered variety of the Marsh Marigold or Kingcup. The golden-yellow flowers provide spring colour in the water garden

Where space is available—the leaves can be as much as 6ft. across—the spectacular *Gunnera manicata* should certainly be considered for waterside planting

the water, particularly that of a placid pool. But the colour should not be overdone, or it will destroy the cool restfulness one associates with water in the garden. The feathery heads of pink, crimson or white astilbes standing above the mass of handsome green foliage will be doubly effective when seen across a pool, and the elegant heads of the Japanese *Iris kaempferi* are serenely beautiful when growing beside water. Generally speaking, planting is best done in early spring and the finest effect is produced with bold clumps of one variety, rather than dotting odd specimens here and there.
Astilbe. These graceful plants thrive in moist soil beside and above water-level in sun or partial shade (see Chapter Five, p. 112, for description of varieties).
Caltha (Marsh Marigold or Kingcup). One of the earliest plants to flower in the water garden, *C. palustris* bears rich golden flowers on 12-in.

Heracleum. Given sufficient space, the Giant Cow Parsnip, *H. mantegazzianum,* with its large leaves and large flat heads of white flowers in summer is a plant to provide interest in the middle distance of streamside plantings. But it is 10ft. tall and is nearly as much across so only suitable for spacious settings. Also, if well suited, it can seed itself freely—so use with discretion.
Hosta (Plantain Lily). These are admirable for planting near water where their large, prominently veined, cool-looking leaves are most effective. (See Chapter Five, p. 115, for further details.)
Iris. A very large genus, some of which are waterside plants. Among the better known are the Japanese *I. kaempferi* with elegant shades of violet, purple, plum, yellow and white, borne in July on erect stems 2ft. or more in height. They require a lime-free soil, as does the rich blue *I. laevigata* which will grow in water about 3in.

deep, or on the water's edge where its roots can go down below water level. The dainty *I. sibirica* is a delight in June with its bluish-purple flowers on slender 3-ft. stems. There are several good named hybrids, the flowers of Perry's Blue being a most attractive sky-blue colour with a touch of white on the falls, while Caesar is a large, violet-purple. Where space permits the Yellow Flag, *Iris pseudacorus,* is worth planting in shallow water or at the edges of pools. This is the iris which may be seen growing wild in ponds and ditches. The flowers open together and continue throughout May and June on stems up to about 3ft. high.

Kingcup, see Caltha

Lysichitum (Skunk Cabbage). The yellow-flowered *L. americanum* has most striking arum-like flowers which it bears in April and these are followed by enormous deep green leaves. The flowers have a foul smell and should never be cut

by boggy soil it thrives luxuriantly. Planting is best done in the spring just when growth starts. It will grow in partial shade or in sun so long as the roots have adequate moisture. It forms a very distinctive and attractive plant.

Plantain Lily, see Hosta

Primula. A very large genus, many of the members of which thrive in bog gardens. Once established some of the primulas will seed themselves happily and soon form colourful colonies. They like moisture at the roots but should not be 'stuck in the mud'. One of the earliest to flower, in April, is *Primula rosea,* with bright pink flowers on 6-in. stems above tufts of pale green leaves. This is a real bog plant. Earlier still is *P. denticulata* (see Chapter Five, p. 121, for details). Taller growing, up to 18in., is *P. beesiana* (also described on p. 121, together with such other fine species as *japonica* and *pulverulenta*). Another fine primula is *P. florindae* with soft yellow, droop-

Primula denticulata, known as the Drumstick Primula, is a splendid plant for waterside planting. The flowers may be of pale purple, lavender or white colouring

Trillium grandiflorum, the attractive Wake Robin, bears its white, three-petalled flowers in spring and needs partial shade and moist, rich soil to give of its best

for use indoors. It is perfectly hardy and revels in thick mud at the water's edge. There is also a white species, *L. camtschatcense,* which flowers a few weeks later. Both are plants for a large garden.

March Marigold, see Caltha

Mimulus (Musk). These gay moisture-loving plants produce a profusion of flower throughout the summer. There are numerous species and varieties, both perennial and annual, the latter being easily raised from seed sown in damp soil at the water's edge. The markings on the little trumpet-like flowers in shades of orange, yellow, red and pink are delightful, and there are self-colour varieties.

Musk, see Mimulus

Osmunda. The Royal Fern, *Osmunda regalis,* is a magnificent spectacle with its large, handsome fronds, perhaps 4ft. in length, swaying beside a pool. When planted on a small mound surrounded

ing flowers carried on 3-ft. stems in July and August.

Royal Fern, see Osmunda

Skunk Cabbage, see Lysichitum

Trillium (Wood Lily). These spring-flowering woodland plants do well in partial shade and moist soil near, but above, the water. They like a rich leafy soil and once planted should be left undisturbed. *T. grandiflorum,* the Wake Robin, has large pure white three-petalled flowers and there are other North American species, some of which are still rare in cultivation.

Trollius

(Globe Flower, see Chapter Five, p. 123)

Wake Robin, see Trillium

Wood Lily, see Trillium

For the control of pests and diseases see Chapter Thirteen (p. 297).

The Greenhouse 10

The popularity of greenhouse gardening today is proof of its pleasures. Even a modestly sized and equipped greenhouse can extend one's gardening horizons in a quite remarkable way. It allows one to grow many plants otherwise beyond one's scope, and the opportunities it provides for home propagation add a new dimension to one's gardening interests.

A greenhouse can easily be given year-round interest but it is never more appreciated than in winter and early spring before the garden has come really to life again and colour is everywhere. Even with a small greenhouse there is so much you can do to brighten the dull days for flowering bulbs like the daffodils, hyacinths, freesias and tulips, and other plants like cinerarias and primulas need only modest heat. And if one is prepared to raise the temperature a little more it is easy enough to have cyclamen and poinsettias in bloom.

Of the summer flowers, my thoughts turn first to the fuchsias and regal pelargoniums and, for autumn, the glorious chrysanthemums, to which a chapter is devoted elsewhere (see pp. 183 to 202).

Perhaps the most fascinating and challenging aspect of greenhouse gardening is that one makes one's own climate and, by manipulating the heating, ventilation and damping down, can provide the conditions required by the plants one chooses to grow. As with everything else experience has to be won the hard way. One learns by one's mistakes. Surprisingly quickly, though, most gardeners learn to know the idiosyncracies of their equipment and the quirks of their plants and soon achieve the success which it is so much desired to have with any hobby.

If one can afford it the heated greenhouse is the choice every time, but I must make the point that an unheated greenhouse also allows one to grow many showy and interesting plants superbly well. In winter bulbous plants like the miniature cyclamen and small narcissi, which are perfectly happy in the open garden, can be flowered without the risk of their flowers being damaged by bad weather. And in summer plants like fuchsias and tuberous-rooted begonias, which are normally brought on in a heated greenhouse, can be flowered a little later in the greenhouse which must make do with the natural heat of the sun.

Types of Greenhouses

When deciding on a greenhouse there are various factors to take into consideration and the most important of these is the kind of plants you want to grow. Are they tall growing or will they be pot plants for growing on staging; will you want to grow, say, a wall-trained peach or nectarine? These and other considerations must be thought about first, and I shall discuss the advantages of the different types of house available. Also there is the question of metal or wood. The former has advantages in that the non-rusting alloy frames are slim, to let in maximum light, and impervious to the weather, but wood has a more sympathetic appearance in many gardener's eyes and in the case of such hardwoods as Western Red Cedar does not need painting or other maintenance.

The Span-roof Greenhouse. This is the most adaptable kind of greenhouse for it is suited to a wide range of plants and there are various modifications of the basic design. For example, the standard span-roof house has low walls of brick or wood with staging on both sides at the level where the glass starts. Alternatively, one can obtain such houses with staging for pot plants and a wall on one side and glass to the ground on the other, to allow plants to be grown in a bed. Another alternative is to have glass on both sides to ground level to grow crops such as lettuce or tomatoes in

Below. Part of a span-roof greenhouse—the most adaptable kind available. It is suited for a wide range of plants and there are various modifications of the basic design
Bottom. Part of a lean-to greenhouse, a type which is economical to buy and provides a house wall—through which some heat is transferred—on which plants can be trained

beds on the floor and to follow these with chrysanthemums in the autumn.

The Lean-to Greenhouse. This type is also popular for it makes use of the house wall on one side; saves money, of course, because the materials needed are less; and provides a wall on which one can train, say, a peach or nectrine or an ornamental climber. The house wall will also transfer some heat.

The Conservatory Greenhouse. This kind of house is once again becoming popular, but in a lighter, airier style than its Victorian forerunner. My own house of this type gives me enormous pleasure, and if such a house is separated from a living room by a glass door the ease of access and the visual enjoyment is obviously far greater than it could otherwise be.

Dutch-Light Greenhouse. This type of greenhouse, constructed basically from standard Dutch Light panels, is excellent for tomatoes, lettuces and other crop plants which benefit greatly from the high light factor associated with its large panes of glass.

Three-quarter Span Greenhouse. This type, which is a cross between the span-roof house and the lean-to, has many advantages but it is not seen as much nowadays as it formerly was. Like the lean-to it makes use of a wall for one of its sides, but the house in this case is higher than the wall and the short span from the ridge to the wall allows more light into the house than does the type just mentioned. Also this short span, or sections of it, can be hinged to allow top ventilation.

Other Considerations. Having outlined the advantages offered by different types of greenhouse let us now briefly consider one or two other points. Height is one of the key factors in successful greenhouse cultivation but it must not be bought at too high a price. Where the glass reaches to the ground the light availability must be better, but the heat loss is also greater. Also with large glass sections, as in a Dutch Light greenhouse, and a southern exposure, the temperature will rise rapidly in summer with sun heat and the consequent temperature variations during the day may not be too easy to control.

Door fittings and ease of access are something I always like to pay close attention to as well. Most important I consider is that the door should allow a wheelbarrow entry without the contortions impossible to make if it is carrying a load of plants or compost.

More important is the amount of ventilation the greenhouse you are interested in provides for. Ventilation must be adequate for the kind of plants you intend to grow. Roof ventilators are essential and side ventilators in addition will give more precise control of the temperature and air circulation. I go into this subject more thoroughly on p. 229

Choosing a Site

As with so many things in gardening, choosing a site for a greenhouse is largely a matter of common-sense. For obvious reasons it should be near the house (think of all that foul winter weather!) and within easy reach of electricity and water supplies, neglecting this point can be a costly business. Equally important it should not be sited so that the house or trees cast unwelcome shade. For most of the year the need is going to be for more light than can be provided so that the site chosen must be the lightest possible, which conforms with the second criterion I mentioned. Try also to avoid draughty sites for cold winds can lower the internal house temperature significantly; or if such a site is especially good for other reasons, do what you can to eliminate the draughts by screening with a hedge or fencing. North and east winds are, of course, the most damaging ones.

Lean-to and three-quarter span greenhouses are usually placed against south-facing walls. North-facing walls are not suitable as the small amount of light they receive limits too severely the number of plants which can be grown successfully.

Heating

As I have already remarked, quite a wide range of plants can be grown in an unheated greenhouse, but as frost cannot be excluded it cannot be used as a permanent house for tender plants. Provide artificial heat so that frost can be excluded in even the coldest weather and the range of plants which can be grown will be increased tremendously. Go one step further and maintain a minimum temperature of 7°C. (45°F.) and the scope is still wider. However, to put things in perspective, let me hasten to add that many tropical plants need even higher minimum temperatures and the cost of supplying these extra degrees of heat rises disproportionately.

Electrical Heating. This is the cleanest, safest (from the plant's point of view), easiest and most adaptable form of heating for greenhouse, but it is also likely to be the most costly. It is important that all installation should be by a qualified electrician.

Tubular heaters have much the same capacity for even distribution of heat as hot water pipes; the usual loading is 60 watts per foot of tube and they can be installed singly or in banks. Compact fan-assisted heaters can be used and these have the advantage of being portable and of circulating the hot air throughout the house. Another form of electrical heating employs an immersion heater to warm the water in ordinary 4-in. water pipes, as with a solid-fuel boiler system but without the boiler. This system incorporates a thermostat.

Special thermostats are available for greenhouse use and the most sensitive of these is the rod-type which should be mounted as nearly as possible in the middle of the house where it will register the mean temperature. If shaded from the direct rays of the sun by silver foil it will record the air temperature more accurately.

Soil warming is another form of greenhouse heating which is especially valuable. It is used to provide early crops when the plants are grown direct in the greenhouse border, and for providing bottom heat in propagating cases to hasten the rooting of cuttings and seed germination. Two kinds of electrical soil warming are available: low voltage current reduced by a transformer can be passed through bare wires, or current at full mains voltage can be passed through insulated soil-warming cable. Both bare wire and insulated cable are buried 4 to 6in. deep. Thermostats can be installed in such units, if desired. The loading is usually enough to provide a temperature of about 16°C. (60°F.).

Solid Fuel Boilers. Boilers which burn anthracite, coke or other suitable fuels are widely used and probably the cheapest form of heating but, on the other hand, they are time consuming and need quite a bit of attention. However, if time is not a problem this does not matter.

Metal hot water pipes of 4in. diameter are used with such systems, the water being circulated by thermosyphon action, and these must have a steady rise of about 1in. in 10ft. to their furthest point from the boiler with a corresponding fall on the return journey. As the boilers for small greenhouses are usually installed in an end wall with the flue nearby, changes in wind strength and direction can affect the rate of burning. It is a good plan to reduce this risk by building a screen for the boiler.

Daffodils, cinerarias and cyclamen, which bring colour to the greenhouse when all is bleak outside

Left. The controlled environment of a greenhouse opens up new fields of enjoyment and experience to the gardener. Even quite a small structure can be made to provide year-round interest. For summer display, begonias and pelargoniums (geraniums) are especially colourful

Left, below. The attractive Zonal pelargonium Dagata. This variety is notable for the size of its flower trusses

Right. The gay, curiously pouched flowers of the calceolarias always attract attention. These plants are easily raised from seed and need a minimum winter temperature of 7°C. (45°F.)

Right. The tuberous begonia
Roy Hartley, a variety of
especially attractive
colouring and quality
Below. The brightly marked,
handsome leaves of coleuses,
or Flame Nettles as they are
called, make them popular
with gardeners who can
maintain a minimum winter
temperature of 13°C. (55°F.)

Oil- and Gas-fired Boilers. Solid fuel boilers can be converted to oil burning quite easily and, with thermostatic control, they then need little attention. Gas-fired boilers are also fully automatic and for the busy home gardener are competitive alternatives to electrical forms of heating; with gas, though, it is important to place the boiler and the flue in a position where gas fumes will not be likely to seep into the greenhouse.

Paraffin Heaters. For simplicity and cheapness of installation nothing can beat the portable paraffin heater, but as most gardeners are aware fumes can be a problem. This possibility will be reduced if a heater intended for greenhouse use is purchased, for these have been designed to reduce the risk of fumes to the minimum, and they are fitted with devices to distribute the heat as evenly as possible. Such heaters must be kept scrupulously clean, never be turned too high and have their wicks trimmed regularly. They must also be kept out of draughts.

The commonly held view that paraffin heaters dry the atmosphere is the reverse of the truth. As paraffin burns water vapour is produced and the atmosphere, therefore, remains humid.

Controlling the Temperature

The four prongs of temperature regulation are insulation, thermostatic controls for certain heat installations, shading and ventilation—all in their different ways of importance to good plant cultivation. The difference which good insulation makes to the heating bills is very considerable, and this is something every greenhouse owner should be aware of. A simple method of insulation is to fit polythene sheeting inside the roof glass but one must then be prepared for a certain loss of light which may or may not be significant depending on the plants being cultivated. Even small cracks allow a great deal of heat to escape and cold air to enter, and wooden houses in particular are in need of extra insulation with glass wool, asbestos packing or other suitable heat-insulating materials.

Electrical heater installations and oil- and gas-fired boilers can be thermostatically controlled to the benefit of one's pocket and the well-being of the plants. A maximum-minimum thermometer is also a very valuable piece of equipment for monitoring what actually happens in one's absence.

Shading can now be done automatically, and one must remember that most greenhouse plants require some shading from direct sunlight in the summer. Automatic shading is, however, still rather expensive and most gardeners will still use more traditional methods—fabric or polythene blinds inside the house, wooden lath or split cane outside; or by coating the glass with white-wash or one of the proprietary shading compounds. The most obvious disadvantage of the last-mentioned type of shading is that it is permanent and in dull spells is a nuisance. That most gardeners use this cheap form of shading is, however, an indication that this drawback is not too much of a liability.

Ventilation is necessary to change the air in the greenhouse and keep the temperature from rising too high. It is one of the most important aspects of greenhouse management. Ventilation must be supplied even during winter, but draughts must be avoided. This is achieved by opening the ventilators on the leeward side of the house.

If hinged ventilators are used there should be some mounted near the ridge to let hot air out and some in the sides to let cool air in, especially when the weather is very hot. The aim should be to maintain a fairly steady temperature, but higher by day than at night.

Electric fans can also be used for ventilation and if controlled by thermostat can be completely automatic in operation.

Greenhouse Accessories

Greenhouse staging, the platforms on which plants are stored, is usually fixed so that the plants are about waist level for an average-sized person. Sometimes, too, if the layout of the greenhouse allows, as in a lean-to or three-quarter span house, they are arranged in tiers to allow attractive plant display and make the best use of the space available. The staging may be either open (with spaces left between wooden slats) or solid as in the case of cement, concrete or corrugated iron sheeting bases, these last being covered with a layer of gravel or small stone chippings to retain the moisture.

In recent years there has been a rapid advance in the use of plastic pots as opposed to the traditional clay pots. Light in weight, unbreakable, easier to store and clean, and cheaper than their clay counterparts, they have indeed much to offer the home gardener. Plants grown in plastic pots have been found to dry out less quickly than those in clay pots. Watering techniques for the two kinds of pots are rather different but this presents no problems.

The most frequently used sizes of pot are $3\frac{1}{2}$in., 5in. and 7 or 9in. (for the final potting of large plants). Two-inch pots are useful for rooting single cuttings and seed pans of 6in. diameter are sometimes preferred to seed boxes or trays for certain jobs. Seed boxes with a standard measurement of $14\frac{1}{2}$in. by $8\frac{1}{2}$in. by 2in., once only available in wood, are now offered in plastic and it is thought by some gardeners that seedlings and pricked out plants come on faster in the latter than in wooden boxes.

Coarse and fine roses are available for watering cans, but these are used only when watering very young plants, seeds or cuttings. Other plants are watered from the spout

Damping down the floor in a greenhouse on a sunny day—an easy and effective way of increasing the atmospheric moisture content

Other equipment which is necessary or useful to have in the greenhouse is as follows:

A watering can with both fine and coarse rose attachments.

A ½-in. sieve for compost and a small perforated zinc sieve for covering seeds with compost.

A sharp knife for taking cuttings and a dibber for inserting them.

A widger (a small tool for lifting young plants from seed boxes etc., loosening old topsoil in pots and so on).

A hand sprayer for insecticide and fungicide application.

A measure.

A bucket (many diverse uses).

A 'patter' (a wooden block fitted with a handle for firming compost after seed sowing).

Fillis, stakes and labels.

Watering and Damping Down

It always seems to be watering which, in greenhouse terms, separates the men from the boys. Watering, correctly done, is a skilful job and its mastery is one of the goals of every newcomer to greenhouse gardening.

Coarse and fine roses are available for watering cans, as I have already mentioned, but these are only used when watering very young plants, seeds and cuttings. All other plants should be watered from the spout, this being held close to the soil so that the latter is not displaced. When water is given it should be given in good measure—sufficient to soak right through the pot.

The most difficult thing for the beginner is to tell when plants need water. With clay pots an aural check is a good indication. Make a wooden 'hammer' by pushing a cotton reel on to a cane and be guided by the tone when the pot is struck on its side: if dry it will give out a ringing sound, if wet the response will be dull and muted.

Some gardeners find that lifting the pots is another good guide for, with practice, they can tell by the weight how wet or dry the compost is. With plastic pots this is the method we must adopt for the aural test does not work.

A form of automatic watering which has become very popular is the capillary bench. Like so many good ideas it is essentially a simple device which makes use of plants' ability to absorb water by capillary attraction from a base of damp sand on which the pots are stood. This bench with its sand covering is kept supplied with water automatically by means of perforated piping laid in the sand and connected to a water tank whose water level matches the height of the bench. The plants simply take up water from the bench according to their needs and their water requirements are therefore more accurately met than is possible by human means. Plants in clay pots must have wicks placed in the drainage holes of the pots to ensure capillary action.

It is important that the atmospheric moisture should be correct for the types of plant being grown. Some plants, succulents for example, like a much dryer atmosphere than ferns or foliage plants. There are various ways in which the moisture in the air can be increased: by damping down the paths and the ground under the staging, by having trays filled with water dotted around the house, and by syringing between the pots on the staging and, where the plants enjoy such treatment, on the leaves.

Cleaning and Fumigating

In winter most plants need all the light it is possible to give them and when the shading (if permanent shading has been applied) is removed at the end

A winter job—washing down the greenhouse and removing dirt from between panes of glass with a strip of metal. All the light available is needed at this time

Painting heating pipes with old sump oil. This acts as a preservative and makes them more efficient at radiating heat

of the season it is a very good plan to wash down the inside glass of the greenhouse as well. Use a long-handled broom, warm water and detergent for this job. The dirt which accumulates between the overlap of the glass can be removed by taking a thin strip of metal and slightly bending it so that it can be inserted between the panes and worked along.

In winter all woodwork and walls need scrubbing down with warm water to which some disinfectant has been added but do not let this come in contact with any plants. If possible remove them before work starts. Brick or concrete walls look neater and are certainly more hygienic if they are lime-washed at this time. The light-reflecting properties of the lime-wash will also be beneficial during the duller months.

If there are cast-iron heating pipes in the greenhouse brush these thoroughly with a wire brush and then paint with old sump oil. This will act as a preservative and make the pipes more efficient at radiating heat.

Composts for Seed Sowing and Potting

Several generations of gardeners have been brought up now on the John Innes composts, which, by standardising compost mixtures into just a few different categories for seed sowing and potting, have made it easy to get good results without remembering or looking up countless compost recipes. There are two basic formulae for the John Innes composts—one for seeds, the other for potting and the three ingredients used in them in bulk are loam, peat and sand.

Medium grade loam is needed for John Innes composts, neither too heavy (clay) nor too light (sand) with a pH of about 6·5. Such loam can be

obtained from turves cut with 2 to 3in. of soil attached from a good meadow or building site. The loam is stacked, grass side down, and left for up to a year if possible so that the grass and soil fibres have a chance to decay. It should be riddled through a $\frac{3}{8}$-in. sieve before being used.

The horticultural grade peat used should be granular or fibrous and reasonably free from dust. It must be well watered and made thoroughly moist before use, the watering being repeated several times, if necessary, to obtain the desired result. The sand needed is defined as coarse and damp with particles grading up to $\frac{1}{8}$in. in diameter. Cornish river sand is ideal for the purpose; never use builder's sand.

This compost can be bought from garden stores and nurseries but it is as well to be sure of the source because some so-called John Innes composts offered by dealers bear little resemblance to the real thing. If you make your own the soil should be sterilised by standing it over boiling water in a saucepan or copper, or placing it in one of the special sterilisers which can be obtained through garden stores, etc. The temperature of the soil should be raised to 93°C. (200°F.) and kept at this level for 20 minutes.

The John Innes Seed Compost is prepared by mixing 2 parts sterilised loam, 1 part peat and 1 part sand, all parts by loose bulk. To each bushel of these combined ingredients is added 1½oz. of superphosphate of lime and ¾oz. of either finely ground chalk or limestone. If the compost is to be used for lime-hating plants substitute flowers of sulphur for the ground chalk or limestone, at the same rate. First, the loam is sieved to remove stones. Then the peat and sand are added and finally the other ingredients, carefully measured out, are scattered over the top. The whole heap must then be turned several times.

The John Innes Potting Compost is prepared

in a similar manner, but the proportions of the main ingredients are different: 7 parts loam, 3 parts peat and 2 parts sand. A base fertiliser is added to this. It can be purchased ready mixed or can be made with 2 parts of hoof and horn meal (13% nitrogen), 2 parts superphosphate of lime (18% phosphoric acid) and 1 part sulphate of potash (48% pure potash), all parts by weight. This is added to the other ingredients at the rate of 4oz. per bushel for No. 1 Potting Compost, 8oz. per bushel for No. 2 Potting Compost, and 12oz. per bushel for No. 3 Potting Compost. To No. 1 add ¾oz. of ground chalk, or limestone (except for lime-hating plants) per bushel of mixture; double and treble this amount for Nos. 2 and 3. The No. 1 compost is used for all ordinary purposes, No. 2 for older plants in pots over 4in. diameter, and the No. 3 for some very strong-growing plants in 10in. pots.

Soilless Composts. The increasing shortage of loam has resulted in a search being made for an ingredient which will take its place, and it has been found that a mixture of peat and coarse sand with fertilisers added provides good growing conditions for plants. These soilless composts are now sold under proprietary names. It is interesting that some plants with a reputation for being tricky to grow are proving easier to manage in these composts. However, at present they are most suitable for propagation purposes and growing smaller plants. Large plants like chrysanthemums are too heavy for these composts to support firmly. On the other hand, some plants, such as African Violets, do particularly well in them.

Propagation

I have dealt with seed sowing, the growing of seedlings under glass and the rooting of softwood cuttings in Chapter Five (Perennials, Annuals and Biennials), pp. 106 to 108. Rooting half-ripe cuttings of trees and shrubs in cold frames and by mist propagation, and the raising of these plants from seed sown under glass, is described in Chapter Three, pp. 44 to 46. The propagation of dahlias is described in Chapter Seven, pp. 174 to 181, and that of chrysanthemums in Chapter Eight pp. 195 to 201.

Hanging Baskets

Perhaps one associates hanging baskets most often with outdoor displays but they are, of course, splendid for greenhouse decoration as well. Fuchsias, pendulous begonias, Ivy-leaved pelargoniums and *Impatiens sultanii* (the Busy Lizzie) are some of the best and most popular plants used for this purpose.

Early spring is the time to prepare such baskets and the moss lining, which should be thick, is the

Hanging baskets planted with such plants as pendulous begonias are admirable for greenhouse display. Regular feeding in summer is essential

framework on which one starts to build. John Innes No. 1 Potting Compost is suitable for this kind of planting, and planting should proceed as the compost is built up. It is then easy to plant through the sides of the basket and end up with an extremely effective display of plants arching over the sides and building up to a well-furnished crown of more upright-growing plants. Some of the plants planted near the edge or top of the basket should be angled so that the new growth hangs over the side.

If polythene film, with drainage holes made in the base, or puddled clay, is placed round the inside rim of the basket this will aid watering, which in hot weather will need to be frequent. With so many plants packed into such a relatively small area (baskets of 12- or 14-in. diameter are usually used) the demands on available soil moisture are heavy. For the same reason regular feeding with a soluble fertiliser is, I consider, necessary every seven to ten days during the summer. This is well worthwhile for given such treatment planted baskets will keep up a fine display well into the autumn and give a great deal of pleasure.

Such baskets look best, of course, if they can be hung from a suitably rigid beam, but this is not always possible and the alternative is to stand them on the staging on large pots which provide a firm base.

Abutilon vitifolium, the Indian Mallow. This evergreen shrub has lilac-blue flowers and there is a white variety, *album.* It will reach a height of 10 to 15ft.

Aechmea rhodocyanea, a handsome bromeliad for growing in a humid greenhouse. Plenty of warmth is needed for the aechmeas to produce flowers

List of Greenhouse Plants

In the list of plants which follows, I have described as wide a range as possible, including foliage plants which provide such a good background for the brighter colours of the flowers. Bulbs, corms and tubers are not included for I have described those suitable for greenhouse cultivation in Chapter Six (pp. 154 to 166).

Abutilon (Indian Mallow). These are evergreen shrubs which have funnel-shaped flowers and handsome foliage. The first two mentioned below can be grown out of doors in sheltered, warm parts of the country, but are much more successful if grown in a greenhouse. *A. megapotamicum* has bell-shaped, red and yellow flowers during spring and summer, and will grow up to 8ft. tall, though 5 to 6ft. is more likely. If grown outdoors it should be planted against a sunny wall. *A. vitifolium* has lilac-blue flowers in summer, and soft grey, rather hairy leaves.

Another well worth growing is *A. striatum thompsonii* with orange-yellow flowers and yellow-variegated leaves.

Use John Innes No. 1 Potting Compost. Give them a sunny position and a temperature of 13° to 18°C. (55° to 65°F.) from March to September and 7° to 10°C. (45° to 50°F.) for the rest of the year. Regular pruning is not essential but overgrown plants can be cut back in March. Increase by seed sown in March in a temperature of 16° to 18°C. (60° to 65°F.), but varieties and hybrids are best increased by cuttings in spring or early summer rooted in a close frame.

Acalypha. Although not often seen, these are fine shrubby plants for a warm greenhouse. Some, such as *A. godseffiana,* 1 to 3ft., have attractive green and white leaves, but *A. hispida* is perhaps the most striking with its long red tassel-like flowers, as much as 12 to 18in. long. To grow them well a minimum winter temperature of 13°C. (55°F.) is needed, but in the summer temperatures of 21° to 27°C. (70° to 80°F.) can be maintained, provided the atmosphere is kept moist by regular syringing and damping. Shade from the sun should also be given in the summer. Use John Innes No. 2 Potting Compost with 1 part extra peat and re-pot when necessary in spring. If pruning is required, this is done in February or March, but only to shape them. Increase by cuttings of young shoots taken in spring and rooted in sandy soil, providing temperatures of 24° to 27°C. (75° to 80°F.) and a close atmosphere. They will root within a few weeks.

Achimenes, see Chapter Six, p. 154

Aechmea. Like most bromeliads the members of this genus have tough, often spiny leaves, arranged in large rosettes, and bizarre, colourful flower heads. Though best grown in a humid greenhouse, and needing plenty of warmth to produce flowers, they are very good house plants, and a plant bought in flower will remain attractive for a long time.

A. fasciata (Billbergia fasciata) is the most commonly grown species with broad grey, silver-mottled leaves and large rosy-pink spiky flower heads in August. Another good species is *A. fulgens*, with scarlet and purple flowers in August and September above recurving, shining green leaves.

Pot in March in a mixture of equal parts peat, leafmould (oak or birch) and coarse sand. Keep the compost moist and maintain water in the cup-like centre, through which the spikes of flowers will emerge. Give a temperature of 21°C. (70°F.) or more in summer, and 16° to 18°C. (60° to 65°F.) in winter, though they will stand 7°C. (45°F.) if not watered too much.

Increase by the suckers produced after flowering, rooted in moist, warm conditions in a temperature of 27°C. (80°F.) if possible, though a lower temperature can be used, but rooting will take longer.

African Violet, see Saintpaulia

Allamanda. This is a climbing plant with evergreen leaves and most attractive, large trumpet-shaped flowers. *A. cathartica* bears golden-yellow flowers in June, and reaches 5 to 10ft. in height. *A. c. hendersonii* has orange-yellow flowers. Grow them in large pots in John Innes Potting Compost or plant in a border of good soil. They should have a minimum temperature of 13°C. (55°F.) in winter, but this can be exceeded in summer. Water freely while growing, but not so much at other times. Prune by cutting back the side shoots to within one or two buds of their base in January or February. Increase by rooting the tips of the shoots removed when pruning; put in a mixture of moist peat and coarse sand in a temperature of 18° to 21°C. (65° to 70°F.).

Amaryllis, see Chapter Six, p. 154

Asparagus. This genus is, of course, best known for its edible member, but it also includes the popular greenhouse species *A. plumosus*, with fine feathery leaves, and *A. sprengeri* with needle-like foliage. Both are easy to grow and can be raised from seed which is sown in pans of light compost in spring and germinated in a temperature of 18°C. (65°F.). The seedlings should be potted singly at an early stage into 3-in. pots and later moved into 5- or 6-in. pots. Grow them in John Innes No. 1 Potting Compost in pots or baskets, preferably in the latter in the case of *A. sprengeri* for it is very handsome when suspended from the greenhouse roof. When pot-grown, *A. plumosus* can be trained up wires to the rafters of the house. Pot in March and syringe with water frequently in the summer. A minimum temperature of 10°C. (50°F.) is required.

Asplenium (Spleenwort). *A. bulbiferum* is a popular greenhouse fern with handsome fronds 1 to 2ft. long. This species produces small plants or bulbils on the veins of the fronds and these can be removed and laid on the surface of a mixture of 2 parts moist peat and 1 part coarse sand in a seed box. Bent pieces of wire can be used to keep the detached bulbils in place. A temperature of 16°C. (60°F.) is required for rooting in a close frame and the resulting plants can be potted into 2-in. pots of John Innes No. 1 Potting Compost when ready for the move. Mature plants grow well in a mixture of 6 parts peat and 1 part each of loam, charcoal and silver sand. Pot in March and water freely in summer. Provide a winter temperature of 10° to 16°C. (50° to 60°F.).

A. nidus, the lovely Bird's Nest Fern, with fresh green undivided fronds, is another good greenhouse plant now also popular as a house plant. Its temperature requirements are similar to those of *A. bulbiferum* but the potting compost most suited to it is a mixture of peat and sphagnum moss in equal proportions. These ferns, as is usual, do not like exposure to direct sunlight.

Azalea. Indian azaleas with their colourful flowers can be seen in most florists' shops in winter and early spring, and if one has a cool greenhouse, and the plants are given little attention, they can be kept for many years. After flowering is over the remains of the withered flowers should be removed and the plants re-potted. They will not tolerate lime or chalk in the soil and so a special potting compost must be used consisting of 2 parts moist peat and 1 part coarse sand, measured by volume. The peaty soil must never be allowed to dry out; if this happens it is extremely difficult to get it properly moist again. New growth soon begins after flowering and it is best to keep the plants in a warm part of the greenhouse; sprays of water over the plants will help to encourage good growth. In districts where the tap water is hard and contains lime it is wise to use rain water for these plants.

From June to September the plants can be plunged outside in their pots. Give them a position in partial shade as they do not like strong sunshine. As the potting soil contains little plant food it is necessary to feed the plants regularly so that they make good growth and flower well. Either use a liquid or soluble fertiliser or provide a dry feed of 2 teaspoonfuls of dried blood to 1 of sulphate of potash for each plant every 14 days in spring and summer.

Watering must be attended to each day when the plants are in the open during the summer, although the compost in the pots will not dry out so rapidly when the pots are plunged in ashes or soil. Overhead sprays of water are also beneficial. As the nights become colder in September, the plants should be returned to a cool greenhouse and sufficient heat should be provided to keep the air fairly dry.

Allamanda cathartica, a climbing evergreen which bears golden-yellow flowers in June

The colourful winter-flowering Indian azaleas are splendid pot plants for the cool greenhouse and can give pleasure over many years

Beloperone guttata, which bears the apt common name of Shrimp Plant, for its decorative feature is the salmon-red bracts which surround the white flowers

The Indian azaleas that are imported into this country from Continental sources, to be forced in time for Christmas, are usually grafted plants, but they can also be increased from cuttings. Choose young shoots that have begun to harden at their base and insert the cuttings made from these in small pots of moist peat and sand. Stand in a warm moist propagating frame until rooted.

Babiana, see Chapter Six, p. 154

Begonia, see Chapter Six, p. 155

Beloperone (Shrimp Plant). The salmon-red bracts which surround the flimsy white flowers of *B. guttata* give it the apt name of Shrimp Plant. In a moderately warm greenhouse they are produced almost continuously, especially in the summer and autumn. The plant reaches a height of $1\frac{1}{2}$ft. Pot in spring in John Innes Potting Compost, and water freely in spring and summer, but less so in autumn and winter. Give it a temperature of 16° to 18°C. (60° to 65°F.) but in winter this can drop to 7°C. (45°F.). If the leaves drop, as they may do with the lower temperature, cut the stem back hard to encourage new growth from the base in spring. Increase by stem cuttings in spring and summer in sandy soil with some bottom heat.

Billbergia. These, like the aechmeas (see p. 233), are members of the bromeliad family; *B. nutans,* with a rosette of long, toothed narrow leaves, up to $1\frac{1}{2}$ft. tall, is almost hardy, but needs a fairly warm greenhouse if it is to produce its curious drooping flowers in yellow, green, red and blue. Give billbergia similar treatment to the aechmeas, and increase similarly by side shoots taken in spring or early summer. Root these in a sandy, peaty compost in a close frame with a temperature of 18° to 24°C. (65° to 75°F.).

Boronia. The most popular species, perhaps because it is fragrant, is *B. megastigma,* a shrubby plant 2ft. or more in height needing a cool greenhouse. It comes from Australia. The brown and yellow flowers appear in spring. John Innes Potting Compost is suitable, and 5- or 6-in. pots are adequate for the final potting. Give them a winter temperature of 7° to 10°C. (45° to 50°F.) and water carefully at all times. Keep the atmosphere rather dry and airy, and put the plants in a sunny position. Re-pot each year when flowering has finished. Increase by cuttings of partly matured shoots in summer, rooting these in a propagating box in a greenhouse.

Browallia. These plants have pretty blue or white flowers, and are not difficult to grow in a frost-proof greenhouse. They reach a height of $1\frac{1}{2}$ to 3ft. *B. demissa* bears blue-violet or white flowers in June and July, and reaches a height of up to $1\frac{1}{2}$ft. *B. speciosa major,* about 2ft. tall, has violet flowers 2in. across in summer. To have plants in flower in winter seed can be sown in July. Young plants can be potted in 3-in. pots but

235

as they grow they should be moved into 5-in. pots, using John Innes Potting Compost. Three plants of *B. demissa* can be grown together in a 6-in. pot to produce large specimen plants. It is wise to pinch out the tips of the shoots of *B. speciosa major* to encourage a bushy habit.

Butterfly Flower, see Schizanthus

Calceolaria (Slipper Flower). The bright and gay colours of the greenhouse calceolarias with their curious pouched flowers always attract attention in the spring. Plants can be easily raised from seed and the time for sowing is May to July to produce plants for flowering the following May. The seed should be sown in pans on the surface of sifted John Innes Seed Compost; no further covering with soil will be required. The pans should be covered with glass and put in a cold frame. The resulting seedlings should be pricked out into boxes and moved when ready into 3-in. pots filled with John Innes No. 1 Potting Com-

Camellia japonica donckelarii, one of many fine camellias for growing in a frost-proof greenhouse. It has white splashes and streaks on its dark red petals

post. In October another move to 5-in. pots can be given. Keep them in a minimum temperature of 7°C. (45°F.) in winter—they do not like high temperatures—but the air must be kept fairly dry, and watering should be sparing otherwise they may damp off. I like to give the plants another move into 6-in. pots in February to get really good specimens and, as the flower buds begin to show, regular feeding with a liquid fertiliser will benefit the plants considerably. Greenfly are often a nuisance on these plants and should be dealt with as soon as they are seen, otherwise they will build up very considerably.

Camellia. Although camellias are hardy enough and they can be grown in sheltered places in the open, they are also fine flowering shrubs for a frost-proof greenhouse, where the blooms are able to develop to perfection without being damaged by the weather. The kinds most often seen are varieties of *C. japonica*, and some particularly good varieties are: Adolphe Audusson, dark red; Contessa Lavinia Maggi, white with rose-cerise stripes; *donckelarii*, dark red with white splashes and streaks; Jupiter, bright red; *magnoliaeflora*, semi-double soft pink; Mars, semi-double dark crimson; and Sode-Gakushi, semi-double, white, with golden stamens. Donation is an excellent hybrid of orchid pink colouring and J. C. Williams is another fine hybrid, blush pink with yellow stamens.

Re-pot the plants in spring, not necessarily every year (once every two or three years will often be found to be all that is required) just before growth begins, and make sure the drainage is good by adding plenty of crocks at the bottom of the pot and a layer of rough peat. Use a lime-free compost.

Stand the plants outside during the summer in a partially shady place, and never allow them to become dry at the roots, otherwise the flower buds may drop off later. Feed regularly through the summer with liquid or soluble fertiliser and give overhead sprays of water. Then, in early autumn, return them to an unheated greenhouse with full ventilation, with a little heat being given in cold weather.

Increase by leaf cuttings, each with a bud and some stem attached. Do this in March, and insert the cuttings round the edges of 3-in. pots in a mixture of 2 parts coarse sand and 1 part moist peat. Water well and put in a warm, moist propagating box. When the cuttings have rooted lift them carefully and pot the young plants singly in 3-in. pots filled with 2 parts peat, 1 part lime-free soil and 1 part coarse sand. All the parts mentioned are by bulk. Leave in a cold frame with shade from strong sunlight in the summer. Use lime-free water always, particularly if rooting the cuttings under a mist propagation unit. Cuttings can also be taken in late July from shoots made in the current season that have begun to harden at the base. Treat as for leaf cuttings.

Cape Primrose, see Streptocarpus

Celosia (Cockscomb, Prince of Wales' Feathers). There are two celosias that are popular for greenhouse decoration in the summer. They are *C. cristata*, the Cockscomb, and *C. c. pyramidalis*. The former have crests of tightly-packed red flowers, and the plants grow 9 to 12in. tall. The other type has attractive plumes of red or yellow flowers about 18in. tall, and there are good strains available of all colours between red and yellow. Treat them as annuals by sowing seeds in pans of well-drained compost in March in a temperature of 18°C. (65°F.). The seedlings are transplanted 2in. apart when 1in. high and are later moved into well-drained 3-in. pots of John Innes No. 1 Potting Compost and kept in a temperature of 16° to 18°C. (60° to 65°F.). The plants are moved again in June into 5-in. pots. Celosias need careful cultivation, and although they must never be allowed to become dry, watering must be done very carefully, particularly when they are young, as overmoist conditions can cause losses. Give an overhead spray twice a day if possible, and feed when the flowers appear.

Chilean Bellflower, see Lapageria

Chlorophytum (Spider Plant). *C. elatum variegatum* is called the Spider Plant because of the rosettes that develop on the ends of the flowering stems. It is an easy plant to grow in a cool greenhouse, and it will also do well as a room plant. It is grown mainly for the attractive green and cream-striped foliage. The flowers, borne on long, arching stems, are insignificant but they are soon followed by the rosettes of leaves which add to the decorative value of the plant. As plants age and decline in vigour the rosettes of leaves can be detached. These are, in fact, tiny plants and root initials are usually already present. If they are placed in small pots of good soil, new roots will soon develop and they can be moved to 5- or 6-in. pots. Chlorophytums have tuberous root systems and can also be increased by separating these and potting them individually. Feed established plants regularly with a liquid or soluble fertiliser as they soon use up all the goodness in their potting soil.

Chrysanthemum, see Chapter Eight

Cineraria. Anyone with a greenhouse with sufficient heat to maintain a minimum temperature in winter of 7°C. (45° F.) can grow cinerarias, and their brightly-coloured daisy flowers give a wonderful splash of colour in late winter and early spring. In small greenhouses it is probably best to grow the Nana Multiflora strains, which grow about 15in. tall. My favourites are the large-flowered Grandiflora hybrids that have large flowers and grow about 2ft. tall. Even taller are the Stellata cinerarias that have a branching habit and masses of small flowers.

Seed can be sown in April, May and June to provide a succession of bloom; plants from an April sowing should flower in December or January. Sow the seeds in John Innes Seed Compost, and when the resulting seedlings are large enough to handle prick them out into boxes of John Innes No. 1 Potting Compost. Before they become overcrowded in the boxes move them individually into 3½-in. pots filled with John Innes No. 1 Potting Compost. The final potting is into 5- to 6-in. pots of John Innes No. 2 Potting Compost. Potting should be done as soon as the roots begin to fill the small pots. At all stages of potting adequate drainage must be provided in the pots with broken pieces of pot and a little rough peat or loam fibre. The final soil level in the pots should be 1in. below the rim to allow for watering.

Cinerarias do not like high temperatures and for the summer they are best kept in a cold frame, preferably standing or plunged in a bed of ashes to prevent rapid drying out. The plants must be given shade from the sun otherwise they will wilt and watering must be attended to carefully. Greenfly are very fond of these plants and at the first signs of attack the plants should be sprayed with a suitable insecticide. The tops of some of the plants can be pinched out if it is desired to delay flowering and so provide a longer display of flowers. Stopping should not be done until the plants are established in their final pots. Feed with a liquid or soluble fertiliser to keep them growing steadily through the late summer and autumn, but this should cease as the flowers appear.

In late September the plants must be taken into the greenhouse and arranged so that each plant has ample room on the staging. Ventilate freely whenever the weather is favourable. When tem-

Celosia cristata pyramidalis in its various strains has showy plumes in colours from brilliant red and orange to yellow. They are treated as annuals

peratures are low watering must be done very carefully. If the soil is kept too moist the stems may rot. The cineraria is a perennial but it is best treated as an annual by raising new plants from seed each year.

Cissus (Kangaroo Vine). Apart from some peculiar succulents the cissus are climbing plants of the vine family. Two of them are tough house plants, *C. antarctica*, the Kangaroo Vine, with large pale green notched leaves, and *C. sicyoides*, which has glossy five-lobed leaves. *C. striata* is a much smaller and more graceful version of the foregoing, but has not the same stamina. The exotic looking *C. discolor* with velvety, silver-mottled leaves, red beneath, will thrive in the really moist, warm greenhouse. *C. adenopodus* is remarkable for the red hairs that cover the young growths and leaves, and *C. gongylodes* for the bright red aerial roots and aerial tubers which it forms in a moist atmosphere. The tropical species such as *C. discolor* need much more warmth and moisture, a winter minimum of 16°C. (60°F.) rising to 21° to 27°C. (70° to 80°F.) in summer, and more water, particularly in summer. *Cissus antarctica*, *C. sicyoides* and *C. striata* prefer cool conditions and shade from full sun, the Kangaroo Vine in particular tending to drop its leaves if it gets too hot. The soil should be allowed to dry out partly between waterings. If a lot of leaves fall, the stem should be cut back to encourage the production of new shoots.

Increase by cuttings rooted in a close, moderately warm propagating frame, in rich, well-drained soil.

Clivia, see Chapter Six, p. 155

Cobaea (Cup and Saucer Vine). A climbing perennial usually grown as an annual, *C. scandens* has violet-greenish, tubular flowers, which are borne in summer. It is also suitable for warm walls outdoors facing south or south-west or for arches and trellises in sheltered, sunny positions. It flowers continuously from June onwards and can climb to as much as 24ft. although in practice it rarely does so.

Grow it in John Innes No. 1 Potting Compost, and give it plenty of water in summer, but much less in winter, where it is kept permanently in the greenhouse, instead of being treated as an annual. The stems should be cut hard back in late winter and when new shoots develop the weakest can be thinned out. The remainder should be allowed to ramble up the greenhouse roof or through a trellis. Re-pot in spring and give a winter temperature of not less than 7° to 10°C. (45° to 50°F.).

Increase by seeds sown in light soil in a temperature of 18°C. (65°F.) in March.

Cockscomb, see Celosia

Codiaeum (Croton). This plant is invariably referred to by its common name, Croton. The greenhouse kinds are all varieties of *C. variegatum*, and, like all crotons, are noted for their foliage which has variegations in various beautiful colours. They need a summer temperature of 16° to 21°C. (60° to 70°F.) and the atmosphere must be kept moist by damping down and syringing. The lighter the position they are given in the greenhouse the better the colour of the foliage will be. Feeding in the growing season with liquid manure or a proprietary soluble feed is beneficial but winter feeding is not necessary and water should also be given sparingly at that time. Pot codiaeums in March in John Innes No. 1 Potting Compost with some extra peat or leafmould added and grow on in a warm greenhouse.

Cuttings from the ends of the shoots can be used to increase stock, these being inserted in sandy soil and rooted in a temperature of 24°C. (75°F.) at any time. Air layering is another form of increase.

Cobaea scandens, the Cup and Saucer Vine. The violet-greenish flowers are borne from June onwards. It needs plenty of water in summer, but much less in winter

Coleus (Flame Nettle). The handsome leaves of *C. blumei* are variegated in many bright colours and it is not surprising that it is so popular with gardeners who can maintain the necessary minimum temperature of 13°C. (55°F.). With lower temperatures than this the plants are liable to drop their leaves. The flowers, of little decorative value, should be pinched out as soon as they appear. The best form of increase is by cuttings of young side shoots, a few inches long, taken from mature plants in spring or summer. These are inserted in small pots of sandy soil and rooted in a propagating box with a temperature of at least 16°C. (60°F.). The young plants are then potted singly into 3-in. pots of John Innes No. 1 Potting Compost and re-potted later into 5- or 6-in. pots. Shade from strong sunshine is needed in summer, and when the temperature is high the atmosphere must be kept humid by damping down the floor and staging. Nip out the tips of the shoots at intervals to induce the desired bushy habit.

Crocosmia, see Chapter Six, p. 156
Crocus, see Chapter Six, p. 156
Croton, see Codiaeum
Cup and Saucer Vine, see Cobaea
Cyclamen, see Chapter Six, p. 156
Euphorbia pulcherrima (Poinsettia). This is a plant known to everybody by its popularity as a florists' plant in mid-winter. Its striking feature, of course, is the large scarlet bracts which surround the insignificant flowers.

In addition to the ordinary poinsettia there are some with even more colourful and larger bracts and others with pink and white bracts. Poinsettias are grown from stem cuttings taken in May and June off plants which were cut back after flowering to induce them to make side shoots suitable for this purpose. The plants are then kept in a warm, moist atmosphere and the young shoots

The well-known Poinsettia *(Euphorbia pulcherrima)* which gives so much pleasure with its large scarlet bracts in mid-winter

atrocaeruleum is more richly coloured. It reaches a height of about 1ft. and is rather bushy. A biennial, it does best when raised from seed sown in August, the resulting plants being grown on in 3-in. pots through the winter, after which they are re-potted into 5-in. pots. Smaller plants can be obtained by sowing in March for flowering in the same year. Use a compost of equal parts peat and loam with enough sand to keep it open. Good drainage and careful watering are essential as the plants are inclined to damp off rather easily, and shade from hot sun is advisable. For plants sown in August a winter temperature of 16°C. (60°F.) is required.

False Castor Oil Plant, see Fatsia

Fatsia (False Castor Oil Plant). The evergreen *F. japonica* is an excellent foliage plant for the cool conservatory or for living rooms. The plants should be potted in spring in John Innes No. 1 Potting Compost, water being given freely between April and September and moderately for the rest of the year. Shade is necessary in summer and the winter temperature should not fall below 4°C. (40°F.). Increase is by firm young cuttings inserted in sandy soil in a propagating frame with bottom heat during late summer.

Flame Nettle, see Coleus

Freesia, see Chapter Six, p. 156

Fuchsia. The fuchsia is one of my favourite greenhouse flowers and there can be few plants that give such a long display of bloom in a cool greenhouse for the attention they need. They are easy to grow and they should be within the capabilities of most amateur gardeners. Apart from growing plants in pots there are numerous varieties with a pendulous habit that are excellent for hanging baskets. Pot-grown standard fuchsias are also extremely decorative. Some vigorous upright varieties, such as Rose of Castille and Duchess of Albany, can be treated as permanent climbers in the greenhouse if they are planted in a border and a framework of branches is built up under the greenhouse roof—a large plant in full bloom is certainly a magnificent sight.

Varieties I like for hanging baskets include the white and deep carmine Cascade, the creamy-white and rosy-cerise Mrs Marshall, the crimson-scarlet Marinka, and the pink and purple Lena. For growing in pots some of my favourites are the purple and red, striped with rose, Peppermint Stick, the red and mauve Tennessee Waltz, the red and white Swingtime, the salmon-orange Lord Lonsdale, the red and purple Winston Churchill, the pink and white White Spider, the white Flying Cloud and the old red and white Ballet Girl—still one of the best varieties.

Training standard fuchsias is an interesting experience. It is best to start with young plants and these should not be stopped until the main stem has reached the desired 3 to 4ft. in height.

removed when they are 3 to 4in. long. These are prepared in the usual way and the cut ends dipped in sand to prevent 'bleeding'. They are then inserted round the edge of pots filled with a mixture of loam, sand and peat. Poinsettias can also be increased by cutting up the stem, made in the previous year, into pieces about 2in. long. Treated like ordinary cuttings they soon form roots.

Plants raised from cuttings in May and June will make plants some 6ft. tall by the end of the year but if smaller plants are required, about 12in. tall, cuttings can be taken in August. Plants of this size can be grown in a 5- or 6-in. pot.

The pots containing the cuttings should be placed in a propagating box or frame with bottom heat of 21°C. (70°F.) or more. When well rooted they are potted individually, first into 3½-in. pots and later into 6- or 8-in. pots, using John Innes No. 1 Potting Compost for the first and No. 2 for the second potting. Grown at first in a warm rather humid atmosphere, they can be gradually accustomed to a lower temperature so that during August and September they can be stood in a frame. They must be returned to the greenhouse before there is any danger of frost and a temperature of 16° to 18°C. (60° to 65°F.) should be maintained. To obtain the finest possible bracts, restrict the plants to one stem each, and feed with a soluble or liquid fertiliser from late September until flowering time.

Water very carefully as dryness at the roots causes the leaves to turn yellow and drop. Temperature fluctuations also have the same effect. After flowering plants can be cut back and re-potted unless it is preferred to rely on fresh stock from cuttings.

Exacum. The species commonly grown, *E. affine*, bears a mass of bluish-violet sweetly-scented flowers in summer and autumn. The variety

Hydrangea macrophylla and its varieties are popular shrubs for garden or greenhouse. To obtain the best blooms, plants should be raised annually from cuttings

This stem should be supported with a cane and as the plant develops it will need potting on into a 5-in. pot and subsequently in one of 6- or 7-in. size. For the final potting John Innes No. 2 Potting Compost should be used, this being especially well firmed. If a main stem of 3ft. is required allow the plant to reach $3\frac{1}{2}$ft. before taking out the tip. Side shoots will form after this is done which must be stopped in turn so that a shapely plant is obtained.

Apart from the hybrid fuchsias there are also some interesting species for the greenhouse. *F. corymbiflora* is a handsome plant when carrying its 3-in. long deep red flowers, and *F. fulgens* is another species with long and slender flowers, in this case of scarlet colouring. Both exceed 4ft. in height.

Fuchsia cuttings are taken in July and August, to provide plants for flowering the following summer. These are rooted in a temperature of 16° to 18°C. (60° to 65°F.). Young side shoots, a few inches long, are removed from the plants and any flowers on them should be taken off. The base of each cutting is trimmed below a node (or joint) with a sharp knife and some of the lower leaves are removed. Use a compost consisting of equal parts moist granulated peat and coarse sand or one consisting of 1 part good loam, 2 parts moist peat and 3 parts sand. Insert the cuttings round the edges of 3-in. pots. Water well and put the pots in a propagating box in a warm, shaded part of the greenhouse. Inspect each day and remove condensation from the glass. Pot on the rooted cuttings into 3-in. pots in John Innes No. 1 Potting Compost and grow slowly throughout the winter in a warm, light part of the greenhouse. As the plants require it pot them on into 5- or 6-in. pots using John Innes No. 2 Potting Compost.

Stop the plants when they are 6in. high to induce a bushy habit, and stop the resultant side shoots in their turn. Remove any flowers that occur at this time so that plenty of growth will be made before the main flowering season begins. Feed each week in summer with a liquid or soluble fertiliser (this will keep the plants in flower well into the autumn), and water freely. Give light shade from strong sunshine and damp down regularly. Partially rest the plants in winter, giving only sufficient water to prevent the soil ball from drying out completely. Prune in early spring by cutting hard back the side shoots made in the previous year to within two or three joints of their base. Then re-pot the plants into pots of the same size as those they were in previously. Some of the old soil must be teased out from among the roots to allow this to be done.

Galanthus, see Chapter Six, p. 156
Gladiolus, see Chapter Six, p. 156
Gloxinia, see Chapter Six, p. 161
Haemanthus, see Chapter Six, p. 161
Hedera (Ivy). There are numerous ivies with ornamental leaves that make excellent pot plants. They derive mainly from the hardy *H. helix*, the English Ivy, and one of my favourites is the variety Glacier, which has grey leaves with white margins. Another interesting variety is *H. h. cristata* with crimped edges to the pale green leaves. *H. canariensis*, the Canary Ivy, is a delightful plant with large, rounded, bright green leaves and is usually grown in one of its variegated forms. These are also hardy. Give these ivies a cool partially shaded position with good ventilation, and spray overhead regularly. The temperature of the greenhouse in which they are grown should not rise above 10°C. (50°F.) in winter. *H. canariensis* likes to be somewhat warmer, with evenly maintained temperatures.

Increase is easy from tip cuttings or leaf bud cuttings. The former are made from the tips of the stems. The lower leaves are removed and each cutting is trimmed below a leaf joint, leaving it just a few inches long. Several such cuttings can be rooted round the edge of a small pot filled with a mixture of 1 part loam, 2 parts moist peat and 3 parts coarse sand. A leaf bud cutting consists of a leaf with a small piece of stem attached. It is cut just above and below the bud in the joint of the leaf and the cuttings are dibbled in pots or boxes containing the same mixture as that mentioned above. Such cuttings are best taken in spring and summer and rooted in a propagating box inside the greenhouse. After rooting pot them singly in 2-in. pots filled with John Innes No. 1 Potting Compost and later move them on into 3½-in. pots. Provided they are well watered and fed they can be left in these pots for some time and only when they are really pot-bound need they be moved on into 5-in. pots.

Hippeastrum, see Chapter Six, p. 161

Hyacinth, see Chapter Six, p. 162

Hydrangea. The popular hydrangeas, known to every gardener, are hardy or slightly tender deciduous shrubby plants with white, pink or blue flowers in summer. The familiar shrubby varieties seen in gardens and greenhouses are varieties of *H. macrophylla* (for further details see also Chapter Three, p. 62).

The plants should be potted in autumn in John Innes No. 2 Potting Compost and placed in a cool greenhouse or cold frame from then until March. They can be in the greenhouse, a room window or on a warm terrace from April to September. Prune after flowering, cutting out weak shoots and shortening flowering stems to side growths or growth buds. Make an application of a liquid feed weekly to plants showing flower. For details of 'blueing' the blooms see p. 62.

The best blooms are obtained from plants raised from cuttings taken annually in March or April. Some gardeners also like to take cuttings in August. These are raised in sandy soil in a propagating frame with a temperature of 13° to 18°C. (55° to 65°F.). These cuttings should be made about 4in. long and if a propagating frame is not available they can be raised in the open, shaded bench covered with polythene. After rooting move the young plants into 3-in. pots of John Innes No. 1 Potting Compost and then into 5-in. pots of John Innes No. 2 Potting Compost. For blue varieties, however, the chalk in these composts should be replaced by a proprietary hydrangea colourant as their colour is usually poor if lime is present in the soil. As soon as the earliest plants are a few inches high pinch them out to encourage the production of side shoots. Cuttings rooted after the end of May are best grown as single-headed plants without removing the terminal shoots.

Iris, see Chapter Six, p. 162

Indian Mallow, see Abutilon

Ivy, see Hedera

Ixia, see Chapter Six, p. 162

Jasmine, see Jasminum

Jasminum (Jasmine). A good jasmine for the cool greenhouse is *J. polyanthum,* with white, pink-

Jasminum polyanthum, a good jasmine for the cool greenhouse. The very fragrant, white, pink-flushed flowers appear in February and March

flushed flowers in February and March which are extremely fragrant. The pale yellow *J. primulinum* is spring flowering. Both species reach a height of about 10ft. the first mentioned climbing up any suitable support but the latter needing tying in. They can be grown in the greenhouse border or in pots, preferably the former as growth is then usually better. Plant them in February or March, in a well-drained compost consisting of loam, peat and a little sand. Prune them moderately after flowering. During March to October they should be watered freely, for the rest of the year only moderately. Syringe daily during spring and summer. Increase by cuttings of firm shoots inserted in sandy soil in a propagating frame in summer, at a temperature of 16° to 21°C. (60° to 70°F.).

Kangaroo Vine, see Cissus

Lachenalia, see Chapter Six, p. 162

Lapageria (Chilean Bellflower). The only species is *L. rosea*, a delightful climber with pink bell-shaped flowers, whose petals almost look as though they are folded over one another. The flowers appear in autumn. There is a white variety. Plant in March in good lime-free loam or compost; it is essential that the drainage should be first class.

Plants do better in the border than in pots. Shade from strong sunshine. During spring and summer water freely and syringe growths. Give little water in winter, and provide at that time a minimum temperature of 7°C. (45°F.). No pruning is required except to remove weak and overcrowded shoots in early spring. Increase by layers in spring or early summer.

Lilium, see Chapter Six, p. 163
Mallow, Indian, see Abutilon
Matthiola (Stock). Few people seem to grow stocks under glass although they are excellent for a cool greenhouse in late winter. The Beauty of Nice type are particularly recommended and there are numerous varieties with rose, violet, pink, salmon and white flowers. There are also varieties which, by selecting only the lighter coloured seedlings, will produce entirely double flowers. Increase by sowing seed in July or August in pots or boxes of John Innes Seed Compost. Place these in a shaded cold frame until germination takes place. Prick out the resulting seedlings into John Innes No. 1 Potting Compost and keep in a frame so that they have cool, airy conditions. Take the plants into the greenhouse before cold weather arrives and keep in a temperature of 7°C. (45°F.). Pot young plants into 3-in. and then 5-in. pots of John Innes No. 2 Potting Compost. Keep in a light place during the winter and ventilate well. Water sparingly and only when the soil is dry, otherwise the plants will damp off.

Mimulus (Bush Musk). Several colourful perennial musks can be grown in the cool or even cold greenhouse to provide an attractive display but the one which is mostly used for this purpose is the related but quite different *M. glutinosus,* a species from California. This is a half-hardy evergreen shrub some 5ft. tall which bears in summer large trumpet-shaped flowers of pale buff to deep orange colouring. All that is required is that the greenhouse should be frost proof. It may either be pinched out to form a bush or allowed to grow more or less as a climber on a suitable trellis. Pot on as required. Plenty of sun and water are needed and even in winter the plants should not be allowed to become too dry. Increase by cuttings which root readily in moist sandy soil in April.

Morning Glory, see Pharbitis
Muscari, see Chapter Six, p. 163
Narcissus, see Chapter Six, p. 163
Nerine, see Chapter Six, p. 164
Nerium (Oleander, Rose Bay). Evergreen shrubs, up to 10ft. or more high with terminal clusters of large, showy flowers and handsome willow-like foliage. The commonly-grown species, *N. oleander* and *N. odorum*, are both available in shades of white to deep pink, with either single or double flowers. Pot in February or March in John Innes No. 1 Potting Compost or plant in loamy soil in a light sunny greenhouse. Pot-grown plants should be taken outdoors from June to September. Prune immediately after flowering or in October, shortening firm shoots to within 3 or 4in. of their base. From September to March maintain a temperature of 7° to 13°C. (45° to 55°F.). Water heavily from March to September, moderately from then to November and keep nearly dry in winter. Apply a liquid feed once or twice weekly from May to September. Remove young shoots which emerge from the base of flower trusses as soon as they appear. Syringe twice daily from March to June.

Increase by cuttings taken in spring or summer. These cuttings must be made of firm young wood, be 3 to 6in. long and rooted in a propagating frame in a temperature of 16°C. (60°F.). Alternatively, take cuttings of mature wood in summer, place these in bottles of water in full sun and pot them very carefully as soon as roots form.

Oleander, see Nerium

Passiflora (Passion Flower). The Passion Flowers are so named because of the fancied resemblance of parts of the flower to certain features of the Crucifixion. *P. caerulea* is the most commonly grown species and its large flowers, of which the most remarkable feature is the prominent 'corona' of blue-tipped filaments, are well-known. The flowers of this lovely climber only last a day or so, but are produced so freely that there is a constant display from June to September. This species is grown on walls outdoors in sheltered positions and in warm summers the flowers are followed by bright orange fruits. A striking variety called Constance Elliot has ivory-white flowers. Pot in February or March in well-drained tubs or large pots using the John Innes No. 1 Potting Compost, or plant in beds of good loamy soil. Train the shoots up supports and prune in February, thinning out weak shoots and removing strong shoots by about one-third. Water heavily from March to September, but moderately apart from this period. Spray overhead daily from April to September. Feed healthy plants occasionally while in flower but not at other times. Increase by seed sown in sandy soil in a temperature of 16° to 18°C. (60° to 65°F.) in spring, or by cuttings of young shoots inserted in sandy soil in a propagating frame with a temperature of 18°C. (65°F.) between April to September.

Pelargonium (Geranium). There are many types of pelargoniums in cultivation, chief among these being the Zonals, which are splendid bedding and pot plants; the Regal pelargoniums with large flowers in more diverse colours, grown as greenhouse pot plants; climbing Ivy-leaved varieties which are much used for outdoor bedding, hanging baskets and for training on sunny walls; and the more subdued but delightful species with fragrant leaves.

The showy Regal pelargoniums are very easy to grow. Magnificent varieties are available and my two favourites are Grand Slam and Capri, both a lovely shade of bright vermilion orange. Cuttings made from non-flowering shoots 3 to 5in. long are usually, for convenience, taken in July when the plants have finished flowering, but they can in fact be taken at any time during spring or summer. They should be cut cleanly at the base just below a leaf, the lowest leaf or the two lowest leaves then being removed and the base of the cutting dipped in hormone rooting powder. They are then inserted round the edge of 5-in. pots or singly in 2½-in. pots filled with a mixture of 1 part medium loam, 2 parts peat and 3 parts coarse sand with a little extra sand spread on the surface. A little of this surface sand will fall into the holes when the cuttings are dibbled in and will aid rooting. Water the cuttings in with a can fitted with a fine rose. Shade from direct sunlight and syringe with water once or twice a day if the cuttings are being rooted on the open bench, but if this close attention cannot be given root the cuttings in a propagating frame.

The young plants, potted individually in 3 or 3½-in. pots of John Innes No. 1 Potting Compost, should be grown on in a temperature of 10° to 13°C. (50° to 55°F.) if possible, although it can be allowed to fall to 7°C. (45°F.) if necessary. The plants must be kept moist but never sodden in autumn and winter but when spring arrives increase the amount of water given considerably. Re-pot into larger pots as soon as the pots are well filled with roots. Young plants will probably need the 6- to 7-in. size by flowering time. Before this time arrives start feeding with a soluble or liquid fertiliser, applied as directed. It is especially important at this stage that the plants' water requirements are fully met. Cut back the plants quite severely after flowering and stand them in a sunny, sheltered place outdoors or in a deep frame until September, when they are returned to the greenhouse. From this time until new growth appears watering should be considerably reduced.

Although mainly grown as summer bedding plants, the Zonal pelargoniums are also fine pot plants for frost-free greenhouses. Most varieties have a zone of darker colouring on the leaves—hence the name given to this type. Favourites of mine are Irene, and Paul Crampel, vermilion, King of Denmark, Salmon; and Prince of Wales, a pink shade. Cuttings are usually taken in spring or late summer from firm young shoots. The lower leaves are removed and each cutting trimmed below a joint at the base. Several cuttings can be placed around the edge of a 5-in. pot or they may be inserted singly in small pots of sandy soil. If they are stood on the greenhouse staging rooting will soon take place and they should then be moved into 3½-in. pots. Those rooted in late summer should be left in their pots for the winter maintaining a minimum temperature of 7° to 13°C. (45° to 55°F.) and be moved into 5- and 6-in. pots in the spring and summer. The top of the shoots should be nipped out to induce a bushy habit and the side shoots may need to be stopped later.

The plants will be in flower in the following summer but if the flower buds are pinched out until September they can be had in flower during the winter. These plants are best stood outside in a sunny spot in the summer. They soon fill their pots with roots and use up all the goodness in the soil. Feeding is therefore necessary every week with a liquid or soluble fertiliser and watering must be done regularly. Before the end of September the plants must be returned to the frost-proof greenhouse which should be ventilated freely whenever the weather allows. It is not necessary to raise new plants each year. The stems of old plants can be cut back in the spring. Some of the old potting soil should be shaken from the roots so that the plants can be re-potted into 5- or 6-in. pots.

Cuttings of Ivy-leaved varieties, so decorative when used in hanging baskets, are taken in later summer and the young plants grown on in a cool greenhouse with a minimum temperature of 7°C. (45°F.). In February or March the plants will be ready to move into 4-in. pots, the points of main shoots being nipped off at this time. In April or May re-pot again into 5-in. pots or plant out in hanging baskets. For all the stages detailed use a similar compost to that recommended for Zonal pelargoniums. If the plants are grown on in 5-in. pots they will need staking. Water freely in summer and sparingly at other seasons. Feed with liquid manure when in flower. Prune back old plants in February or March. Climbers need only have the laterals cut back. Plants in hanging baskets should be pruned severely to encourage fresh growth. Re-pot the plants as soon as growth re-starts.

Scented-leaved pelargoniums are grown in the same way as Zonal varieties raised for summer flowering, but they do not need a period of rest outdoors in summer as the latter should have.

Pharbitis. This genus includes the well-known and very pretty Morning Glory, *P. tricolor*, which is still usually referred to by its old name of *Ipomoea rubro-caerulea (syn I. tricolor.)*. This half-hardy annual is often grown to flower outdoors but it is also a fine pot plant for the cool greenhouse, flowering from June to September or even later. The brilliant sky blue flowers each last only part of a day, normally closing up during the early afternoon, but if the plant is shaded from mid-day sunlight they may last until evening. A twining perennial species, *P. learii,* has deeper coloured bluish-purple flowers which are borne in late summer.

Pharbitis tricolor is raised from seeds sown in late March or early April, preferably singly in small pots, in as much heat as possible (at least 13°C. [55°F.]). They germinate best if dipped or soaked in water for 24 hours first. Acclimatise the seedlings to full light as soon as possible as spindly seedlings are prone to damping off. Pot them on in rich soil to flower one in a 6-in. or three in an 8-in. pot. Provide twigs, light trellis or wire for support. Well-grown plants can reach 8ft.

Pharbitis learii should be potted or planted in the greenhouse border in spring in a mixture of fibrous loam, peat and sand. It is best grown in a border or bed if this is practicable, but otherwise use large pots or tubs or containers. The shoots should be trained up the roof or on trellis. Shorten over-long growths in February. This species needs a temperature of 7° to 13°C. (45° to 55°F.) between September and March with the natural rise in temperature in summer.

Platycerium (Stag's-horn Fern). The platycerium are tender evergreen epiphytic ferns and it is the divided fronds which gives them their common name, Stag's-horn Ferns. Most need really warm treatment but one, *P. bifurcatum,* does best in a cool greenhouse where its fertile fronds, forked and divided, make an outstanding feature.

Grow the plants on blocks of wood suspended from the roof or sides of the greenhouse. Cover the roots with a layer of sphagnum moss and fibrous peat and secure by means of copper wire. Top-dress annually with fresh peat and moss in February or March. Water freely from April to September and moderately from September to March. Unlike most ferns they should be given full light and be allowed to almost dry out between waterings, which should then be very thorough. Shade from the sun. A winter temperature of 7° to 13°C. (45° to 55°F.) is needed and 16° to 21°C.

Fuchsia Swingtime, a striking variety well suited for cool greenhouse cultivation

Above. *Clivia miniata,* a colourful bulbous plant for spring and early summer display. It is worth remembering that this plant only flowers well when potbound

Left. The intricately formed floral parts of the Passion Flower, *Passiflora caerulea,* make this a most fascinating climber for greenhouse cultivation

Right. Schizanthus, popularly called the Poor Man's Orchid or Butterfly Flower, is among the best of all plants for the cool greenhouse or even for the cold greenhouse in summer

Right. Lachenalias, or Cape
Cowslips, are splendid bulbous
plants for the cool greenhouse.
Natives of South Africa, there
are many named hybrids available
in a lovely range of colours
Right, below. Another South
African native, this time a tender
herbaceous perennial, is
streptocarpus, the Cape Primrose.
Hybrids are available in a colour
range which includes red, pink,
purple, blue and white. Shown
here is the lovely variety Constant
Nymph

(60° to 70° F.) from March to October. Increase is by division.

Poinsettia, see Euphorbia
Poor Man's Orchid, see Schizanthus
Primula. The primula family is enormous and those members of it grown in greenhouses are exotic relatives of our native primrose and cowslip. They are mainly winter flowering and are ideal for the amateur gardener as they do not need much heat. The three most popular kinds are *P. obconica*, *P. malacoides* and *P. sinensis*.

Primula obconica often flowers intermittently throughout the year, but it is as a winter and early spring-flowering plant that it is chiefly valued. The colours available are pink, salmon, red, blue and white. Some people are, however, allergic to the plant which can cause a skin rash. *P. malacoides* has smaller, more numerous flowers borne in elegant sprays, one whorl of flowers above the other. The colour range is carmine, rose, salmon-scarlet, pink, blue, violet and lilac and this is certainly one of the most valuable winter-flowering plants for the amateur. *P. sinensis* has larger flowers and a greater range of colour and flower forms, one type with more star-shaped flowers being distinguished as *P. sinensis stellata*. Another form with fringed flowers is called *P. s. fimbriata*. *P. sinensis* in all its forms is a little more difficult to grow than either *P. obconica* or *P. malacoides*.

All three kinds are cultivated in a similar way but the sowing times differ. Seed of *P. obconica* and *P. sinensis* is best sown in March or April and *P. malacoides* in late June or early July. All need a temperature of 16°C. (60°F.) for quick germination. The seed, which is small, should be covered only very lightly. Sometimes germination is erratic and it is then wise to prick out seedlings a few at a time as they become large enough to handle. At both these stages the John Innes Seed Compost is suitable. Cover the seed pans or pots with both glass and paper until germination takes place but then the seedlings must be given full light to keep them sturdy. At all times the soil must be kept moist.

Prick the seedlings out into seed pans or boxes about 1½in. apart, give them the same temperature and shade from strong sunshine. Alternatively, seedlings of *P. obconica* may be pricked out 2½in. apart and one potting omitted. The first potting should be done when the leaves are touching in the boxes and before the young plants get checked by starvation—into 3-in. pots if the seedlings were pricked off closely and into 5-in. pots if they were widely spaced. Lift and separate very carefully and use the John Innes No. 1 Potting Compost for this first potting. Firm the compost only lightly with the fingers.

In high summer it is best to move the plants to a frame for a greenhouse is liable to get too hot, but if no frame is available shade the greenhouse glass permanently on the sunny side. Give plenty of ventilation and maintain a moist atmosphere by frequent damping of the paths and staging. Plants in frames should, if possible, be plunged in sand, ashes or peat. Plants potted into 3-in. pots should be moved to 5-in. pots as soon as they have comfortably filled their pots with roots. Use John Innes No. 2 Potting Compost.

The plants should be moved back to the greenhouse (if in frames) by the end of September and careful watering is now needed. Avoid getting water on the leaves in autumn and winter as this encourages grey mould. This trouble is less likely to occur at temperatures above 13°C. (55°F.) but perfectly good primulas can be grown at a temperature of 7°C. (45°F.) if care is taken to prevent damp. Any leaves showing signs of decay should be removed immediately. If yellowing of the leaves occurs sprinkle calcined sulphate of iron around the base of the plant. When the plants have filled the 5-in. pots with roots start to feed, especially in the case of *P. malacoides*. Use a good soluble or liquid fertiliser mixed as directed by the manufacturer's and repeat every week or ten days up to and during the time that the plants are in flower.

Pteris. *P. cretica* with simply divided light green fronds up to a foot in length is a popular fern for the cool greenhouse or home. It should be potted in March into a mixture of equal parts peat and loam with ½ part sharp sand added. Water freely in spring and summer and moderately in autumn and winter. A winter temperature of 7°C. (45°F.) is needed with a natural rise in spring and summer. Shade in summer. Increase by division at potting time.

Rose Bay, see Nerium
Saintpaulia (African Violet). The saintpaulia grown usually is *S. ionantha*. This has the familiar violet-blue flowers and velvety leaves, but there are many named varieties in different shades of violet, pink, rose and deep purple-blue. These include double-flowered varieties. Flowering is almost continuous. Although grown as house plants they prefer warm, humid conditions such as are usually found in greenhouses; in the home they seem to prefer kitchens, provided there is no gas, or the bathroom. They should be potted between February and May into 3-in. pots for small plants and 4½-in. pots for large plants, using the John Innes No. 1 Potting Compost with extra peat added, or a soilless peat/sand compost can be used with liquid feeding. Between October and April give a temperature of 13° to 18°C. (55° to 65°F.) and allow the natural temperature increase in summer. Careful watering is necessary at all times, avoiding splashing of the leaves; and provide ample shade and humidity in summer. Keep on the dry side when flower buds show, and also in winter.

The saintpaulias, or African Violets, are plants with fine decorative qualities, available in colours from violet, pink and rose to deep purple-blue

Increase from leaf stalk cuttings in mid-summer, placed in a peaty-sandy mixture with about 1 in. of the stalk buried. Plantlets will appear after about four weeks if the cuttings are put in a propagating case with a temperature of 18°C. (65°F.). Pot these singly into 3½-in. pots of John Innes No. 1 Potting Compost.

Saintpaulias can also be increased by seed. Spring sowing will provide autumn-flowering plants and August sowing plants for flowering in summer the following year. Sow the seeds thinly on the surface of John Innes Seed Compost and provide a temperature of about 21°C. (70°F.). When the resulting seedlings are large enough prick them out into boxes of John Innes No. 1 Potting Compost and finally thereafter to 3½-in. pots using the same compost.

Schizanthus (Poor Man's Orchid, Butterfly Flower). There are several fine seed strains of this lovely half-hardy annual, colours ranging from red, pink and crimson, to mauve and purple. These are among the best of all plants for the cool greenhouse or even for the cold greenhouse in summer. The large-flowered heads make a magnificent show on plants 2 to 4ft. tall, but the dwarf Bouquet strains are also excellent.

Seed can be sown at different times according to when one wishes to have flowers. It is, however, usually sown in August to provide plants for flowering in the following May, but seed can also be sown in February, March or April, if flowers are required in the summer. Use the John Innes Seed Compost and space out the seed to avoid overcrowding of the seedlings. For the September sowing place the pots in a cold frame. As soon as the seedlings can be handled they should be moved individually to 3-in. pots of John Innes No. 1 Potting Compost and be kept cool and close to the glass so that they form sturdy plants. Move to 5-in. pots in November, and use John Innes No. 2

Potting Compost. The plants are likely to damp off in winter if the soil is kept too wet when the weather is cold, so water sparingly until growth accelerates in the spring. Keep at a temperature of 4° to 7°C. (40° to 45°F.) and ventilate well. Stop when about 6 in. tall to encourage side shoots to develop. Adequate support is required, first with a split cane and with twiggy sticks as side growths develop. In February the plants from an early autumn sowing should be ready for their final pots of 6- or 7-in. size. Use John Innes No. 2 Potting Compost and firm it evenly. When the plants are well established in their final pots the final staking can be given. I like to use three canes spaced round the side of the pot to which raffia can then be fixed to keep the growth in place.

Feeding should be carried out as necessary from early spring, using a liquid or soluble fertiliser. Yellowing of the foliage indicates that the plants need extra nourishment.

Scilla, see Chapter Six, p. 164
Shrimp Plant, see Beloperone
Slipper Flower, see Calceolaria
Solanum. *Solanum capsicastrum,* the Winter Cherry, is a popular pot plant with bright red berries. These, as the common name indicates, are borne in winter. It is not difficult to grow, needing a minimum temperature of 7°C. (45°F.). Although plants can be kept from year to year new ones can be raised from seeds found inside the berries. These should be removed carefully ready for sowing in February or early March in small pots filled with John Innes Seed Compost. Space the seeds out individually and cover them with fine soil and firm. Water well and stand in a propagating box with a temperature of 16° to 18°C. (60° to 65°F.). As soon as germination takes place (it is rapid) remove the glass.

When the seedlings are about 1in. tall move them into 3-in. pots of John Innes No. 1 Potting Compost and move them on to 5- or 6-in. pots as necessary. Pinch out the tip of each plant when it is a few inches tall to induce a bushy habit, and in the summer pinch the side shoots for the same reason. The plants may be put in a frame from early June to September. Alternatively, they can be planted out in a warm border. In September the plants must be returned to a light position in the greenhouse, and if they were planted out re-potted into 6-in. pots and kept warm and close until they have settled down. Plants kept in pots in the summer need feeding with a liquid or soluble fertiliser and copious supplies of water.

To ensure a good set of berries I spray my plants with water each day when they are in flower.
Sparaxis, see Chapter Six, p. 164
Spider Plant, see Chlorophytum
Spleenwort, see Asplenium
Stag's-horn Fern, see Platycerium
Stocks, see Matthiola

Streptocarpus (Cape Primrose). These plants are tender herbaceous perennials, and the most useful for greenhouse cultivation are the hybrids, of which seed is offered by every seedsman, usually in a range of colours including red, pink, purple, blue and white. The flowers are trumpet shaped and freely produced in clusters on stems about 18in. high in the hybrids.

Seed should be sown in January or February to provide plants to flower in the autumn, or in July to provide flowers early in the following summer. The seed is small and should be germinated in a temperature of 18°C. (65°F.). Seedlings are potted first into 3-in., later into 4-in., and finally into 5- or 6-in. pots for flowering, John Innes No. 1 Potting Compost being used throughout. Water should be given freely throughout this period and the plants fed with weak liquid manure as soon as the final pots are comfortably filled with roots. Shade from May onwards, and spray overhead frequently during this period to maintain a moist atmosphere. Throughout the summer a temperature of 16° to 21°C. (60° to 70°F.) should be maintained. After flowering the plants can be either discarded or they may be overwintered, if kept rather dry in the greenhouse in a temperature of 10° to 13°C. (50° to 55°F.). In addition to raising plants from seed, increase can be by leaf cuttings. These are prepared so that the leaf stalk remains attached and the cuttings are pressed into the compost in a close frame with bottom heat. This latter method is often used for the species or for specially selected forms.

Streptosolen. A shrubby plant with bright orange flowers in June. There is only one species cultivated, *S. jamesonii*, a handsome climber, making a good pot plant in a 6- or 7-in. pot; it can also be planted directly into the greenhouse border. Potting or planting should be done between February and April, in the first case in John Innes No. 1 Potting Compost and in the second in good loamy soil. The plants need a light, sunny position. Water freely during spring and summer but only keep the plants just moist in autumn and winter. Do not allow the temperature to fall below 10°C. (50°F.) during the winter months. The shoots can be trained up wires or canes and should be pruned back fairly hard after flowering. Apply weak liquid fertiliser occasionally during the summer. Increase is by cuttings inserted in light sandy soil in a propagating frame in a temperature of 16° to 21°C. (60° to 70°F.) in spring or summer.

Tigridia, see Chapter Six, p. 164

Trachelium. A very pretty plant, *T. caeruleum* is usually grown in the greenhouse as an annual. The small blue flowers are produced in close heads on 18-in. stems during June and July. These plants succeed best in a sunny greenhouse with plenty of ventilation and no artificial heat after about the beginning of May. They should be

Solanum capsicastrum, the Winter Cherry. The berries are bright red and a good set is ensured by spraying the plants with water each day when they are in flower

watered freely throughout and staked when placed in their final pots as the growth is rather fragile. The plants are discarded after flowering. Increase by seed sown in February or March in a temperature of 16° to 18°C. (60° to 65°F.) and pot the seedlings into 3-in. pots and later in 4- or 5-in. pots in John Innes No. 1 Potting Compost.

Tritonia, see Chapter Six, p. 164
Tulipa, see Chapter Six, p. 165
Vallota, see Chapter Six, p. 165
Zantedeschia, see Chapter Six, p. 165
Zebrina (Wandering Jew). Often confused with the tradescantias, and like them called Wandering Jew, *Z. pendula* has slightly larger leaves, usually about 3in. long and about 1½in. wide, striped with silver and backed with purple. Its variety *quadricolor* has the upper sides of the leaves striped with white, rosy-purple, dark green and silvery-green, and the underside in varying shades of purple. Any good light soil suits zebrina, and it should be watered well in summer, but less in winter when it needs protection from frost. Keeping the plant on the dry side will emphasise the purple colour, but do not give too much light as it then turns brown. The variety *quadricolor* should also be kept on the dry side because of the danger of rotting, and unlike the species should be given the brightest light available.

Zephyranthes, see Chapter Six, p. 165

Fruits and Vegetables Under Glass

Information on fruits and vegetables to grow under glass is given in Chapters Eleven and Twelve respectively. See Index for page references to individual kinds.

Pests and Diseases

For pest and disease control see pp. 301 and 302.

The Fruit Garden

11

Every garden should have its fruit section for even where top fruits such as apples, pears, plums and cherries cannot be grown in tree form because of space limitations, room can usually be found for such soft fruits as raspberries, say, or currants and gooseberries. Blackberries, loganberries and other hybrid berries can often be found a place in the corner of the garden, or be trained on poles or fences. In many gardens, too, there are walls and fences which will provide opportunities to grow trained fruit trees such as peaches and nectarines which will benefit from the shelter and extra warmth; or such positions can be used for pears, plums or sour cherries trained as fans. Fan-trained Morello cherries, which belong to the last-mentioned section, will fruit well on north walls. Apples and pears trained in cordon or espalier form will also provide one with a decorative fruit 'hedge', and quite a few gardeners like to make use of such a feature.

When considering apples and pears remember that varieties are available grafted on to dwarfing rootstocks and these are the sort which, trained as cordons, espaliers, dwarf pyramids (usually apples only) or dwarf bushes are most suitable for small gardens. The apple rootstock known as Malling IX is very dwarfing, but if the soil is poor the semi-dwarfing Malling VII will be needed. Another for fairly small standard trees is Malling II. Additional to these is the stock M.26; trees on this dwarfing stock do not need staking. A quite recently introduced range of rootstocks is the Malling-Merton type of which MM.106 is semi-dwarfing, MM.104 rath vigorous, and MM.109 and MM.111 vigorous. These new stocks are resistant to woolly aphis.

Pears for restricted forms of training like cordons and espaliers are worked on to Quince rootstocks of which Quince C is the most dwarfing and Quince B and Quince A progressively more vigorous. Those on pear rootstocks grow much too large for average-sized gardens.

Plum trees need a lot of room and the amateur gardener will usually be more interested in bush- or fan-trained specimens, the latter for growing against a warm, sunny wall or fence. These should be worked on to Common Mussel or Brompton rootstocks. There is, unfortunately, no dwarfing rootstock for cherries and standard or half-standard trees grow very large. However, they can be trained in fan-formation against walls and the birds, which are always a problem at fruiting time, can at least be more easily countered by netting the trees when they are grown in this way. Peaches and nectarines are usually grown as fan-trained specimens, also against warm, sunny walls.

Of the soft fruits I personally would give priority to raspberries as they will produce fruit from June until late July, a time when fruit is really appreciated. Raspberries are also good for deep freezing. Next I would place black currants for this also is a fruit which crops well, is free from most troubles and is comparatively easy to grow. Third in my assessment come strawberries and fourth gooseberries. Strawberries are not a permanent crop, of course, for three years is the maximum any planting of strawberries should be left in a bed and many gardeners prefer to grow their plants for one year only. They are then grown in the vegetable garden and included in the normal crop rotation. Gooseberries are very popular, I find, particularly in certain parts of the country, and they do not take up a lot of room, nor are they a lot of trouble.

Flavour is of prime importance with home-grown fruits and it is as well to remember that this is not necessarily synonymous with good colour and general appearance. This applies particularly to apples.

Obtaining Trees. There are nurserymen who

specialise in the production and supply of all kinds of fruit; there are also those who offer general hardy nursery stock, which includes some fruit, usually the more popular kinds and varieties; and there are the garden centres. At the last-mentioned one can, of course, see the actual trees or bushes, decide on the best for one's purpose, take them away and plant them immediately, straight from the containers. This can be done at almost any time of the year, but with summer planting they must be carefully watched to make sure they do not dry out.

From the other sources, however, it is necessary to order some months in advance of the planting season. The best time of all to plant is early November but it may be done at any time until March provided the weather is suitable. One can specify the form of tree wanted, so that when it arrives it will already have been started and trained in the way required; after this it is up to you to continue the good work.

It is perhaps wise to point out that money spent on buying cheap fruit trees is liable to be money wasted—not to mention time, for several years may well be lost before you realise your mistake.

Pollination. Another point to remember is that when ordering top fruit varieties should be bought which are either self-fertile or will pollinate one another. Some varieties are self-sterile and must have a cross-pollinator; some, even with a special pollinator, do not set a good crop, and need a third variety to ensure a heavy set. This lack of a suitable pollinator is often the reason why fruit trees do not bear any fruit. Specialist nurserymen will always give advice on this matter and their catalogues often give the necessary information.

When considering this aspect of cultivation remember that it is only the currants, raspberries and gooseberries which are entirely self-fertile. Nearly all strawberries are also self-fertile.

Very few apples are fully self-fertile and all are better for the presence of a suitable cross-pollinator. Pears must have a compatible variety growing nearby to provide cross-pollination. All sweet cherries are self-sterile and a few of the sour cherries also fall in this category. Plums include both self-fertile and self-sterile varieties and others which fall in between, only setting poor crops of fruit without cross-pollination.

Planting. A very important point to be considered before planting is whether the favoured site is a frost pocket. If it suffers from late frost the blossom will be ruined, and poor crops and cracked fruits will result. This is particularly important with peaches, which flower early in the year, in March, when frosts and cold winds are still a factor to reckon with. A windy position is not good either, as it tends to discourage pollinating insects and the trees get blown into a bad shape if the wind is prevailing. A south-facing, sheltered slope is ideal.

Having decided on the site, ordered the plants, and prepared the soil, a final dressing of a compound fertiliser (containing nitrogen, phosphorus and potash) should be added about 10 days before planting. (The nursery will advise you when your order is likely to arrive). The fertiliser should be forked lightly into the top few inches of the soil.

When the plants arrive, they can be planted immediately if the weather is suitable, but if there is frost and snow about or the soil is very wet leave them in their wrappings for a few days until conditions improve.

If the bad weather is persistent take out a trench and heel the trees in, i.e. lay them in it at an angle and cover the roots with soil. They will be perfectly safe until conditions improve and planting can take place.

When planting fruit trees the usual rules of good husbandry apply. One must be sure, for example, that the drainage is good and the soil is up to standard. With bushes and dwarf pyramids the planting method is as described for trees and shrubs on p. 41, but espaliers and cordons are most easily planted by opening up a trench rather than digging individual holes. With all fruits the planting hole (or width of trench) must be sufficiently broad to take the plants' roots at their full spread.

Staking is a necessity with all the larger fruits and these should be set in position in the planting hole before planting actually takes place. In this way damage to roots is avoided.

Planting at the correct depth is most important and a good guide in this respect is the soil mark on the stem which indicates the depth at which the plant was growing in the nursery. Where top fruits are grafted on to rootstocks to ensure that they assume certain characteristics, dwarf habit and fruitfulness etc., it is essential to make sure

Growing apples and pears by the cordon method of training is space-saving and productive. The apple variety shown in this picture is Blenheim Orange

that the union of stock and scion is kept above soil level. Otherwise scion rooting will take place and the benefit of the rootstock will be lost.

If rabbits and hares are likely to be a problem provide fruit trees with wire-netting or black polythene protection to a height of about 4ft. Mice and voles, which nest in long grass and weeds, will also gnaw bark and can cause severe damage to a tree and even its death. These also should be watched for. Make sure, too, that there are no depressions round the trunks of apple trees where water can collect for this can lead to trouble from root rotting and the possibility of fatalities if it is not attended to promptly. Such depressions are equally undesirable with other fruits.

Pest and Disease Control. Although there are a great many pests and diseases which can attack fruit, only a few may, in fact, do so. It is unnecessary to take elaborate precautions to safe-

guard one's fruit from every known trouble, in the same way that one cannot always be taking precautions against every illness which might attack one.

However, there are routine precautions which should be taken to avoid the most likely attacks. First, try to grow your fruit as well as possible; give it the food, soil and management that will produce healthy, strong-growing trees and bushes right from the start. If this is done, any pests or diseases which may beset them will be much less damaging than would be the case with weakly, stunted plants. Secondly, carry out routine spraying with one or two fungicides and pesticides to keep those pests and diseases at bay which always attack fruit trees. In winter, when the top fruit trees are completely dormant (late December and January), spray them thoroughly with tar-oil winter wash. This will kill the eggs of many overwintering pests and get rid at the same time of lichen and moss. Then in spring and summer individual kinds of fruit are sprayed as necessary to counter specific pests and diseases. Details are given in Chapter Thirteen.

Do not spray fruit trees when these are in flower for this will kill bees and other pollinating insects, which is highly undesirable. Be careful also to note any time restrictions which apply to the harvesting of fruit when insecticides and fungicides have been applied. Indeed, read all the manufacturer's instructions with special care.

Non-production of Fruit. Some gardeners who grow fruit run up against the unhappy fact that for no apparent reason their trees fail to produce a crop. There are quite a number of reasons for this. One very obvious one, which nevertheless often gets overlooked, is frost damage. If the blossom gets frosted when it is full out the crop may fail completely because the stamens and stigmas are killed; damage can even be done to the unopened buds, and this is still more likely to be overlooked. This is why it is so important not to plant in a frost pocket, i.e. at the bottom of a slope, or in an area enclosed by hedges on a slope, or any sort of situation where cold air can be trapped. Cold air flows downhill like water, being heavier than warm air, and builds up at the lowest point, exactly as water does, and it is often possible to see a 'tide level', a high water mark so to speak, to which the frost has risen; above this the fruit and blossom will be untouched. Peaches are particularly subject to this type of trouble, as they bloom so early in the year.

Another cause is buying varieties which are not self-fertile and for which no suitable pollinating variety has been planted as companion (see p.253). Strong, cold winds at blossom time often also result in a poor or non-existent crop, as bees are very much less likely to be out in such conditions.

Another factor is birds. These are becoming more and more of a menace in that they start pecking at fruit buds during the winter, sometimes as early as November and often completely strip the trees, so that they come into leaf in the spring, but have no flowers. One of the signs of these attacks is that the tree produces an excessive amount of leaf and shoot when it breaks in the spring, particularly where the blossom should have been.

Sometimes trees are grafted on to the wrong kind of stocks which make them so vigorous and strong-growing that they take many years to come into fruit, and in some cases never do so. Seedling trees may also take a very long time to come into fruit.

If the soil is very fertile certain sorts of trees will again produce vegetative growth at the expense of fruit bud, and one way of counteracting this is to bend the shoots and tie them down in a curve. This tends to make them produce fruit buds rather than vegetative buds.

Lack of fruit can, in some cases, be due to starvation. If the tree is growing in very poor, stony, 'hot' soil it may produce blossom, but it sets badly or sets and drops off before it matures. Sometimes the blossom gets infected with maggots, and this also results in the fruit dropping off too soon.

The wrong sort of pruning leads to poor fruiting, and to a condition known as biennial bearing, in which the trees bear heavily in one year, and practically not at all the next.

Fruit Storage. Unless one has a deep-freeze unit apples and pears are really the only fruits that can be stored, and these must be of varieties known to have good keeping qualities. The fruits should be picked just before they are fully ripe, and placed on slatted shelves or trays in tiers, so that they can 'breathe'. They will also store very well if placed in closed chests, where they generate carbon dioxide which builds up and helps to preserve them. Another way is to wrap them individually in oiled paper and place them on shelves; and yet another is to place small quantities, say five or six, in closed polythene bags. They should never be stored in a dry atmosphere, as they shrivel and dry up; this particularly applies to pears. Sheds, cellars and similar buildings with brick or stone walls, provided they are cool and have a moist atmosphere, are very suitable. Only store those fruits that are free from disease or mechanical damage of any kind to the skin and flesh, for such damage provides disease spores with an ideal form of entry.

Picking apples for storage two or three days either side of the optimum date for the task will not affect their keeping qualities. The timing is much more crucial though in the case of pears, and it is very easy to pick them too late when they will simply go 'sleepy' and never really ripen evenly. On the other hand, if picked too early they shrivel; the right time can only be learnt by experience. In general, fruit is ready to be picked, though not necessarily to be eaten, when the stalk breaks away from the tree easily after the fruit has been lifted gently upwards in the palm of the hand. A change in the colour also indicates that picking time is close.

APPLES

Apples will grow in almost any reasonably good garden soil although the best results are obtained in a rather deep loam with good drainage. On heavy waterlogged soil some varieties are apt to suffer from canker, while on poor, sandy or gravelly soil the fruits are likely to be of small size and inferior quality. Well-rotted manure or compost can be used freely in the preparation of the ground, which should be dug as deeply as possible.

When apples or pears are stored they must be inspected at regular intervals, any showing signs of decay being removed immediately

Apples can be trained in a great variety of shapes from standard trees, which will probably take up too much valuable room in smaller modern gardens, to bushes and dwarf pyramids and restricted forms of training like espaliers or cordons, which may be single or double stemmed.

Pruning of trained trees should be carried out in both summer and winter. During July and the early part of August the new side shoots should be shortened to three or four leaves each, not counting the basal rosette of small leaves. Then in November these laterals can be further shortened to within two or three dormant buds of the main stem to encourage the production of fruiting spurs. At the same time the leading shoots should be shortened by about a third. It is not essential to carry out such restrictive pruning with well-established standard and bush apple trees, which after a while can be left to go their own way, with little beyond thinning of badly-placed branches in November, to prevent overcrowding and the removal of any cankered or otherwise diseased branches or shoots.

When trees are cropping freely it is necessary to reduce the number of fruits to not more than two at each spur, or if exhibition fruits of culinary varieties are required, only one should be left at each spur. The thinning should be done when the fruits are about the size of marbles, the 'king' fruit (the one at the centre of the cluster) being removed first and badly-shaped specimens as necessary.

Propagation. Increase by grafting in March or April, or by budding in July and August. Although seedling crab apples are sometimes used as rootstocks it is best to use the much more reliable Malling or more recently introduced Malling-Merton rootstocks. The latter are resistant to woolly aphis in so far as the rootstocks themselves are concerned but not the variety grafted on to them. I have already referred to some of the most useful rootstocks for those with smaller gardens on p. 252. M.IX will provide dwarf bushes to about 7ft. in height and is also suitable for cordons, espaliers and pyramids. A slightly more vigorous and equally useful stock in M.26. M.VII and M.II are good for small standards and the average bush and MM.106 is their equivalent.

Some Varieties (in order of ripening). The very large selection available includes Scarlet Pimpernel, dessert, August. Grenadier, culinary, August–October. James Grieve, dessert, September–October. Egremont Russet, dessert, October–December. Bramley's Seeding, culinary, November–March (too large for small gardens). Cox's Orange Pippin, dessert, November–January (I like to grow James Grieve with it as a pollinator, superb flavour). Lane's Prince Albert, culinary, November–February. Sunset, dessert (not very good colour on some soils, but can be a good substitute for Cox's Orange Pippin), November–December. Rosemary Russet, dessert, December–March. Winston, dessert, January–April (crops best as a bush form).

APRICOTS

In many respects the apricot stands midway between the plum and the peach as regards cultivation. Apricots can be grown in the open, but they rarely fruit satisfactorily unless given wall protection. A sunny position must be chosen and the trees trained as fans (see peaches p. 260). The soil should be prepared as for peaches and nectarines, and planting carried out in November.

The apricot bears on spurs produced on the old wood as well as on one-year-old shoots. In consequence, it is not necessary to disbud so drastically as with peaches. Instead, unwanted or badly placed shoots can be pinched out from May to July, and in November young growth made during

Each year, as soon as the fruit has been picked, the old canes of loganberries are cut back to ground level and the new canes trained in their place

the previous summer can be trained in at almost full length where there is room for it, but to prevent overcrowding some of the new growths may have to be cut back to within two dormant buds of the main stem.

Propagation. Increase by budding during the summer months on to plum rootstocks.

Recommended Variety. The most widely grown variety is Moorpark, which is ready for picking in late August.

BLACKBERRIES AND LOGANBERRIES

The cultivated blackberry is a great improvement on the wild one in size, yet it still retains the characteristic flavour. Since it is a native of this country, it is very easy to grow and, indeed, the main trouble with blackberry cultivation is to keep the plants under control. Plant them during autumn or winter; they will grow in practically any soil except a light sandy one, on which they do not produce satisfactory fruit. They form a good barrier between different parts of the garden. What I have done with mine is to plant them against a hedge and let them grow up through it.

Dig in plenty of organic matter before planting, and put them at least 10ft. apart. Train them against horizontal wires fastened to really strong posts, or against walls or trellis or on poles. After planting little general cultivation is needed, beyond mulching each year fairly heavily, and applying a general fertiliser in spring if the soil is on the light side.

Pruning and training are not complicated. Each year, as soon as the fruit has been picked, the old canes which have been fruited are cut back to ground level, and the new canes trained in their

Mulching round a young black currant bush. This is either done after fruiting in summer or in late winter. After fruiting also apply a light potash dressing

place. It may be necessary to lay all the new canes on the ground so that they can be disentangled and fastened in place tidily, and a good strong pair of gloves is a 'must' when pruning. During the season, as the canes grow, they are trained up above the centre of the plant and along the wires over the old canes. When the latter are removed at the end of the season the new canes are untied and brought down lower to take their place, they are spread in a fan shape, and arranged so that the centre of the fan is left empty for the next year's new shoots.

Propagation. Increase by bending down the tips of strong new growths and fixing them to the soil in May or June, when they will root. The tips can then be cut off in September, and the new young plants lifted and planted in their new positions in October or November.

Varieties. These include Bedford Giant, early but short season; vigorous grower. John Innes, ripens mid-August, and goes on until the first frosts; good for a small garden. Cut-leaved Blackberry, sometimes called the Parsley-leaved Blackberry, late season, also suitable for small gardens; leaves change to pleasant colours in autumn. Merton Thornless, ripens August, high-quality berries; its thornlessness is an additional attraction.

Loganberries. These are treated in the same way as blackberries. They ripen from mid-July onwards, and should be planted in soils which are on the medium to light side; they are not so amenable to heavy soil, however, as blackberries. In addition to the Loganberry itself there is the Thornless Loganberry. The LY59 strain is the loganberry usually offered by nurserymen.

CHERRIES

Cherries grown for their fruits must be considered in two groups—the sweet cherries, usually grown as standards and bearing their fruit mainly on spurs on the older wood, and the sour cherries, used for cooking purposes (and including the Morello cherry), which are often grown as fan-trained trees against walls and bear their fruits mainly on young wood made during the previous year. Both kinds thrive in fairly rich loamy soils, preferably well supplied with lime.

Plant from November to March, in an open sunny place for the sweet cherries, and in sun or shade for the Morello kinds. Unfortunately, there are no dwarfing rootstocks for sweet cherries and they make large trees which do not like being pruned. They are not really suitable for small gardens as they should be planted at least 25ft. apart. Another disadvantage is that they are self-sterile so that more than one variety is needed to obtain fruit. Give them a frost- and wind-free site.

Sweet cherries require little cultural care once established; an occasional dressing with compost and a sprinkling of Nitro-chalk each year will be sufficient. Pruning consists only of cutting out crowded, broken, crossing or diseased shoots and branches, this being done in April to lessen the chance of Silver Leaf disease and bacterial canker gaining entry through the cuts, for cherries are susceptible to these troubles.

Sour cherries are grown 15 to 20ft. apart as fan-trained specimens against walls at least 10ft. high, keeping three or four main shoots to form the skeleton of the fan, and leaving the centre empty until the side shoots are strong and can be used to fill it up. I find it best to leave in as many young branches as possible and cut out some of the old wood each year. The Morello cherry, like the peach, bears its fruit on the young branches produced the previous year.

The sour cherries can also be grown as trees, of course, and are considerably smaller than the sweet varieties.

Propagation. Increase by budding or grafting on to rootstocks (the main one being F12/1) which has some resistance to bacterial canker.

Varieties. These include—Sweet cherries: Early Rivers, mid to late June (fine flavour). Governor Wood, July (also excellent flavour, but suitable only for drier areas). White Heart, late July (one of my favourites). Merton Bigarreau, late July. Sour cherries: Morello, August and September (excellent variety), self-fertile. Kentish Red, July, self-fertile. Both of these varieties are good for preserving.

CURRANTS, BLACK

The black currant is a particularly rewarding crop for the amateur gardener, especially if space is at a premium. The fruit has high food value and its flavour usually appeals to most members of the family; it is comparatively easy to grow; and the fruit is excellent for deep freezing. As smaller and cheaper freezers become available this is giving more and more families the opportunity to use their home-grown produce more effectively as well, of course, as to economise by buying produce in larger quantities.

Black currants are grown in bush form, with the main shoots arising from ground level instead of being borne on a short main stem (or 'leg' as it is called) as is the case with red currants and gooseberries.

Plant the bushes between November and March in rich, well-cultivated rather moist soil in a sunny or partially shaded position. Avoid sites which are subjected to frequent winds. Plant two-year-old bushes 6ft. apart each way, and immediately after planting cut them down to within 4 to 6in. of soil level. This is essential to ensure the production of plenty of good strong new shoots in the season following, and it is on these that fruit will be produced a year later. There will not be any fruit in the first season after planting.

Mulch the plants heavily each year with organic matter, either in the late winter or after fruiting in summer, and apply a light potash dressing after fruiting in summer to help ripen the wood and encourage future fruit production. Putting straw down all over the soil round the bushes is a very good way of keeping weeds under control, and it also supplies some organic matter. It can be left down permanently and renewed as required.

Black currants do not need support or training, but if they are unavoidably planted in a windy position, protect them from the prevailing wind with hessian, sacking or some sort of barrier, otherwise they will suffer from what is known as 'run-off'. In such conditions the bees will not work the blossom and only one or two flowers near the top end of the 'strig' (the bunch of fruit) are likely to set, the flowers getting progressively less well set the nearer they are to the free end of the strig.

Pruning can be carried out in summer after fruiting, or as soon as the leaves have fallen in November. The black currant fruits best on new wood, and hardly at all on the older wood, so fruiting shoots should be cut right out. This encourages the production of new shoots the following season. It will mean removing at least one-third of the wood, cutting it back to the point of origin. Some varieties are inclined to produce strong new shoots on older shoots, and then the older ones should be cut back only as far as the new.

Propagation. Black currants can be increased from both hardwood and softwood cuttings and by layering. The hardwood cuttings, made from the current season's wood and some 10in. long, are planted in a prepared trench in October so that only the top three or four buds remain above soil level. The other buds, however, are not removed. Space the cuttings 9in. apart and when returning the soil to the trench firm well with the feet. Soft-wood cuttings, made from side growths of which the top 3in. is used, are preferred by some gardeners as they are taken in July when buds affected by Big Bud (Reversion), the most serious trouble of this fruit, can be seen and affected plants avoided for propagation purposes. These cuttings are rooted in a shaded cold frame and they must be well supplied with moisture to make sure that they do not dry out. The rooted cuttings will be ready for lining out in nursery rows in autumn. The third method of increase, by layering, is carried out in June and July.

Varieties. These include Mendip Cross, early. Wellington XXX, mid-season, heavy cropper (rather straggling grower and needs severe upward-tending pruning). Westwick Choice, mid-season to late, fine flavoured (makes a neat bush and grows strongly). Daniel's September, late, heavy cropper (compact growing).

CURRANTS, RED AND WHITE

The treatment necessary is similar to that given to black currants except that they are grown on a leg, and are thus pruned differently. Also, grow the early croppers in a sunny position and the later ones in a shady position. Plant 4 to 6ft. apart in rows the same distance apart, and cut back after planting to leave about 6in. of stem on the main side shoots, of which there should be four or five. Any others should be cut right out. The cuts should be made to buds pointing outwards, and any pointing inwards should be rubbed off. Thereafter prune in winter, when dormant, removing the leading shoots so as to leave about 6in. of the new season's wood, and cutting back the side shoots to leave one bud; this helps to build up the spur system which is the best way of making the red currant produce fruit. Cutting back the leading shoot will encourage the growth of side shoots, and thus the production of more fruit. Keep the centre of the bush open.

Propagation. Red currants are increased from hardwood cuttings in a similar manner to black currants but only the top four buds are left on each cutting and the trenches in which they are planted are much shallower.

Varieties. These include—Red currants: Earliest of Fourlands, early (strong upward-growing variety). Laxton's No. 1, early (heavy cropper, needs wind protection). Red Lake, an American variety, mid-season (a heavy cropper, strong growing). Wilson's Long Bunch, late (smallish berries but good cropper). White currants: White Versailles and White Dutch.

FIGS

These can be grown outdoors in the milder parts of the country, or under glass anywhere. The position chosen outdoors should be as sunny and sheltered as possible—for example, against walls facing south or south-west. Plant in March or April in a border 2ft. deep and 3ft. wide, enclosed with a brick or concrete wall. The planting mixture for figs should consist of 2 parts fibrous loam and 1 part brick rubble. This restriction of the roots is to curb the production of top growth which is otherwise liable to be made excessively at the expense of fruiting.

Fruits may be borne along the entire length of previous year's shoots, but only one crop at the base of the young shoots is borne outdoors in England. Prune in April or October, simply removing damaged, dead or weak branches. Pinch the points of vigorous young shoots in July. Apply liquid manure once in August to trees bearing heavily. Figlets the size of filberts should be picked off in September or October as these are unlikely to produce usable fruits. Protect the branches in December with straw or mats, removing these in April.

Figs Under Glass. When growing figs under glass, the compost, position and time of planting are the same as for planting and growing outdoors. The branches should be trained up the roof or against the wall. The fruits are borne on shoots of the previous year's growth for the first crop and those of the current year for the second crop. Prune and pinch as described for outdoor cultivation, and disbud young shoots when too many buds are forming. Water and syringe freely in summer, and apply liquid manure occasionally.

Propagation. Increase by cuttings of firm young shoots in autumn, inserted singly in pots filled with sandy soil and plunged in a frame or cool greenhouse.

Varieties. Of those available Brown Turkey is, perhaps, the best for outdoor and indoor cultivation.

GOOSEBERRIES

Gooseberries are grown in the form of a bush with a short leg like the red currant. They are one of the earliest fruits to ripen, coming into harvest in early June. Two-year-old bushes should be planted in the autumn or early winter, 5 to 6ft. apart, in well-drained soil which has been manured beforehand. After planting cut all the shoots back, removing all but four or five strong shoots, and cutting away about three-quarters of each of these. This will encourage the bush to develop strong new shoots. In spring, or earlier in February, each year mulch heavily with organic matter and give a light dressing of potash.

Pruning is done in autumn and need only be fairly light. Cut out shoots growing into the centre of the bush as well as those which are crowded, crossing or diseased. The aim is to have an open-centred bush to which light and air have free access and to facilitate fruit picking. Just tip the rest of the leaders. Leave most of the side shoots at full length, say about three-quarters, and cut the rest back to one bud from the base. Some varieties have a drooping habit and this can be corrected by cutting back to an upward pointing bud at the apex of the curve of the drooping shoots.

Propagation. Increase by hardwood cuttings of one-year-old wood prepared in early October. These should be about 9in. long and be inserted to two-thirds their length in light soil. Rub off the lower buds on the cuttings leaving four or five at the top to provide each with a 'leg' of about 6in. Gooseberries can also be increased by layering in summer.

Varieties. There are dessert and culinary varieties, and the fruits of dessert varieties can be thinned in late May and early June, using the thinnings for cooking and leaving the remainder to increase in size and ripen. They include Careless, culinary, mid-season (very liable to suffer from potash deficiency). Keepsake, dessert or culinary, early. Golden Drop, dessert, mid-season (very suitable for the small garden). Lancashire Lad, culinary, mid-season (not one of the best varieties but makes good jam). Leveller, dessert or culinary, mid-season (sulpher shy). Whinham's Industry, dessert or culinary, mid-season. Lancer, dessert or culinary (very good flavour).

GRAPE VINES

Grape vines can be grown outdoors, trained against sunny walls or over trellis, or wires, or under glass in heated or unheated glasshouses. Outdoors, ripening tends to be uncertain and selection is limited to certain very hardy varieties.

Vines need good, rich soil and good drainage. Greenhouse borders specially prepared for them should be at least 3ft. deep and the full width of the house. Hard rubble should be placed in the bottom for drainage, a layer of turves on top of this and then the border should be filled with good loamy soil, well sprinkled with wood ashes and coarse bonemeal, and with a little well-rotted manure all thoroughly mixed in.

Vines are best planted in late February or early March, and the roots should be well spread out and made thoroughly firm. Either one vine can be trained to fill a whole house with numerous stems or else—the more usual method—each vine can be restricted to one main rod, when the vines should be 5 to 6ft. apart. The first year the newly planted vine is only allowed to make one main growth and this is trained upwards towards the ridge of the house for single rod cultivation or horizontally along the eaves for multiple-stem cultivation.

In subsequent years side growths are allowed to

form every 15 to 18in. on each side of each main rod of single-stem vines. Multiple-rod vines are permitted to form a main growth every 5ft. along the horizontally-trained stems and these are trained to the ridge, each being subsequently treated like the main rod of a single-stem vine.

The side growths carry the flowers and bunches of grapes, not more than one to each side growth which is stopped two leaves beyond the flower cluster or when it is 2 to 3ft. long. All secondary growths which appear are stopped at the first leaf. All these side growths are carefully tied to wires strained horizontally about 9in. below the glass.

Each winter the side growths are cut back to within one or two dormant growth buds of the main rods or the woody 'spurs' which in time form on them. In January the main rods are un-tied and allowed to hang down to check the uprush of sap but, when growth has started, are re-tied in their former positions. Allow only one side growth to form at each spur and rub off any others. Vines respond to fairly generous feeding. In winter the top 2in. of soil can be carefully removed and replaced by a mixture of equal parts well-rotted manure, and good, loamy soil. Immediately after flowering the border can be further enriched by a top dressing of a well-balanced general fertiliser at the rate recommended by the manufacturer.

Vines are started into growth at any time from January to March by closing the ventilators and allowing the temperature to rise to 10 to 16°C. (50 to 60°F.). The border should be thoroughly soaked at this time. Daily syringing with clear water is necessary now but should be discontinued while the vines are in flower, when the temperature may rise a little and the atmosphere be kept dry.

Most grapes are self-fertile and good sets of berries are usually obtained under glass without outside help. Some gardeners like to further aid pollination, however, by tapping the rods sharply at about mid-day during this period to further distribute the pollen. Outdoors, pollination should be done with a camel hair brush.

In June, when the berries are well formed, they must be carefully thinned with narrow, pointed scissors. The smaller seedless berries are cut out first and then the remainder of the surplus berries can be removed. To avoid touching the berries a small forked stick is useful. It is best to start at the bottom of the bunch and gradually work upwards.

Water is necessary to keep the soil moist, and keep the paths and walls damp to maintain a fairly humid atmosphere, but allow the air to become much drier and the temperature to rise a little as the grapes ripen.

Out of doors, vines can be treated in much the same way as those indoors, but it is best to plant them against a warm and sheltered wall.

Propagation. Increase by taking 'eyes', or dor-

Spraying a dormant grape vine—untied and carefully spread out—with tar-oil winter wash. Vines are started into growth at any time from January to March

mant growth buds, cut from well-ripened side growths in autumn with a short piece of stem attached. These 'eyes' are laid horizontally, but uppermost, in John Innes No. 1 Potting Compost in small flower pots, one per pot, and are started into growth in a propagating frame with a temperature of 16°C. (60°F.).

Varieties. These include—Under glass: Black Hamburgh, Madresfield Court, and Muscat of Alexandria; the fruits of the last-mentioned are not black as are all the others, but yellow. For warm, sunny walls outdoors: Buckland Sweetwater and Royal Muscadine are white grapes which ripen early. Brandt is an early-maturing black grape and the leaves sometimes have attractive autumn colour.

PEACHES AND NECTARINES

Peaches, and their smooth-skinned forms, the nectarines, have an exotic appeal many gardeners, including myself, find hard to resist. Both are cultivated in the same way. The best way to grow them outdoors is on a warm south-facing wall as fan-trained specimens, but they are also grown as bushes in southern parts of the country. In the latter case, though, only a few varieties such as Hale's Early, Peregrine and Early Rivers will fruit satisfactorily.

Autumn or early winter planting is best, as they come into flower early in the spring, and

If peaches are grown in a cool greenhouse—as here on the wall of a lean-to—it will be necessary to pollinate them artificially with the help of a rabbit's tail

require as much time as possible in the preceding season to get established. Manure the ground well before planting and dress it at this time with lime and a liberal sprinkling of bonemeal. Fan-trained trees should be planted 10 to 12ft. apart, and free-standing bushes 15 to 20ft. each way. Thereafter mulch heavily with organic matter in February or June each year, and water the plants freely in dry weather, particularly if grown against walls. An annual sprinkling of Nitro-chalk will also help good growth. The fruit will want thinning, starting when each is about the size of a walnut, and doing it gradually over several weeks, so that finally there is one left to every 9in. of shoot growth.

When the trees are in bud or in flower they must be protected against frost and cold winds. Provide the necessary protection by erecting a plastic covered framework or a hessian screen if you prefer that material.

Peach-tree pruning and training, where a fan-trained tree is involved, can appear to be very complicated. It is not, although it is spread over a comparatively long time, compared with other fruits. Once the framework of the fan has been formed, by taking the two strongest side shoots and fanning them out one to each side, then the strongest laterals off these are retained and tied in each year, gradually filling the centre as well as the sides.

When new shoots start to grow up in spring, there will be far too many of them, and it is necessary to rub them out before they grow to any length. On each fruit-bearing shoot remove all the new young laterals, a few at a time, until only two are left, one at the tip and one at or near the base of the shoot. Where shoots appear with the fruit, reduce these to one and pinch this back to leave one leaf, which will help to feed the fruit. This de-shooting is spread out over several weeks, just as fruit thinning is. The leader is allowed to grow on. After disbudding, tie in the shoots so that they fill the space evenly. Careful disbudding makes fan-training of the tree much easier.

In November, cut out the old fruiting growths and tie in the basal shoot which grew during the summer to take its place. Prune back the leading shoots if more extension growth is required, otherwise prune right back to the replacement shoot. It is often easier to do this pruning if the whole tree is untied and then re-tied in position so that even spacing can be arranged. This is because the shoots will have grown unevenly, some being stronger than others.

Propagation. Increase by budding in summer on to plum stocks, such as Brompton. Seedling peach rootstocks can be used where they are to be grown as bush trees. The stones can also be used to provide trees, but will not come true to variety.

Varieties. These include—Peaches: Noblesse, end August. Peregrine, early August. Barrington, mid-September. Hale's Early, late July-early August. Nectarines: Early Rivers, end of July. Violette Hâtive, mid-August. Pine Apple, early September.

PEARS

Pears are treated in a very similar manner to apples with regard to planting and general cultivation. They do better if given a rather warmer and more sheltered site than apples—they do not stand up to wind well—and need a slightly lighter soil (they resent cold, wet clay soil), though they do not produce good fruit on poor, sandy or gravelly soils. The choicest varieties should, for preference, be trained against a sunny wall or fence, but there are many kinds which can be grown quite satisfactorily in the open, either as bushes or standards. Trained trees are usually grown as cordons or espaliers. If the soil is naturally heavy it is advisable to improve the drainage. Grow them in soil which has been well manured beforehand and plant at any time between November and March, but preferably before Christmas. Thereafter mulch every spring, after applying a slow-acting nitrogenous fertiliser such as hoof and horn or dried blood; this will not be necessary in all cases and, as with many plants, feeding depends on the growth and general health of the plant and the soil in which it is growing.

Doyenné du Comice, the finest of all pears for flavour. A price has to be paid though for this excellence for it is shy cropping

Summer pruning a pear trained against a sunny wall. The laterals are shortened a few at a time during July and early August to five leaves

Pruning is practically the same as that advised for apples, but pears form spurs even more readily than apples and so are particularly well adapted to the more restricted forms of training such as single-stem cordons or horizontally-trained trees (espaliers). These should be pruned in summer shortening laterals a few at a time during July and early August to five leaves. This should be supplemented by winter pruning, at which season the laterals may be further shortened to two buds each, leading growths being cut back by about a third. The fruits require thinning but this must not be done too drastically as some will fall before they are fully grown.

Propagation. Increase by grafting varieties on to the following rootstocks: Quince A for larger trees, Quince C for cordons, espaliers, etc. Do not obtain pears grafted on to unselected seedling stocks as they take a very long time to come into bearing. Sometimes a variety is incompatible with quince, so double grafting has to be carried out, using another suitable pear variety for the intermediate graft. Do make sure when planting that the union of stock and scion is above ground; it is very easy for the scion to root if the union is at or below ground level, and then the tree will be tall and unfruitful.

Varieties. Pears will pollinate one another without difficulty, but the more different varieties that are grown the more heavily they will crop. The following dessert varieties—a small selection of those available—are given in order of season. Doyenné d'Eté, August (very good flavour). Williams' Bon Chrétien, September (can be picked a few days before ripe, kept in a cool store and used as required; very good flavour). Dr Jules Guyot, early September (can also be picked and kept a few days before being eaten). Louise Bonne of Jersey, October. Beurré Hardy, October. Conference, October–November (nearly self-

fertile, needs to be picked and kept a few weeks before eating). Doyenné du Comice, November (a shy cropper but the pear *par excellence* for flavour; do not spray with lime-sulphur). Packham's Triumph, November–December (good for bottling). Joséphine de Malines, December–January (very good flavour, stores well and makes a good pear for Christmas). Two good cooking varieties are Bellissime d'Hiver, which keeps from December until March, and Catillac, with a season from December to April.

PLUMS

This fruit succeeds best on fairly rich soils supplied with lime and not subject to waterlogging in winter. The best time for planting is in early November, but the work can be continued at any time until March, provided the weather is open.

Choose a site that is not exposed to wind, as bees and other pollinating insects will not work the trees when there are cold winds blowing. Birds unfortunately can be a problem, ruining the crop by stripping the trees of fruit buds; this is an especial hazard if there are woods nearby. Half-standard and bush forms of training are to be preferred for the open garden and it is worth while growing especially good varieties as fan-trained specimens against walls. Half-standards should be planted 18ft. apart and bushes 15ft. apart.

Plant in soil that has previously been manured, and thereafter mulch each spring, and feed with a well-balanced general fertiliser in February. Plums are inclined to throw suckers, and these should be pulled off at the roots. They are more surface-rooting than apples or pears, and so it is never wise to carry out deep cultivation around established trees. Thin plums to 2in. apart when they have finished stoning, doing it over a period of about a week; this also helps to check the biennial bearing for which plums are notorious.

Thinning the fruits on a wall-trained Victoria plum. This results in better fruit and helps to check biennial bearing. This variety is susceptible to Silver Leaf disease

Pruning established raspberries consists of cutting out to ground level all wood which has fruited immediately after the crop has been gathered

The pruning of plums grown in the open should be very light, this being carried out in late summer after the crop has been picked. Cut out some of the side shoots completely each year and just lightly tip back vertical shoots. Cut out more wood before a heavy year for fruiting and prune lightly before an off year; and keep the trees well manured. This will help to ensure regular annual cropping.

Fan-trained trees must be pruned more drastically. Unwanted side shoots can be cut back to within two dormant buds of the main branches, but where possible young laterals should be trained in at practically full length.

Propagation. Increase by budding or grafting on to rootstocks such as Common Mussel or Common Plum for moderate-sized trees; Pershore, for small trees, particularly if the variety is naturally strong growing and space is limited; Brompton for varieties which lack vigour and where the soil is poor.

Varieties. Plums fall into three groups with regard to pollination and further details are given on p. 253. Varieties available include Rivers' Early Prolific, culinary, late July (partially self-fertile, makes very good jam and is very suitable for small gardens). Early Laxton, dessert, late July (partially self-fertile, good flavour). Czar, culinary, early August (self-fertile, good for small gardens, except that it is liable to be attacked by Silver Leaf). Early Transparent Gage, dessert, mid-August (self-fertile). Victoria, dessert and culinary, mid to late August (self-fertile, inclined to contract Silver Leaf Disease). Pershore (Yellow Egg), culinary, late August (self-fertile, makes good jam, as well as cooking well). Monarch, culinary, end of September (self-fertile).

RASPBERRIES

Raspberries are an easy crop to grow, needing comparatively little attention, but to obtain good crops they should be given a deep, rich, well-drained soil and an open, sunny position. Their main fruiting season is summer but there are also autumn-fruiting varieties. They are the best fruit of all for deep freezing. Raspberries, unfortunately, are prone to virus infection and it is wise to buy only stock certified free from this trouble, for which there is no cure.

Planting can be done at any time between October and March when the weather is suitable, but for preference choose a date near the beginning or end of this period. Firm planting is essential and the plants should be spaced 2ft. apart in rows 5ft. apart, the canes being trained to wires strained between posts at the end of the rows. Two rows of wires is sufficient, $2\frac{1}{2}$ft. and 5ft. above the ground.

After planting cut back the canes to about 9in. above soil level and do not let the summer-fruiting varieties bear any fruit the first year. Mulch the plants in spring with decayed manure or compost and be careful to avoid damaging the roots, which lie near the surface when keeping weeds under control by hoeing. Water the plants freely if the weather is dry during the bearing period.

Pruning for established plants consists of cutting out to ground level all wood which has fruited, immediately after the crop has been gathered. At the same time cut out weak, broken or diseased canes. Then reduce the number of new canes at each root or stool to the six strongest and tie these in to the wires, spacing them about 6in. apart to give an even coverage of the area available. These canes should be tipped in February or March.

Propagation. Increase by taking suckers with roots from the parent plant in autumn. Cut them off cleanly. There is never any shortage of such suckers. I have already remarked on this fruit's liability to suffer from virus infection and only plants free from disease should be perpetuated.

Varieties. These include Malling Promise, vigorous, grows to 7ft. and more on heavy soils; pick early July. Lloyd George, an old variety, early-mid July, very good flavour, and will fruit on new shoots in the autumn; be certain to obtain stock free from virus. Malling Jewel, vigorous, good cropper, pick mid-July. Norfolk Giant, heavy cropper, pick end July, very vigorous; virus-free stocks must be obtained. Hailsham, vigorous, autumn-fruiting. September, a new American variety, pick early September; good cropper, strong grower.

STRAWBERRIES

This is a fruit which is usually popular with all the family and it is worth growing not only for dessert but also for jam making and deep freezing.

A sunny slope from which cold air can drain freely is the ideal, for frost damage can be a problem, but most of us have to make do with less satisfactory sites. The best results, so far as soil is concerned, are obtained on rather rich loam that has been deeply dug and enriched with well-rotted stable or farmyard manure. Planting can be done in late summer or early autumn or in March, but spring-planted strawberries should not be allowed to fruit during the first year. Space the plants 2ft. apart in rows 2½ft. apart. The best results are obtained from one-year-old plants and after three years they should be discarded. Strawberries are shallow-rooting plants and do not take kindly to weed competition; so perennial weeds should be eliminated from the ground, as far as possible, during the propagation stages.

The planting operation must be done with care for the crown of each plant must be kept at surface level and the roots spread out to their fullest extent. Firming the soil round the roots is also very important.

Spread black polythene sheeting or clean straw around the plants and under the leaves in May to keep the fruits clean. At this time, too, cover the bed with fish netting or other suitable protective material to avoid losses from birds. If there are signs of mildew or other diseases and straw has been used, burn this when all the fruit has been gathered. This will burn off the old foliage but new, disease-free leaves will soon appear. Remove all runners unless these are wanted for propagation purposes.

If the plants are covered with cloches in March they will produce ripe fruit in late May in sheltered areas, but make sure that the soil does not become dry when giving this protection. By using cloches and choosing one's varieties carefully the season can be spread over six or seven weeks in summer, with more fruit in the autumn if perpetual-fruiting varieties are grown.

The perpetual-fruiting varieties flower continuously from May onwards, and produce good-sized and well-flavoured fruit in the autumn in abundance. However, they will only do this if the first blossoms are removed, and only the later ones allowed to fruit.

Alpine and perpetual-fruiting varieties can be planted closer together (about 9in. apart) and be allowed to form a matted bed. They throw runners which start to flower and fruit in the same season if allowed, but the flowers from these runners should be taken off if strong plants are required for the following year. Strawing is said not to be necessary because they hold their fruit high, but in practice I find that it is still more satisfactory to protect them in this way. Providing the plants with the protection of cloches in the autumn will result in larger fruits of better colour.

Propagation. Increase strawberries from the plantlets formed on runners, choosing the best and planting out the young plants in late summer or early autumn. The plantlet on each runner nearest the plant should be chosen in each case, the rest being removed. It is better to remove the unwanted runners before they grow to any size and deplete the parent plant of its energy. Between four and six should then be left on each parent plant. The runners are pressed down with a bent piece of wire. The best months for this job are June and July and the plantlets should be well rooted by the end of August and can then be severed from the parent plants. They should be left in position for about a week longer and then transferred to their permanent quarters.

Some gardeners are reluctant to propagate strawberries themselves because of the proneness of these plants to virus diseases, and the decision is not made any easier by the difficulty of positively recognising some forms of virus attack.

Some of the perpetual-fruiting strawberries do not make runners freely and must be increased by division at planting time. Alpine strawberries, which produce their small fruits over a long season, are raised from seed sown in a warm greenhouse in February or in a frame in March or April, the seedlings being planted out after a few weeks of hardening-off in a frame.

Varieties. These include Cambridge Favourite, early to mid-season (good for cloches). Royal Sovereign, early to mid-season (strong grower and runs to leaf on heavy soils; first-class flavour). Talisman, late mid-season (produces a second crop in a mild autumn with cloche protection). Redgauntlet, mid-season. Hummi Grande, mid-season (a new variety from Germany which produces exceptionally large berries of good flavour). Alpine variety: Baron Solemacher. Perpetual-fruiting varieties: Sans Rival and St Claud.

Pests and Diseases

Advice on the more common pests and diseases is given in Chapter Thirteen (p. 302 to p. 305).

Right. The late-maturing dessert apple Rosemary Russet. This excellent variety has a season lasting from December to March. It makes a tree of reasonably small dimensions

Right, below. The October-maturing pear Louise Bonne of Jersey, a variety noted for the superb quality of its fruits which are usually borne freely. The blossom on this variety is highly decorative—a factor not to be overlooked when planning the garden

Far left. The fruits of the raspberry
Malling Jewel, a vigorous,
good-cropping variety for picking
in mid-July. Raspberries are an
easy crop to grow and need
comparatively little attention.
They are the best fruit of all for
deep freezing

Left. The strawberry Redgauntlet
which matures its fruit in
mid-season. Strawberries are
grown most of all, of course, for
dessert but their usefulness for jam
making and deep freezing should not
be underestimated

Right. The Morello cherry, an
excellent fruit for growing as a
fan-trained tree against a north wall,
as well as growing as a free-standing
tree or bush. It bears its fruit in August
and September and is self-fertile

Left, below. No plum is better
known than Victoria, shown here.
This very free-cropping plum is
suitable for dessert or culinary use,
matures from mid to late
August and is self-fertile. It is,
however, inclined to contract Silver
Leaf Disease, a factor which must
be weighed against its many
excellent qualities

Right, below. The strong-growing
red currant variety Red Lake. Of
American origin, this heavy
cropping, mid-season variety bears
large fruits and has much to
commend it

Below. The author casts a critical eye over his Ailsa Craig tomatoes. This well-known variety has a natural appeal for it crops extremely well and the fruits are of medium size and good shape

Far right. The wise gardener makes every inch of space count in the vegetable garden by catch-cropping, intercropping and successional cropping

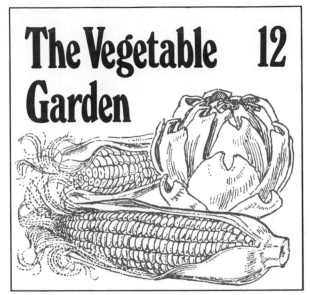

The Vegetable Garden 12

My own garden would not be complete without an area for growing vegetables. It might be thought that vegetables are hardly worth growing nowadays, but I would not agree with this contention for several resons. Take the question of freshness; there can be no doubt at all that home-grown produce, freshly picked, has a quite different flavour to shop-bought produce, and when the family numbers, say, four or more there can be quite a big saving on the greengrocer's bill if only some of one's requirements are grown in the garden. Most important of all, it really can be enormous fun to plan for and produce a regular supply of vegetables. A well-cared-for vegetable plot has its own visual appeal.

What to Grow

Of course, one wants to consider vegetable growing objectively, and it is only common sense to try and work out a plan of action which takes economics into account. This means growing some of the vegetables which are always rather more expensive and stealing a march by advancing the maturing date of some crops through the use of cloches. Plastic cloches are freely available nowadays at very reasonable prices and with these it is possible, for instance, to advance the planting dates for early potatoes and the sowing dates for broad beans, onions, lettuce, peas, carrots etc., by two to three weeks, which is really advantageous. I agree that in a small garden it is not worth growing main-crop potatoes, for they take up too much space for too long, but a few rows of earlies are certainly worth growing, for these are cleared in time to allow the ground to be used for a follow-up crop in the same season. The area set aside for vegetables will depend on the size of the garden and it may not be sufficient to supply fresh vegetables for the family throughout the year. To get things in perspective I would expect a plot of

Some vegetables, like the beetroots
below, can be stored for consumption
as and when required. Growing
vegetables which can be used over
an extended period in this way makes
home-grown produce even more useful

already indicated and all things being equal—to grow the less usual crops which are never cheap in the shops, if obtainable at all. For example, such salad crops as endive, which is much more than a substitute for lettuce in salads, and if correctly blanched, introduces a new flavour to dishes. Chicory and seakale are good vegetables which are always a little expensive to buy but easy enough to grow and blanch for use in winter and spring, especially if a greenhouse or warm cellar is available.

Whatever part of the country one lives in there will be certain vegetables which do better in that district than other kinds. Both crop yields and flavour vary with different soils and it is always worth enquiring which kinds do best in your district. For this and many other reasons it is well worth joining the local horticultural society. In regard to varieties, it is also worth remembering that those varieties which are favoured by commercial growers, because they give high yields with a minimum of trouble, need not necessarily be the best for the home gardener to grow. The commercially-favoured varieties and strains are not always the kinds with the best flavour, and this should be a primary consideration when you are growing for your own table. So, talk to local gardeners and make use of their experience, as well as doing some experimenting yourself. Study the seedsmen's catalogues closely—it is an entertaining pastime—and get off your order early so that there are no unnecessary delays.

Catch-cropping, Intercropping and Successional Cropping. As one becomes more experienced in vegetable growing, it is possible to increase crop yields by catch-cropping, intercropping and successional cropping. Catch-cropping consists of growing a quick-maturing crop between the timings of the main growing period for which the ground is reserved. For example, lettuce, summer spinach, radishes, mustard and cress and spring onions are some of the crops one can use for this purpose. They can be grown, for instance, between the trenches prepared for celery provided they are off the ground before the celery needs earthing-up, or they can be grown on a plot prepared for winter greens.

Ground can also be exploited by intercropping—growing quick-maturing vegetables between others which remain in the ground for much longer periods. For example, radishes can be grown between parsnips or onions, lettuces between peas and so on. Successional sowing of individual vegetables is extremely useful as it ensures a continuity of supply in quantities bearing some relation to one's needs rather than having a glut of supplies over a much shorter period. Lettuce, spinach, carrots and radishes are examples of vegetables which are best cropped in this way thus preventing unnecessary wastage.

about 60ft. by 30ft. to supply a family of four with fresh vegetables throughout most of the year, apart from potatoes.

The most common fault is to have too much ready at once so that lettuces run to seed, beans hang unpicked and radishes grow large and uneatable. If you have a deep freezer this problem is largely solved but if you have not the answer is to sow a little and often. Successional sowings can be made of many salad and other vegetable crops at two- or three-week intervals from March or April onwards to provide sufficient for the family's requirements without having a glut. Some vegetables such as round or globe beet, parsnips, main-crop carrots and onions can be grown also for storage during the winter and consumption at times when they are more expensive if not actually scarce.

Where space is limited it is best—as I have

Essential Preliminaries

The best time to start making a vegetable garden is in the autumn for it is then that one must plan for the year ahead. To grow vegetables well you must have good soil and I feel strongly that at least one-third of the vegetable garden should have manure, compost, peat or other humus-forming material dug into it once a year (remembering, of course, that carrots and parsnips should not be grown in recently manured soil as this causes the roots to fork). Vegetables, on the whole, grow fast and any check to growth, such as shortage of water or lack of food, results in small stringy roots on root crops, running to seed in the case of celery, small and yellow leaves on brassicas and similar crops and poor quality pods in inadequate numbers on the legumes. So, the soil must be well-cultivated, well-manured and well-fed, it must be left sufficiently moist and—very important—free of weeds.

If the soil is of medium or heavy structure I am a great believer in winter digging. As I have stressed elsewhere in this book, with the heavier soils Nature is of great service. The frost, snow and rain will break down the large lumps and make the task of seed bed preparation in spring a comparatively easy chore. If farmyard manure, garden compost or leaves are dug in at this time make sure that they are well rotted. Spent mushroom compost and spent hops are both good for this purpose, and so are fish manure, sewage sludge and seaweed manure.

If your soil is light and quick-draining the manure is best applied only about a month to six weeks before seed sowing. If it is applied too early the food will be washed down into lower levels of the soil and so be unavailable to the vegetables when the time comes.

While the humus-forming materials I have mentioned do much to improve the physical condition of the soil, they are extremely unreliable in the amount of food they provide and with some crops it will be necessary to apply concentrated fertilisers supplying one or more of the three main plant foods—nitrogen, phosphorus and potash.

Bonemeal and basic slag (this is one of the cheapest fertilisers available and it is ideal for the vegetable garden although one must remember that it contains a percentage of lime) are two fertilisers we cannot afford to overlook. Both are slow acting, releasing their food over an extended period and both supply phosphate, especially beneficial to root vegetables, and the formation of root growth in general, including the rootlets of newly-germinated seeds. As I have already indicated, basic slag contains lime and is therefore useful for making acid soils more alkaline. The best time to apply it is during the autumn or winter, and at least a month after any manures are added, but not later than mid-February.

At least a third of the vegetable garden should, if possible, have manure, compost, peat or other humus-forming material dug into it each year. Heavy soils should be worked in winter so that frost, snow and rain break down the lumps

Compound proprietary fertilisers containing the three main plant foods are best applied as a preliminary dressing before seed sowing, being worked into the top few inches of soil about seven to ten days previously. Apply the dressing evenly whether it is broadcast or spread in strips where the rows will be.

Fertilisers are also used as topdressings during the growing season, and I like to stick to the policy of a little and often whether it be a compound fertiliser or a 'straight' fertiliser such as sulphate of ammonia or dried blood (both nitrogenous foods) or a solution of nutrients. Where dry fertilisers are used they must be sprinkled on the soil carefully, avoiding the leaves of the plants which could be scorched; and if the weather is dry they should be watered in.

Vegetables, with one or two exceptions, are grown from seed, and successful germination and

Globe Artichoke, a vegetable grown for its flower heads, which are cooked and the scale leaves eaten as a delicacy. The flowers can be used in flower arrangements

subsequent growth are greatly influenced by the state of the seed bed and the type of soil; both should be first class. Buy the best quality seeds as it is pointless to go to so much trouble to experience disappointment later because the seed was old or otherwise inferior. Store it until sowing time in an air-tight tin or tins, so that mice cannot cause losses.

A lot of gardeners attempt sowing too early. More important than the date is the temperature of the soil and its condition. An old country practice is to watch the hawthorn hedges, for when they start to come into leaf, it is a pretty good indication that the soil temperature is rising and seed sowing can be started. As a general guide it has much to commend it. Again, as a generalisation, one can say that sowing times for the first crops like parsnips, broad beans, onions and lettuces will be early March in the South, the end of March in the Midlands and the beginning of April in the North.

The first preparation of the seed bed should be done when the soil is moist but not wet, a point of especial importance on clay soils where compaction will add to one's difficulties. As recommended earlier, heavier soils should have been dug and left rough for the winter so that the elements could play their part in breaking down the large lumps. Any large lumps remaining should be broken down with the back of a fork and the whole plot lightly turned over, removing any large stones and, of course, weeds. It is then firmed by treading and afterwards worked to a really fine tilth by raking. If you have a hay rake this is a splendid tool for a first raking, following this with another rake-over, at right angles to the first path of progress, with an iron garden rake. This should provide the fine tilth and firm seed bed which is needed.

Some seeds are tiny, and in germination their roots and shoots will have a tough struggle if the seed bed is not up to standard. A firm and level surface is required, but 'panning' must be avoided. This is when the surface is absolutely smooth, like the top of a table, but with the surface soil particles so close together that water lies on the surface when it rains and the soil develops cracks in dry weather.

Seeds can be sown broadcast over the seed bed, as with mustard and cress, or in lines or 'drills', which is more trouble but gives better results. A garden line is then needed and a drill can be taken out with the corner of a hoe or rake or the top of a pointed stick. Most seeds are sown at depths between $\frac{1}{4}$ and $\frac{1}{2}$in., and some gardeners like to line the drill with a thin layer of peat.

Sow seeds evenly and thinly in the drills for to do otherwise is wasteful, and thinning is more difficult later. If the soil is dry, water the drills beforehand, and if peat is put in the drills make sure that it has been well moistened. Cover the seeds by raking soil over them.

Crop Rotation. Where space allows, vegetables should always be grown on a rotation system so that specific vegetables are not grown in the same ground for two successive years. The reason for this is that certain crops tend to exhaust the soil in the same way or to suffer from the same diseases, and it is therefore advisable to avoid growing them one after the other on the same plot of ground. Cabbages, Brussels sprouts, broccoli, cauliflowers and kales form one such group, while peas and beans form another. A common method of rotational cropping is to divide the ground into three approximately equal sections and to grow potatoes, root crops and celery on one, peas and beans, onions, leeks and lettuces on another, and cabbages, Brussels sprouts, cauliflowers and other green crops on a third. Then the next year the groups are shifted round one plot—potatoes etc. going to plot 2, peas and beans etc. going to plot 3 and cabbages to plot 1. A similar change is made in the third year, and in the fourth year the crops are back to their starting quarters again and the whole cycle can then be repeated.

ARTICHOKES, GLOBE

A vegetable grown for its flower heads, which are cooked and the scale leaves eaten as a delicacy. Plants can be grown from seed sown in a frame in March or outdoors in April, but a better method is to detach offsets in early April from plants known to produce good crops. These offsets are planted 3ft. apart each way in fairly rich soil that has been well dug, and in an open position. Mulch with manure in May and remove all flower stems the first year. Best results are obtained in the second and third years when flower heads should be gathered regularly as soon as they are plump and before they start to expand. Remove dying leaves in autumn and protect the crowns from frost with dry straw or bracken.

Varieties. These include Green Globe and Purple Globe.

ARTICHOKES, JERUSALEM

These are grown for their tuberous roots which are cooked and served in a similar manner to potatoes. They are very hardy and easily grown in any ordinary soil and an open position. Plant the tubers in February, 6in. deep and 15in. apart, in rows 2½ft. apart. Hoe them occasionally, drawing the soil towards the rows, and lift them as required for use in autumn and winter. The stems grow to a height of 6ft. or so and make a good temporary screen for, say, a compost heap or anything else best hidden.

ASPARAGUS

When growing this vegetable it is important to make sure that the drainage is really good, and except on light land it is advisable to raise the beds a foot above the surrounding soil level. The ground should be deeply dug and be given a good dressing of well-rotted manure. The planting season is April, and the plants should be placed 15in. apart with 18in. between the rows. Set the corms 2 to 3in. deep. Beds 4ft. wide accommodating three rows are most convenient. One-year-old plants are best, but no shoots should be cut until the third year after planting. During the first year allow the plants to grow and give them a little fertiliser. In the second year pull up a little soil along each side of the row, as though one was earthing-up potatoes, doing this at any time before growth starts. Feed at least twice a year at this stage and then start to cut in the third year. Ample supplies of water must be given during dry weather.

After cutting down the growth, about the first week in November, a dressing of well-decayed manure should be applied. Cutting should never be continued beyond mid-June, and sufficient growth must also be allowed to remain to keep the plants growing strongly. Increase by sowing seed in light soil in rows 1in. deep in early April. Thin the resulting seedlings to a foot apart and plant in their permanent quarters the following April.

Varieties. These include Connover's Colossal and Martha Washington.

BEETROOT

There are three principal types of beetroot: globe-rooted, long-rooted, and tankard, which is intermediate between the other two. Globe beet is favoured for early crops, and long-rooted and tankard for main-crop and storing.

Seed should be sown in April, May and June in groups of two or three, 6 to 8in. apart in drills 1in. deep and 15in. apart. Later, the seedlings

The dwarf broad bean, The Sutton. It provides a long succession in the garden and early crops under cloches. The rows are set 12in. apart for this variety

should be thinned to 6 to 8in. apart. Well-worked ground, manured for a previous crop and dressed with a good general or all-purpose fertiliser before sowing, is best. Keep well hoed, lift the roots from July to October when of the desired size for kitchen use, and store in sand or fine ashes in a shed or other sheltered place.

Varieties. These include Crimson Globe and Sutton's Globe. A new variety said to be resistant to bolting is Boltardy.

BROAD BEANS

Sow the early long-pod varieties in November, or February to April, and the Windsor varieties in March or April, in rich, well-manured soil in drills 3in. deep and 2ft. apart, with the seeds 4 to 6in. apart in the drills. If more than one row is grown, leave 3ft. between the rows. The growing tips of the plants should be nipped out when a reasonable number of pods has formed. Water late crops well in dry weather, especially to prevent attack by blackfly on light soils. The taller varieties may require supporting with string which is strained between stakes placed at each end of the row.

Varieties. I like to grow such varieties as Exhibition Longpod, White Windsor, and Aquadulce Claudia (for autumn and early winter sowing). The Sutton is a dwarf variety.

BROCCOLI

There is no real distinction between broccoli and cauliflower, both of which are grown for the close white heads (or the numerous purple or white shoots of the sprouting broccoli). However, in gardens the term broccoli is usually applied to the hardy autumn and spring kinds, and the term cauliflower reserved for the more delicate summer varieties. All are brassicas and have the same general requirements as other brassicas. There are numerous varieties of broccoli differing in the time at which they produce their heads or curds, and in addition there are sprouting varieties which produce a succession of shoots in spring with white or purple flower buds in close clusters. These are cut as required and, when cooked, make very good eating.

Sow in March, April or May outdoors and transplant in May or June to good, rich, well-worked but firm soil. It is most important that the soil should be firm. Plant 3ft. apart in rows 3ft. apart. Feed during summer with small topdressings of a compound vegetable fertiliser. Draw a little soil around the stems in autumn to provide better anchorage. Break some leaves over the curds as they form to protect them from frost. Cut the curds as soon as they are well grown or pick shoots of sprouting kinds before the flower buds start to open.

Varieties. These include Calabrese (Italian Green Sprouting) (August–September); Veitch's Self-Protecting Autumn (October–November); Early Purple Sprouting (January–February), Purple Sprouting (March); Knight's Protecting and Leamington (April); Whitsuntide and Late Queen (May); and Midsummer (June).

BRUSSELS SPROUTS

These belong to the brassica family and have the same general requirements as other brassicas (see Cabbages, p. 274). Varieties are available to crop from September to February or later according to season.

Sow the seeds in February in a frame or cool greenhouse, or in March outdoors. Plant in April or May, 3ft. apart each way in good rich soil, which should be well-worked and, most important of all, firm. Feed occasionally in summer with small dressings of a compound fertiliser and water freely in dry weather. If the plants get very large, they should be individually staked. The sprouts are picked a few at a time, starting from the bottom of the stems, where they are most mature. The yellowing leaves can be removed at any time, but the tops of the plants should not be cut off until all the sprouts have been picked.

Varieties. These include Cambridge No. 1 and Cambridge No. 5, Exhibition, Early Button (exceptionally good) and Jade Cross (an F1, hybrid which crops heavily and early).

CABBAGES

These belong to the brassica family which includes broccoli, Brussels sprouts, mustard, savoy, seakale and turnip. There are many points of cultivation which this family has in common. First and foremost amongst them is a demand for a rich but well-firmed soil. In loose soil they are never satisfactory. Every endeavour should therefore be made to have sites for them prepared some considerable time in advance of planting, or simply to rake down ground vacated by early short-season crops, in order that the soil may have a moderate degree of solidity. Given firm soil and a reasonable supply of plant food, none of the cabbage tribe is difficult to grow, but as they take a good deal out of the ground it is advisable to make sure that they are crop-rotated, so as to lessen the danger of disease such as club-root building up. The soil should be limed as all brassicas grow best in such soils and it also acts as a deterrent to the disease just referred to.

There are quick-growing varieties of cabbage such as Primo, Velocity and Greyhound for summer use, slower growing kinds such as Rearguard and Winnigstadt for autumn use, very late and hardy kinds such as January King which can be used in winter, yet others such as Ellams Early and Harbinger to be sown in summer and cut the following spring, and red cabbages grown specially for pickling.

The May-maturing broccoli Late Queen. This vegetable needs to be grown in good, rich, well-worked but firm soil. Firming is of special importance

Sow the summer cabbages in a frame or greenhouse in February, or outdoors in March, and plant out in April or May. Sow the autumn cabbages outdoors in March, April or early May and plant out from May to early July. Sow the spring cabbages outdoors between mid-July and mid-August and plant out in September or October.

All require good, well-worked but firm soil. Most varieties should be spaced 18in. apart in rows 2ft. apart, but the big drumhead and pickling varieties need a little more room and the small spring varieties can have a little less. The summer, autumn and winter varieties should be fed occasionally in summer with small dressings of a general or all-purpose fertiliser. Spring cabbage should not be fed in this way until danger of prolonged frost is over, say in April and May.

Varieties. These include the following – Summer maturing: Greyhound, Primo and Velocity.

Autumn maturing: Rearguard and Winnigstadt.
Winter maturing: January King. Spring maturing: Ellams Early and Harbinger.

CARROTS

There are four types of carrot in general cultivation: the shorthorn, stump-rooted, intermediate and long-rooted. The first is used mostly for forcing in frames; the stump-rooted varieties are quick-maturing and will supply roots in early summer, while the intermediate and long-rooted kinds are best for storing during the winter.

Light soil suits carrots best and heavy soils should be improved by thorough cultivation. No fresh manure must be used, or the roots may fork badly. Deep digging in autumn or winter followed just before sowing by a dressing of a general or all-purpose fertiliser is all the preparation required. March is soon enough to sow outdoors, starting with the shorthorn varieties. These are

Blanching celery. Paper is wrapped round each plant and soil drawn up around the plants at intervals until only the final tufts of leaves are exposed

best sown in small successional batches. Main-crop sowing should be made in April; 12in. must be allowed between the rows with long-rooted and intermediate varieties, but 8in. is enough for the shorthorn and stump-rooted kinds. Only a very light covering of soil is needed and thinning to 4 to 8in. apart must be undertaken as soon as the seedlings are large enough to handle. Pull small-rooted varieties as soon as they are large enough to use.

Apart from hoeing, the main objective in the summer months is to keep the carrot fly at bay by the methods given in Chapter Thirteen on pests and diseases (see p. 305). Main-crop and intermediate varieties lifted in October can be stored in sand or ashes in a cool shed or they may be put in clamps of the same types as those used for storing potatoes.

Varieties. These include Early Nantes, Red-cored Early Market and James Intermediate, Chantenay, Stump-rooted Intermediate.

CAULIFLOWER

These need to be grown quickly in good, rich, well-manured soil. First sowings can be made in February, in a frame or greenhouse, to be followed by a further frame sowing in March and an outdoor sowing in April. An outdoor sowing can also be made in early September, the seedlings being transplanted to a frame at the end of October. In this they will spend the winter and will be planted out the following April for an early crop. Plant at least 2ft. apart in rows 2½ft. apart, water freely in dry weather and feed occasionally with small dressings of a general or all-purpose fertiliser. When curds start to form break in some inner leaves over them to keep them white. Cut as soon as well grown (see also Broccoli, p. 273).

Varieties. These include All the Year Round, Snowball, Snow King (an F1, hybrid) and Veitch's Autumn Giant.

CELERIAC

This vegetable is closely allied to celery, but it is the bulbous growth between the roots and the leaf stems that is the serviceable part of the plant. It is often used as a substitute for celery for flavouring soups and stews, as well as a vegetable for use in its own right. Seeds and seedlings are treated in exactly the same way as those of celery. The soil must be well dug and well manured but no earthing-up is necessary. Plants are put outdoors early in May, 9in. apart in rows 1ft. apart. Subsequently, the only attention required is regular hoeing and liberal watering in dry weather. When mature in late summer, the plants can be lifted as required, or they can be left in the ground for the winter in warmer parts of the country. Elsewhere, the roots should be lifted before the frosts arrive and stored in an airy shed in sand.

CELERY

There are three principal types of celery: white, pink or red, and self-blanching. All must be grown in deeply dug and well-manured soil. A common practice is to prepare trenches 18in. to 2ft. in width and as much in depth, throwing out the soil from these and then returning most of it, mixing well-rotted farmyard or stable manure, or if these are not available, garden compost with it. The top 5 or 6in. of the trench is not refilled and the surplus soil is built up into ridges on either side of each trench, to be used later in earthing-up the stems. Many gardeners go wrong by making the trench too deep. Sprinkle a little all-purpose fertilizer along the trench at the rate of 1½oz. per yard run. Lightly fork this into the soil. During the early summer months catch-crops such as lettuces and radishes can be grown on these ridges.

Chicory can be forced in covered pots in complete darkness in autumn or winter, accommodation being provided in a cellar, shed or space under the greenhouse staging

Seed is sown in a warm greenhouse in March, or for an early crop at the end of February. The seedlings are pricked off into deep boxes and are hardened off for planting outdoors 12in. apart in May or early June. During the growing season it is hardly possible to over-water celery. Once the plants are established, feed regularly with very weak liquid manure.

Approximately six to eight weeks is required for blanching. Paper is wrapped round each plant to prevent soil getting into the heart—not more than 2- to 3-in. bands at a time—and then soil is drawn up around the plants at intervals of a week or so until only the final tuft of leaves is exposed. It may be necessary to make a loose raffia tie round the stems to keep them together. Never make the tie tight for the stick is developing all the time and the heart is coming up through the centre. Slugs are a major pest of this crop and slug pellets should be sprinkled along the trench each time the plants are earthed-up.

When celery runs to seed this is due to the plants receiving a check at an early stage of growth, being allowed to starve in the boxes, suffering from dryness, or being left too long before planting out.

During severe weather the tops of the ridges may require some protection, with straw or bracken laid along them. Self-blanching celery is grown in a similar manner except that it is planted in blocks, the plants being set 9in. apart each way, and no earthing-up is required. It is suitable for summer and early autumn use. Next, the white celery should be used, the pink or red varieties being left until last as they are the hardiest.

Varieties. I still prefer the kind which needs earthing-up and varieties of this type which I like to grow are Clayworth Prize Pink and Solid White. Self-blanched varieties include Golden Self-blanching.

CHICORY

This vegetable is grown for the young growths, or 'chicons', which are blanched and either eaten cooked or raw. Seed is sown in early June in good, well-drained soil and drills ½in. deep and 15in. apart. The seedlings are thinned to 1ft. apart. The flowering stems are removed if any appear, and the roots lifted as required for forcing in autumn or winter. The tops should be cut off about 1in. above the crowns and the crowns should then be placed close together, right way up, in large pots or deep boxes with any fairly light soil such as old potting or seed soil. Five can be accommodated in a 9-in. pot, the bottoms of the top roots being cut off. Another pot of the same size is inserted over the pot containing the plants. Forcing should be in complete darkness and in a temperature of 10 to 13°C. (50 to 55°F.). Cellars, sheds or the space under the staging in the greenhouse may be used.

Outdoors, chicory can be blanched where it is grown by covering each plant with an inverted flower pot, or by drawing soil up in a ridge along the rows as when earthing celery. Whatever method is used the blanched growths are broken off close to the crowns when about 9in. high.

Variety. Witloof de Brussels.

CHIVES

These are relatives of the onion and the leaves are used in salads and for flavouring. Any ordinary garden soil will serve for this plant which should be given a sunny position. Plant in March 6in. apart in rows 6in. apart. No subsequent attention beyond hoeing is required, and the plants can be left undisturbed for three or four years, after which they should be lifted, divided and replanted in February, March or early April.

CRESS

Cress, usually grown for consumption with mustard (see p. 279) is an easy crop to grow. Seeds can be sown in the open garden from April to the end of August and under glass during the rest of the year, provided a temperature of about 13°C. (55°F.) can be maintained. Make indoor sowings in shallow boxes filled with light soil and do not cover the seeds with soil—just cover the box with paper until germination takes place. As cress takes longer to mature than mustard, sow the former three days before the mustard if they are wanted for cutting at the same time.

Varieties. These include Curled and Plain.

CUCUMBERS

These can be grown in either heated greenhouses or in frames on a hot bed, but the best and biggest crops are obtained from greenhouses. For this purpose seed should be sown singly in small pots in a temperature of 18 to 21°C. (65 to 70°F.) at intervals from January until the end of April. Water rather freely throughout.

The fruiting beds should be prepared either on the floor of a low span-roofed greenhouse or on a flat staging in a taller structure. In either case the rooting medium should be rich compost of fibrous loam, leafmould and well-rotted manure. A ridge of compost about 15in. wide and 7in. deep in the centre is sufficient for young plants. When roots appear on the surface they should be covered with a thin layer of rich, light compost similar to that used in the preparation of the ridge. The plants should be from 3 to 5ft. apart, and the shoots must be tied to wires strained lengthwise along the house.

Main growths may be allowed to run until they reach the apex of the roof. Side growths are trained horizontally and are pinched at the second leaf joint. The plants should be syringed daily with tepid water and must be shaded from strong sunshine. Weak liquid manure can be given as soon as the first fruits begin to develop. In heated frames cucumbers can be planted in April, and in unheated frames at the end of May or early in June. For this purpose seed should be sown as for indoor cucumbers but in March.

Ridge cucumbers, which are much hardier than any other kind, can be planted outdoors, in early June, on ridges of good soil built up in a sunny, sheltered position. They should be watered freely during the hot weather. The seed of this type of cucumber should be sown in April.

Varieties. These include Butcher's Disease-resisting (a sound old variety); Improved Telegraph (especially recommended for frame cultivation); and Simex (one of the new completely, non-bitter, all female, F1 hybrids). A good ridge variety is Baton Vert, also an F1 hybrid).

ENDIVE

A vegetable like lettuce in appearance, endive is grown for use as a salad. Seed is sown in ½in. deep drills at intervals from April to mid-August in good, rich, well-dug soil and an open situation. The seedlings are thinned to 9in. apart. When the plants are well-grown each is covered with an inverted flower pot, the hole in the base being covered with a plate or piece of wood to exclude all light and blanch the leaves. This will take six

A side growth of a cucumber being stopped. Main growths are allowed to reach the apex of the greenhouse roof, before they are stopped

weeks. Late-sown endive is best protected with a frame or cloches in autumn and for the last few weeks the glass can be white-washed to secure a sufficient measure of blanching.

FRENCH BEANS

The ground should be well prepared by digging, the addition of rotted manure or compost, and a good sprinkling of a general or all-purpose fertiliser a week or so before sowing. Sow in late April and early May, 1in. deep in drills 18in. apart, spacing the seeds singly 6in. apart in the drills. Gather the beans regularly as soon as they attain usable size, and do not allow any to mature and produce seed as this checks further cropping. Early crops can be produced by sowing in January or February, five or six seeds in a 6- or 7-in. pot in John Innes No. 1 Potting Compost and keeping in a greenhouse or frame in which a temperature of 13 to 18°C. (55 to 65°F.) is maintained. Only two-thirds fill the pots at first and then top dress with more John Innes compost as the plants grow.

Climbing French beans are grown in exactly the same way except that the rows should be at least 3ft. apart, and pea sticks should be stuck in firmly along the rows for support. There is a purple-podded variety of good flavour (named Purple Podded) which cooks green and is easy to grow. It also crops well; bean sticks are needed for support.

Varieties. Varieties of dwarf French bean include The Prince, an early forcing variety; Masterpiece, main-crop; and Canadian Wonder, a late variety. In addition to the climbing French bean already referred to, a good variety is Tender and True.

GARLIC

A well-drained, rather light, but reasonably rich soil, and an open, sunny position should be chosen. The bulbs are planted in February 2in. deep and 6in. apart in rows 8in. apart. Beyond regular hoeing, no further attention is required until July or August, when the bulbs are lifted and stored in a cool, airy place. They are often tied in small bunches and suspended from a nail or beam in a shed.

LEEKS

A deep, rich, well-manured soil is necessary to produce the best results. Seed should be sown in January in a warm greenhouse for an early crop, or outdoors in March in rows 1ft. apart. Seedlings raised under glass must be pricked out into boxes as soon as they can be handled and any required for exhibition may have a further move singly into small pots. Hardening-off should be completed in time for planting in May. Outdoor-sown leeks are usually planted in June. In both cases the plants should be 8 or 9in. apart in rows about 18in. apart. Hoe frequently during the summer and water freely in dry weather. Feed as advised for onions (see p. 279). As growth proceeds the stems must be earthed-up to ensure blanching.

An alternative method is to plant in holes bored with a stout bar so that only the tips of the leaves show above the rim, but this is not suitable for exhibition purposes. A third method, suitable where small leeks are preferred, is to thin the outdoor-sown leeks in their seed bed instead of transplanting them. Then soil is gradually drawn up around them to blanch the stems.

Varieties. These include Musselburgh, The Lyon and Holborn Model.

LETTUCES

There are two main types of lettuce, the cabbage and the cos, and a great many varieties of each differing in colour, crispness, size and season.

Lettuces like a well-dug, rather rich, well-manured soil. They are frequently treated purely as a catch-crop, and good yields can be obtained from sowings made on the ridges of celery trenches or between rows of peas. For earliest crops a sowing can be made in pots or boxes of sandy soil under glass in February. Prick out the seedlings as soon as they can be handled and plant out in a sheltered place in late March or early April when thoroughly hardened. Outdoor sowings can be made at fortnightly intervals from early March onwards until mid-August (see below for details of a later sowing). Seedlings can be thinned where they stand and thinnings can be transplanted elsewhere to give a slightly later supply. Space the large varieties 1ft. apart in rows 1ft. apart; small varieties such as Tom Thumb 8in. apart in rows 1ft. apart.

Selected varieties are available for sowing in late August or early September in the open garden to reach maturity in early spring. Plant the seedlings out 9in. apart in rows 12in. apart in early October. These should be grown in a warm, sheltered border. Slug pellets will need to be put down the rows in late autumn and early spring. If some of the plants are covered with cloches, cuttings can be advanced. Other varieties can be sown at the same time for winter cultivation in greenhouse or frame.

Varieties. These include – Cabbage lettuces: All the Year Round; Attraction and Cheshunt Early Giant (both varieties for greenhouse and frame); Continuity; Imperial (a variety for autumn sowing outdoors); May Queen (for greenhouse and frame); Tom Thumb (very small hearts and very early maturing); and Webb's Wonderful (a large leafy lettuce). Cos lettuces: Buttercrunch, Little Gem (this variety, my favourite, is sometimes listed as Sugar Cos, very early and of especial value as it takes up little space and there is little waste); and Winter Density (for spring, summer and autumn sowing).

MELONS

Melons may be grown in greenhouses or frames. For greenhouse cultivation, seeds are sown singly in small pots in a temperature of 18 to 21°C. (65 to 70°F.) at any time from January until the end of May. The plants are grown on ridges of rich soil either on the floor of a low greenhouse or on the staging of a taller structure. The bed should be about $2\frac{1}{2}$ to 3ft. in width with a 6-in. depth of soil spread over it, and, in addition, a narrow ridge, a further 6in. deep, should be made towards the back. On this ridge the plants will be set 2ft. apart. The compost should consist of fibrous loam mixed with a little good leafmould, and well-rotted manure added. The plants are trained as single stems to wire strained from one end of the house to the other, 6in. from the roof glass. When each plant is about 30in. in height pinch off the top. Side shoots will soon grow and four to six per plant should be retained. On each, female flowers will form, distinguishable by the small embryo fruit immediately behind the flower. These must be fertilised with pollen from the male flowers, and all the female flowers of one plant should be fertilised at one time. Usually four fruits per plant are sufficient.

When the young melons begin to swell all

sub-lateral shoots should be pinched out. Water should be given freely at all times until the melons are nearly ripe and begin to emit their characteristic smell, and the atmosphere must be kept humid. Melons can be fed freely with weak liquid or soluble fertiliser from the time the fruits commence to swell.

For frame cultivation, seed should be sown early in April or May and the plants set out in June. Two plants can be accommodated in a frame measuring 6ft. by 4ft. The plants should be stopped at the fourth rough leaf and four side growths retained for flowering and fruiting. Subsequent culture is the same as for melons in greenhouses. It helps if the frames can be stood on a hot bed or if they can be soil-warmed with electric cables.

Varieties. These include Dutch Net (an early, large fruited variety with orange-pink flesh especially suited for frame cultivation), Hero of Lockinge (this has white flesh and is excellent for the greenhouse or frames), Emerald Gem (excellent flavour), and Superlative (with scarlet flesh).

MUSTARD

Outdoor sowings can be made at intervals from early April until the end of August in an open border. The seeds may either be covered with light soil or with mats or boards until germinated. Indoor sowings should be made on the surface of shallow boxes filled with light soil and covered with a sheet of paper (until germination occurs), and, provided a temperature of 13°C. (55°F.) or thereabouts can be maintained, can be made at practically any time of the year. Mustard is ready more quickly than cress, and the cress should therefore be sown three days in advance if these two crops are required simultaneously.

Varieties. Brown and White.

ONIONS

The soil for onions should be deep, well-worked and crumbly. It should be prepared as early as possible in advance, preferably in the autumn, so that it may be broken up thoroughly by winter frost. Well-rotted farmyard or stable manure, worked in deeply, will improve the quality of the crop. Wood ashes can be mixed generously with the top soil. Seed can be sown in January or February under glass in a temperature of 16°C. (60°F.); in March outdoors, where the plants are to mature; or in late August or early September in sheltered nursery beds outdoors.

Seedlings from early sowings under glass are pricked out when 1½in. tall and later may be placed singly in small pots. In these they are hardened off for planting out about the middle of April. Seedlings from April sowings are thinned out where they stand. Seedlings from autumn

As leeks grow so the stems must be earthed up gradually to ensure blanching. This vegetable needs a deep, rich, well-manured soil for good results

For early lettuces, seeds are sown in February under glass, the seedlings being pricked out, later hardened off and planted out in late March to early April

The tops of onions are bent over when growth slackens in summer to encourage swelling of the bulbs

sowings are transplanted in March. In all cases the bulbs should be 6in. apart in rows 1ft. apart unless they are required for exhibition, when a little more space should be allowed.

An alternative to sowing seed is to plant onion sets in March or April. The soil is prepared in exactly the same way as for seed sowing and the bulbs are planted 6in. apart in rows 1ft. apart. For the amateur gardener who has difficulty in raising plants from seed this is the answer to his problem.

Hoe frequently during the summer, water freely during dry weather, and during June and July feed every 10 days or so, either with very weak liquid manure or with a general all-purpose fertiliser.

Bend over the tops when growth slackens in summer if the plants do not do so of their own accord. This will encourage the swelling of the bulbs. Lift when fully developed and lay the bulbs out to dry in the sun, preferably where they can be protected from rain. I space mine out in a garden frame and cover this with a light. When they have ripened off, store them in a cool, airy, frost-proof place.

Varieties. These include Ailsa Craig, Autumn Queen, Giant Zittau and Sutton's Long-Keeping.

PARSLEY

For a continuous supply three sowings should be made annually, one in early March, a second towards the end of May, and a third in August. The position should be open and the soil reasonably good and well dug. The seed should be sown in drills $\frac{1}{4}$in. deep and 1ft. apart or as an edging to a bed. Thin the seedlings to 6in., and gather the leaves a few at a time so that the plants are not weakened unduly. Some seedlings from August sowings can be transferred in October to a frame for winter use, or alternatively plants may be covered where they grow with cloches.

Varieties. These include Imperial and Perennial Moss Curled.

PARSNIP

This vegetable needs deeply-worked soil that is in good condition but has not been freshly manured. A general or all-purpose fertiliser should be scattered over the ground at the rate of 4oz. per square yard before sowing. The seed is then sown in March, April or early May, in drills 1in. deep and 18in. apart. The seeds are dropped in, two or three at a time, 6in. apart and later thinned to one at each cluster. If they are being grown for exhibition, a little more space should be allowed. Parsnips are hardy and can be left in the ground all winter, to be lifted as required, but it is usually convenient to lift some roots in November and bury them in sand or peat in a shed or sheltered place as it may be difficult to dig parsnips when the ground is frozen.

Parsley is a useful crop and it also has its decorative value if grown as a path edging. For a continuous supply make three sowings annually

Varieties. These include Hollow Crown, Tender and True and The Student.

PEAS

Peas require a good rich and well-manured and well-dug soil, and the ground should be prepared during autumn or winter. Make the first outdoor sowings as early in March as soil and weather will permit. Sowing can commence in February, with the protection of cloches. Subsequently, sowing may be continued every fortnight or so until early June to provide a succession. Seed may either be sown in drills 2in. deep or in shallow flat-bottomed trenches about 2in. deep, scooped out with a spade, in which case two or three lines of seed may be sown in each trench. Space the seeds 2 or 3in. apart.

There are many varieties of peas, and these may be classified in various ways: as early, second-early and main-crop; as tall, medium and dwarf, and as round-seeded or wrinkle-seeded (marrow-fat). The round-seeded peas are hardier, but the wrinkle-seeded peas are sweeter. Dwarf peas need not be supported though they are better for a few short, bushy sticks. Medium and tall peas must always be supported with pea-sticks (usually hazel branches) or netting. Early peas take about 12 weeks from sowing to first gathering; second-early peas 14 weeks; main-crop peas 16 weeks or more.

Successive rows of peas should be spaced roughly according to the height of the peas; 2-ft. tall peas in rows 2ft. apart; 4-ft. tall peas in rows 4ft. apart, and so on. Picking should be done a little at a time as the pods fill up; it is the lower pods that fill first.

Varieties. These include Kelvedon Wonder and Little Marvel (earlies), Onward and Phenomenon (second-early) and Lord Chancellor and Senator (main-crop).

Seeds of parsnips are sown in 1-in.-deep drills at 6-in. intervals, two or three at each station. The resulting seedlings are later thinned to one per station, as above

Rhubarb forced under the staging in a warm greenhouse. The roots are placed close together in deep boxes, kept moist and in the dark

POTATOES

It is not worth growing a main-crop of these vegetables if space is limited as they can be bought quite cheaply and they take up too much valuable ground for too long in the average garden. However, I like to grow a few rows of early potatoes as the flavour is so good. They require a good, rich, well-manured and well-dug soil which can with advantage be dressed with animal manure in the autumn or winter preceding planting. In addition, a general or all-purpose fertiliser should be forked in at the rate of about 4oz. per square yard before planting.

Potatoes are grown from tubers known as 'seed', which should be certified free of virus. It is an advantage if the 'seed' is sprouted before planting. This is done by standing the potatoes eye ends uppermost, in trays in a light but frost-proof place in January.

Plant them in a very sheltered place in late February or early March and in ordinary places in late March, at 15 to 18in. apart in rows 2½ to 3ft. apart, covering them with 2 to 3in. of soil. In my Shropshire garden I plant at least two rows in early March – I am prepared to take a risk – and one or two more rows later that month or in early April. After planting, sprinkle an all-purpose fertiliser over the surface. When the shoots appear draw the soil over them as protection from frost and continue this earthing-up until the potatoes are growing in ridges. The tubers can be lifted in June or July as soon as they are large enough.

Varieties. Early varieties include Home Guard, Midlothian Early (also listed as Duke of York), Sharpe's Express (my favourite) and Ulster Chieftain.

RADISH

The seed of this salad crop is sown thinly in drills ½in. deep and about 5 or 6in. apart in good rich soil or as a catch-crop in celery trenches. Sowings can be made every fortnight from March to mid-August and winter radishes, such as Black Spanish, can be sown in a heated frame or greenhouse in early autumn. All radishes need to grow fast if they are to be crisp and mild-flavoured. They should be watered freely in dry weather.

Varieties. These include French Breakfast, Icicle, Red White-tipped and Black Spanish.

RHUBARB

Single roots with crowns should be planted 2in. below the surface of the soil 3ft. apart in autumn or early spring in rich, well-cultivated soil in sunny, open positions. Topdress with manure each February, forking it into the soil surface. The plants should be lifted, divided and replanted every four years. No stalks should be gathered the first year. It is necessary to remove all flower stems directly they appear.

Rhubarb can be raised from seed sown in a frame in March or in a prepared seed bed outdoors in April. Sow the seeds 1in. deep in rows 1ft. apart and thin the seedlings to 6in. apart as soon as they can be handled. They will be ready to move and be planted in their permanent quarters in the autumn.

When forcing rhubarb, the crowns should be covered with pots or tubs in January or February, these being covered in turn with fresh manure mixed with leaves. Strong roots can also be lifted and placed close together in deep boxes underneath the staging in a warm greenhouse, a shed or cellar. They should be kept moist and dark in a temperature of 13 to 24°C. (55 to 75°F.). The forcing season is from November–February, and roots two to five years old are the best for this purpose. Roots force more readily if exposed on the surface to frost for a few days before being brought inside.

Varieties. These include Glaskin's Perpetual and Victoria.

RUNNER BEANS

These are more tender than French beans and even in the south of England it is seldom safe to sow outdoors before the end of April. Mid-May is more generally a suitable time, and for a late crop another sowing can be made during the last fortnight in June. Early crops can be had by sowing under cloches or in pots or boxes in a greenhouse or frame, the seedlings being hardened off for planting out in late May or early June.

Runner beans like a fairly rich soil with plenty of moisture in summer. In February or March a trench should be dug out at least 18in. wide and 1ft. deep and some well-rotted manure or compost worked into the bottom. The soil is then returned, adding a little more manure or compost and a sprinkling of a general fertiliser. The seeds are sown singly in a double row with 12in. between the two rows of seeds and 9 to 12in. between the seeds in each row. They are then covered with 2in. of soil. If more than one double row is required the successive pairs should be at least 6ft. apart. They must be supported with long bean poles, one to each plant, lashed to a cross bar near the top for additional stability, as a row of runner beans offers considerable resistance to the wind.

The plants should be watered freely in dry weather and the beans sprayed with clear water when in flower to assist setting. The top of each plant is pinched out when it reaches the top of the bean pole. The beans should be picked regularly as soon as they become of usable size; letting them age on the plants checks cropping.

The varieties Hammond's Dwarf Scarlet and Hammond's Dwarf White grow only 18in. high and need no staking. The plants are spaced 1ft. apart in rows 2ft. apart. It is also possible to grow ordinary runner beans without stakes by frequently pinching out the tops of all young shoots or 'runners'. For this kind of cultivation the plants should be spaced 3ft. apart each way.

Varieties. These include Kelvedon Wonder (very early variety), Streamline, (this has perhaps the longest pods of all), Twenty-one (also long podded and good for deep freezing), Hammond's Dwarf Scarlet and Hammond's Dwarf White.

SAVOY

This is the cabbage that has the very wrinkled, hard leaves, and it is grown in exactly the same way as autumn or winter cabbage. There are varieties to cover the period from autumn to spring, those which mature late being particularly valuable.

Varieties. These include Best of All (September), Autumn Green (October–November), Winter King (November–December), Omskirk late

Hammond's Dwarf Scarlet, a dwarf runner bean which grows only 18in. high and needs no staking. The plants are spaced 1ft. apart in rows 2ft. apart

(January–March), Omega (February–April) and Rearguard (December–April).

SEAKALE

This vegetable is grown for its young shoots which must be blanched in complete darkness. It can be raised from seed sown in April outdoors, but a better method is to grow it from root cuttings 6 to 8in. long, planted the right way up in March in well-dug and well-manured soil. Drop the cuttings into dibber holes sufficiently deep to allow the tops of the cuttings to be $\frac{1}{2}$in. beneath the surface. Space them 1ft. apart in rows 18in. apart. Keep them well hoed during the summer and lift the plants in November. Trim off the side roots, tie up in bundles and lay in sand in a sheltered place to provide cuttings for re-planting the next spring. The crowns can also be laid in sand and potted a few at a time, bringing them into a warm greenhouse or shed to force into growth. They should be kept in complete darkness throughout this forcing.

Variety. Lily White.

SHALLOTS

A deeply-dug and well-drained soil is essential for shallots but manure should not be used just before planting. The ideal is a sunny spot that has been

Sweet Corn is a crop for a sunny site. The seedlings are hardened off and planted out of doors in June. The plants need plenty of water in dry weather

In a category of its own is New Zealand spinach, a vegetable with rather a trailing habit and thick, fleshy leaves which it produces in abundance during the summer months. The seed is sown under glass in mid-April or in May outdoors. Seedlings raised under glass should not be planted out until danger of frost has passed. It will grow in hot, dry soils which other spinach would not succeed in.

Varieties. These include the following—Summer Spinach: Long-standing Round, Monarch Long-standing and Victoria. Winter Spinach: Long-standing Prickly.

SPINACH BEET

A form of beetroot grown for its leaves, which are picked and used like spinach. The seed is sown in early April, and again in early August, in good soil and an open situation in drills 1in. deep and 18in. apart. The young plants are thinned to 9in. and the leaves picked as required.

SWEET CORN

The partly-ripened cobs of this vegetable, a favourite of mine, are cooked and eaten. The seeds are sown singly $\frac{1}{8}$in. deep in light soil in well-drained pots in a temperature of 13 to 16°C. (55 to 60° F.) in late April or early May. The seedlings are hardened-off and planted out of doors in June in good well-manured soil and a sunny situation. Space the plants 15in. apart in rows 3ft. apart, but making short rows to produce blocks of plants rather than long single rows, as in blocks the female flowers which produce the cobs are more effectively fertilised by pollen from the flowers or 'tassels'. Alternatively, the seeds can be sown two or three together at a similar spacing outdoors in early May where the plants are to mature, thinning the seedlings to one at each station. Make sure they have plenty of water in dry weather. The cobs are gathered when the seeds emit a milky juice if punctured with the finger nail or the point of a knife—from early August right through to the end of October.

Varieties. These include Golden Bantam and John Innes Hybrid.

TOMATOES

Tomatoes can be grown under glass or outdoors in sheltered, sunny places. There are a great many varieties, some of which, such as Outdoor Girl and Harbinger, are specially suitable for outdoor cultivation. Most tomatoes are restricted to a single stem but the so-called 'bush' varieties are allowed to grow naturally without removal of side shoots and they then produce quite low, freely-branched plants which need little or no support. They, too, are popular for outdoor cultivation. One of the most popular bush tomatoes is Amateur.

well-manured for a previous crop. The bulbs are planted 8in. apart in rows 1ft. apart in February or early March. Frequent hoeing is practically the only aftercare required. In June a little of the soil should be removed from around the bulbs to assist in ripening. As soon as the foliage dies down the bulbs should be lifted, dried and stored in an airy, cool but frost-proof place. This vegetable can be increased by division of the bulb clusters (cloves) and by seed sown thinly in drills 8in. apart in March.

SPINACH

Two types of spinach are commonly grown, the Summer Spinach or Round-Seeded and the Winter Spinach or Prickly-Seeded. Both are cultivated solely for their leaves, which are picked a few at a time as they reach sufficient size. Sowings should be made about once a fortnight from mid-March to mid-July of round-seeded spinach, and a further sowing in mid-August of prickly-seeded spinach. Sow in well-dug and manured ground in sun or partial shade, in drills 1in. deep and 1ft. apart. The plants should be kept well-watered in dry weather as, if allowed to get dry, spinach soon runs to seed and becomes useless. Thin to 3in. and gather leaves as soon as they are of usable size.

Greenhouse Cultivation. For growing under glass, the seeds are sown in January, February or March, in a temperature of 16 to 18°C. (60 to 65°F.). The seedlings are pricked off 2in. apart as soon as possible into boxes filled with John Innes No. 1 Potting Compost and are potted individually in similar compost, in 3-in. pots as soon as the leaves touch in the boxes. They are then potted into 9-in. pots or boxes filled with John Innes No. 2 Potting Compost or with good, well-drained loamy soil. Another method is to plant in beds of good loamy soil, spacing the plants 18in. apart in rows 3ft. apart. Each plant is restricted to one stem by removing all the side shoots as they appear. The main stem is tied to a cane or supported by soft string secured to the rafters in the greenhouse or to wire strained below the rafters. The growing point of each plant is pinched out when it reaches the glass. The plants should be watered moderately at first and more freely as they become established and grow quickly. They can be fed once a week with fertiliser—liquid or dry—from the time the first fruits are set. A temperature of 13°C. (55°F.) should be maintained as a minimum, but it can rise to 24°C. (75°F.) or more with sun heat.

Outdoor Cultivation. As far as outdoor cultivation is concerned, the seeds are sown in late March or April as for indoor cultivation, and the seedlings are treated in the same way, but they are removed to a frame in early May and hardened-off for planting outdoors at the end of May or in early June. The plants are set out 18in. apart in rows which are 2½ft. apart in a well-drained loamy soil and a sunny, sheltered position. Again each plant is kept to a single stem and this tied to a cane or stake. The growing tip is pinched out when four flower trusses have been produced. The fruit is picked as soon as it starts to colour and ripening can then be completed indoors. During dry weather they should be given plenty of water.

Ring Culture of Tomatoes. There is another method of cultivation which has become popular for growing tomatoes in recent years, under glass and outdoors and that is the ring culture system. Its chief advantages are that the amount of soil needed is relatively small and that there is less risk of soil infections. The general principles of the system are that the feeding roots are largely limited to a suitable compost placed in the containers, while the water supply is mainly obtained by the lower roots from a moist bed of some sterile material (known as the aggregate) such as sand, ashes or peat on which the containers stand. The chief drawback is that unless this bed is kept consistently moist failure can very easily occur. The treatment to final potting is as already described, but at this stage the plants are then placed in special bottomless rings, usually of whalehide (bituminised paper) filled with John Innes No. 2

Potting Compost, and stood on a 6-in. layer of gravel or old, sifted cinders. After the first few weeks water is given freely to the gravel or cinders only and liquid food is applied once a week to the soil in the rings. The plants are trained and stopped in the ordinary way.

Varieties. These include Ailsa Craig, Harbinger, Best of All, Moneymaker, Ware Cross (for greenhouse cultivation only), Eurocross, The Amateur and Golden Amateur (bush varieties) and Outdoor Girl (for outdoor cultivation only).

TURNIPS

To be good, turnips must be grown quickly and without check in rich, well-worked soil. The seed is sown successively from March to July in shallow drills 12 to 15in. apart and the seedlings are thinned to 4 to 6in. apart. They should be fed occasionally with small topdressings of a general or all-purpose fertiliser, and should be kept well hoed. When of reasonable size they can be pulled; if required for winter use they should be pulled in October and stored in sand or peat in a shed or cellar. Turnip tops for use as 'greens' in spring are produced by sowing a hardy variety in early September and leaving unthinned.

Varieties. These include Early Snowball, Early Six Weeks and Golden Ball.

VEGETABLE MARROWS

The seeds should be sown singly in small pots in a temperature of 16 to 18°C. (60 to 65°F.) during April. It is also possible to sow outside where the plants are to grow, but this should not be attempted before the middle of May. Rich loamy soil or old turves mixed with a little well-rotted manure make the best compost and this may either be built up into a heap, or a wide trench excavated and filled with the compost, leaving the surface a little below ground level so that water can be flooded round the growing plants during dry weather. Plant early in June and pinch the ends of the long trailing growths occasionally to encourage the formation of laterals. If these become too plentiful they can be thinned out to allow the strongest branches more room. There are also bush marrows which need no pinching or thinning. Water freely at all times and feed with weak liquid manure as soon as the first marrows start to swell. The marrows should be cut for use while still young and tender.

Varieties. These include the following—Trailing: Long Green, Long White and Table Dainty. Bush: Courgette, Green Bush, White Bush and Superlative.

Pests and Diseases

Advice on the pests and diseases of vegetables most likely to be enountered is given in Chapter Thirteen (p. 305 to p. 307).

The melon Hero of Lockinge, which is excellent for growing in greenhouses or frames

Above. Cabbage lettuces grown under cloches, a method of cultivation which has attractions for many gardeners interested in early-maturing crops

Right. Celeriac, a vegetable closely allied to celery, but it is the bulbous growth between the roots and the leaf stems that is the serviceable part of the plant. It is often used as a substitute for celery for flavouring soups and stews

Left. Purple Sprouting broccoli, which matures in March. This vegetable produces a succession of shoots at that time
Below. Masterpiece, a dwarf French bean with long pods which crops heavily and is widely popular. It is used for forcing and outdoor cultivation

Runner beans give a high yield for the amount of space they occupy. To assist setting, the author sprays a row of the variety Streamline while the plants are in flower. This variety has especially long pods of excellent quality

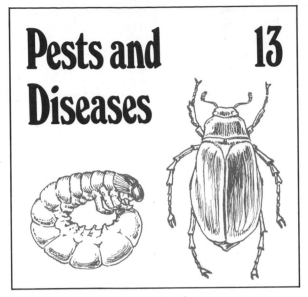

Pests and Diseases

13

Happy is the gardener who has come to terms with pests and diseases. One does not want to be complacent about possible damage, or so conscious of the problem that it detracts from one's enjoyment of the plants. As every professional gardener knows, and every amateur gardener would do well to remember, it is the less well-grown plants which succumb first to attacks from pests and diseases. To obtain strong, healthy plants with high resistance to attack by outside agencies one must necessarily provide good growing conditions, and this is an important factor in pest and disease control. Another is to be one jump ahead of your enemy by taking preventive action as often as possible, before an expected attack develops.

At first it may seem rather a burden to remember which foes attack particular plants at which season, but surprisingly quickly one finds that all this becomes almost second nature. Chemical manufacturers and the makers of spraying equipment have between them done much to make things easier for the gardener. I am thinking now of the multi-purpose sprays and dusts which are available in packages which make mixing and other preparation relatively foolproof—discounting those, of course, who never read instructions carefully. The equipment manufacturers, for their part, have not been slow to make use of lightweight plastic in their garden sprayers and these are available in a range to suit all gardeners' needs and their pockets.

If I may summarise my views on pest and disease control, these are as follows:
1. Grow plants well to lengthen the odds against pest and disease attack.
2. Look ahead and take preventive measures whenever this is possible. When walking round the garden, one should also develop the habit of looking for early signs of attack on plants.

3. Apply garden chemicals at the right time and in the right way, which means reading the manufacturers' instructions carefully.
4. Last but by no means least pay attention to garden hygiene. Many gardeners pile trouble on their heads because they do not bother to destroy immediately diseased plant material. If one has stored bulbs or fruit, for instance, periodic inspections are a necessity.

Lawns

Lawn grasses are usually reasonably trouble-free but, as with other garden plants, the better the health and vigour of the grass the less likely is it to succumb to pest and disease attack. Regular aerating, topdressing and food, and correct mowing all do much to help lawn grass withstand any possible attack.

PESTS

Leatherjackets. Perhaps the worst pest of grass is the larvae of the Cranefly (Daddy-long-legs), known by the familiar name of Leatherjacket; anyone who has tried to kill one of these larvae by stamping on it will know how well-named it is—the greyish skin is extremely tough and resilient. These larvae are an inch or so long.

Leatherjackets can do a great deal of damage to a lawn by feeding on the roots of the grass, just below soil level, with the result that the grass dies off, creating large brown patches. The adult craneflies appear in late summer and lays its eggs in the soil in early autumn, and these hatch in about a fortnight to produce grubs which feed on the grass from then onwards and usually reach a peak of activity in early spring. It is then that the patches begin to be noticed. To confirm that it is leatherjacket damage, flood the area with water, and leave overnight, covered with sacking.

By the morning some of the leatherjackets present will have come to the surface, and can easily be swept up with a besom. Alternatively, BHC or DDT applied as a spray or dust will be found an effective control.

Moles. These animals do much good by devouring leatherjackets and other soil pests, but unfortunately they do a great deal of damage to lawns at the same time by burrowing just beneath the surface of the soil and throwing up loose mounds of soil (molehills) at frequent intervals. Special steel traps are available from garden stores for setting in the runs. Gloves should be worn when this is being done as moles have a keen sense of smell, or mole smoke cartridges can be obtained. Also a few lumps of calcium carbide can be placed in the runs, the smell from these driving the moles out.

Worms. Some gardeners dislike having worms in their lawns because of the casts they produce, but it should be remembered that on the whole they do far more good than harm, aerating the soil very effectively. Where there are large numbers of them present, however, and sweeping does not provide the answer, worm killers of various kinds can be used. Derris will kill most of them in the soil, and has a lasting effect of several weeks; it is harmless to human beings and animals, but will poison fish so should be kept well away from pools containing the latter. Chlordane is another control which kills them in the soil. A good old-fashioned remedy is mowrah meal; this is an expellant and should be watered into the turf under strong pressure to make it froth. Worms will come to the surface within a few seconds, and can be brushed off or not as required. Mowrah meal also is poisonous to fish. It is most effective if used in spring and autumn, when the soil is moist and the weather mild, at which time the worms are likely to be near the surface. The treatment will need to be repeated.

DISEASES

There are several fungal diseases which affect lawns; the four main ones are Red Thread Disease *(Corticium fuciforme)*, Fusarium Patch Disease *(Fusarium nivale)*, Fairy Rings, caused by various fungi, and Damping-off, a trouble of seedling grasses, also caused by various diseases, in particular fusarium and pythium species.

Corticium Disease. This trouble, known as Red Thread, is not a killing disease, but it does weaken the grass considerably and render it susceptible to die off through some other trouble such as waterlogging, or invasion by weeds. It appears most frequently during late summer and autumn, in the form of small circular patches of brownish-buff-coloured grass, which can coalesce if close together. The tips of the grass wither, and this gradually extends down the leaf blade, and a small red or pink thread-like growth about ¼in. long appears, growing out from the blade tip. This is the part of the fungus which gives it its common name and passes the disease on from plant to plant. These threads are rather difficult to see, and a very careful scrutiny of the grass has to be made in order to pick them out. It can be controlled by watering in a mercury-based fungicide, but as it is often found on lawns short of food, the addition of a quick-acting nitrogenous fertiliser as soon as the patches start to appear will do much to enable the grass to grow more strongly and overcome the attack.

Damping-off. Damping-off can be diagnosed when the grass seedlings collapse, with rotting at soil level; they turn brown and the trouble can occur in patches several inches across. It is more likely to give trouble on damp soil or where the grass has been sown without putting on a well-balanced fertiliser beforehand. As soon as the trouble is seen apply a solution of Cheshunt Compound, as directed by the makers.

Fairy Rings. One often see rings of toadstools in fields and on lawns and although some may not harm turf, there are others, *Marasmius oreades* in particular, which will kill the grass and gradually extend outwards so that more and more is killed each year. The grass at the edge of the ring is likely to be a rich dark green. Getting rid of this trouble is rather involved. The simplest method is to skim off the turf from within the ring and to a distance of 2ft. outside it, and then to remove the soil for the same distance and to such a depth as is well outside that to which the white thread-like growth of the fungus has penetrated. For instance, if the white threads extend downwards for 6in. then remove at least 9in. of soil, and so on. Take the soil and turf completely away from the lawn, making sure that none of it spills on to the healthy grass—sacking placed over the full wheelbarrow will help to avoid spillage—to be the cause of a new infection and fill in the hole with fresh soil from another part of the garden.

Another method of control is to treat the grass with sulphate of iron mixed at the rate of 4oz. of sulphate of iron to 1gall. of water and repeating this at half strength one week later. Further applications may also be necessary. This solution should be watered on after rain or after the grass has been well soaked with water.

Fusarium Patch. This disease is rather more serious than Corticium, in that it will kill the grass and can spread very rapidly, causing a lot of damage before any treatment can be undertaken. It is most likely to appear during the autumn and spring, though sometimes mild attacks may occur during the summer, when the weather is cool and damp; however, these attacks are not serious and usually disappear as soon as the weather improves. When seen at other times the grass turns brown, and collapses. If grass is fed with a fertiliser having a high nitrogen content this is conducive to the production of soft growth which is liable to attack by this disease. The patches may be anything from an inch to a foot in diameter and sometimes round the edges a whitish fluffy growth is seen at the base of the grass. As with Corticium, a mercury-based fungicide can be used to control it; thereafter improve the drainage.

Trees and Shrubs

Most gardeners who grow trees and shrubs find that these plants suffer from very few pests and diseases. In my experience, I have found that a large amount of spraying and dusting is unnecessary. However, I am always on the look-out for an attack, and, should one occur, I take the necessary control measures in the early stages before the pest or disease becomes established. In order to prevent attacks of the very common pests and diseases I spray the plants regularly throughout the growing season. With widespread troubles I always feel that prevention is better than cure.

I have listed the pests and diseases most likely to cause trouble and although the list may seem rather long, it by no means implies that you will encounter all of them. Some you may never see at all, but it is best to be aware of them.

PESTS

Aphids. Greenfly and Blackfly are the most common aphids and attack many plants. They are barely $\frac{1}{16}$ in. long and quickly colonise leaves and young shoots, which they seriously weaken by sucking the sap. *Euonymus europaeus* is a winter host of blackfly and may be really smothered with the pest in a bad season. Woolly Aphids are easily identified as they are covered with a white cotton-wool-like substance. A large colony is very conspicuous on branches of beech, hawthorn, pyracantha, *Cotoneaster horizontalis* and numerous other host plants.

I control aphids by spraying regularly during spring and summer with gamma-BHC (lindane) or derris. Woolly Aphids can also be controlled with menazon, BHC, malathion or nicotine and soft soap. The spray must be applied forcibly so that it penetrates the insect's protective covering.

Capsid Bugs. These small, yellowish-green bugs are extremely active and attack a wide range of plants during spring and summer. They suck the sap and cause young shoots and leaves to become distorted. Deciduous trees and shrubs can be sprayed with DDT in petroleum oil just before bud burst, about mid-March. General spraying can be undertaken in spring and summer, using gamma-BHC or DDT, at the first sign of damage.

Caterpillars. The larvae of moths or butterflies may eat the leaves of numerous trees and shrubs and will therefore render them most unsightly. Torrix Moth larvae of various species often do a great deal of damage to rhododendrons, roses and many other trees and shrubs and usually spin webs, which bind the leaves together. Spray the plants with derris or DDT, starting in April or May, as and when necessary. In the case of Tortrix, spray before the larvae bind the leaves.

Cockchafers. The large fat white grubs, which are the larvae of the Cockchafer Beetle, live in the soil and feed off the fibrous roots of plants. This may severely check a young tree or shrub. If a plant seems to be in trouble it will pay to water the ground around it thoroughly with DDT solution. Grubs noticed while the soil is being cultivated can be picked up and destroyed.

Froghoppers. The larvae of this insect are probably best known as Cuckoo Spit, as they are covered with a mass of protective froth. They are yellowish-green in colour and suck the sap of plants such as roses and lavender. They are seen around the young shoots only. The adults are small, yellowish and jump when disturbed. Spray the larvae forcibly with either gamma-BHC, nicotine or malathion.

Leaf Miners. These are the larvae of several kinds of fly and as the name suggests they tunnel inside the leaves, making irregular and unsightly silvery channels. Holly and lilac are, in my experience, the most commonly attacked shrubs. Pick off and burn badly infested leaves and spray frequently with gamma-BHC or nicotine.

Red Spider Mites. These minute creatures can only just be seen with the naked eye, but in spite of their smallness a large colony of them can do a good deal of damage. They are reddish in colour and suck the sap of plants; they are normally found on the undersides of leaves. The leaves gradually take on a yellowish, finely-mottled appearance and fall prematurely. Among the numerous plants attacked are ornamental peaches, plums and crab apples. Plants should be sprayed frequently during spring and summer with either derris, malathion or dimethoate. Ornamental peaches, plums and crab apples can be sprayed with DNC in late winter.

Scale. These small insects are oval in shape and attach themselves firmly to leaves and stems and suck the sap from the tissues. This seriously weakens the shoots. The adult scale insects are covered in a hard, protective shell. There are numerous species, but the greyish Mussel Scale is possibly the most common and attacks many plants. The Brown Scale is also fairly common and is very often seen on yew. Plants may be sprayed at any time of the year with malathion or diazinon, while small numbers of scale insects can be removed with a knife. I also find white-oil emulsion an effective spray.

Weevils. The most troublesome is the Clay-coloured Weevil, a small, greyish-brown insect with a typical, elongated snout. The adult chews the leaves and shoots of many plants, including rhododendrons, roses, azaleas and clematis. The small, white larvae eat the roots of plants. The Vine Weevil does similar damage to that of the Clay-coloured Weevil, and the adults are dull black in colour. Spray plants with DDT or BHC and to kill the soil-borne larvae dust the ground around the plants with DDT or BHC and fork it well in.

DISEASES

Armillaria. Many trees and shrubs, particularly lilac, privet, hawthorns, cherries, pyrus and rhododendrons, may be attacked by *Armillaria mellea*, the Bootlace or Honey Fungus. The tree or shrub may die suddenly for no apparent reason, and inspection of the roots will show the black, bootlace-like threads of the fungus which grow through them. White fungal threads can also be seen beneath the bark at soil level and honey-coloured toadstools appear on the soil surface around the infected plants.

The infected plants, together with their roots, must be dug out and burnt. If ever my shrubs are attacked I dig out as much soil as possible in the infected area and replace it with fresh soil. I usually take the risk of planting another specimen in the fresh soil. I have known of a recurrence, but not often.

Bacterial Canker. Cherries, plums and peaches are prone to this disease. The bark is killed and usually complete branches wither and die. Gum will ooze from the affected bark. I cut the branches right back to healthy wood and seal the cuts with a bituminous tree paint. This disease will eventually kill a complete tree if it is not treated in time. You can also spray heavily with Bordeaux Mixture in late summer and early autumn.

Chlorosis. This is not a disease but a physiological disorder and occurs when plants are grown on very limy or chalky soils. In alkaline soils iron becomes unavailable to plants, which results in the leaves becoming pale yellow, or covered in yellow patches, and in some cases stunting of the growth occurs. Roses and peaches are very often affected.

To rectify this trouble avoid using lime or chalk and incorporate plenty of peat in the soil. Water the ground around the plants—and also spray the leaves—with iron sequestrol—this is a readily available form of iron and is not 'locked up' in the soil by the action of lime.

Gumming. Gum very often oozes from the branches and stems of cherries, plums, peaches and related species and may be a sign that the branch has been damaged in some way. This could be due to Bacterial Canker, therefore this

Branches and even complete trees can be killed by Silver Leaf disease, and fungal outgrowths form on the dead wood. First, however, the leaves take on a silvery sheen

disease must be controlled. The actual gumming is a physiological disorder and usually is no cause for alarm.

Leaf Curl. This fungal disease affects peaches and almonds and causes the young leaves to become thickened and distorted. They appear yellowish at first but rapidly turn bright red. Shoots may die back from the tips and young trees may even be killed by repeated attacks. Spray with Bordeaux Mixture or a copper fungicide, or captan in late February or early March, just before the buds begin to swell. Remove infected leaves and twigs. After serious infections, repeat the spray at leaf-fall.

Leaf Spot. There are various fungi which cause brown or black spots on the leaves of numerous trees and shrubs. Remove those worst affected and spray the plants with either Bordeaux Mixture, colloidal copper, colloidal sulphur or zineb.

Mildew. There are a number of different sorts of mildew, some of which attack specific plants, for example, the Rose Mildew. All of them are seen as white powdery or mealy patches on leaves, stems and buds. They thrive in a damp atmosphere and so are most prevalent in a rainy season. Plants may be weakened in severe, prolonged attacks and shoots and buds crippled.

Plants must be sprayed frequently with colloidal sulphur, thiram or dinocap, or dusted with flowers of sulphur. Start spraying in the spring and spray whenever there are signs of attack, especially after damp spells.

Rust. The most serious rust disease as far as trees and shrubs are concerned is the one that attacks roses. It is seen as raised orange spots, which later turn black, on the undersides of leaves and also stems. An attack will cause stunting and distortion of growth. This fungus can be controlled by spraying at frequent intervals during spring and summer with colloidal copper, zineb or thiram.

The characteristic appearance of rose leaves attacked by the Leaf-cutter Bee. This pest is rarely seen on the plants and control is difficult

All badly affected leaves and stems should be cut off and burnt.

Silver Leaf. This fungal disease will attack ornamental peaches, plums, almonds, cherries, hawthorn, roses and Portugal Laurel (*Prunus lusitanica*). The leaves take on a silvery sheen after the wood has been infected by the disease. Branches and even complete trees can be killed in time and fungal outgrowths form on the dead wood. Sometimes these are flattish, and purplish-mauve in colour, or they may be of the bracket type, arranged in ridges one above the other. The bracket fungus is purplish beneath and brownish above. Infected branches also have internal dark brown staining of the wood.

Infected wood should be cut out in June, July or August. Infection is least likely between these times. Cut back to unstained wood and seal the cuts with white lead or bituminous tree paint. The disease will only enter through wounds, particularly pruning cuts, so paint these as soon after pruning as possible.

Burn all affected branches which have been cut out—never leave diseased wood on the ground.

Roses

Although the pests and diseases which may attack roses are fairly numerous, prompt action when trouble is noticed, or better still early preventive spraying or dusting where this is appropriate, will keep the plants clean and healthy.

PESTS

Aphids. The kind of aphids popularly known as greenfly may prove troublesome on roses from the beginning of May onwards, and will be first detected on the youngest and most tender leaves and shoot tips. The small green 'lice' can multiply very rapidly under favourable conditions so that shoots, leaves and buds become covered with them. These greenflies suck sap from the plant so weakening it and causing distortion. The bushes must be immediately syringed with BHC, derris, malathion or menazon.

During winter, a spray of 3 per cent. tar oil should be applied to bushes which are known to be severely infested.

Caterpillars. Various caterpillars attack roses during the summer months. Though they vary in character they are all to be dealt with in the same way—poisoned with an insecticide such as DDT, BHC, derris or trichlorphon. These sprays should be repeated from time to time so that the young foliage is coated with the poison. Hand picking should also be carried out, and all curled leaves examined minutely for any caterpillars hidden within.

Froghoppers, see p. 291

Leaf-cutter bees. While other rose pests can usually be found on the damaged parts, these bees are rarely seen by gardeners and yet the damage they cause is among the most spectacular. The bees do not actually feed on the leaves, but cut out regular oval or circular portions from them. These portions are used for lining their nests. Other than trying to follow the bees to their nests and destroying it, or waiting by the plant to net them, little can be done to stop the damage which, I must say, reduces the foliage to comic shapes!

Leafhoppers. These are a common pest of the rose and other plants. They received their popular name on account of the habit of the adults of leaping when disturbed, but it is the much less active larvae that do most of the damage, sucking sap from the leaves and causing them to become pale and mottled. These insects change their skins as they grow and, apart from the foliage mottling, the most certain indications of their presence are the white moult skins attached to the under surface of the leaves.

Occasional spraying with BHC or DDT is the best safeguard against injury.

Sawflies. The larvae of these are very destructive. The damage done by one kind is very distinctive, as only the surface of the leaf is eaten, a thin membranous 'skeleton' being left. The larvae are small, black, and have a superficial resemblance to slugs and so are often known as slugworms. Another kind of sawfly causes the leaves to roll up tightly. All may be destroyed by spraying the foliage with DDT or BHC but this must be done early with the leaf-rolling kind or it will be completely protected from the spray by the leaf tightly curled round it.

Thrips. Thrips are minute and rapidly moving pests which frequently attack the flower buds, hiding themselves between the closely folded petals and causing browning of the buds. As they are completely hidden by the petals they are not easy to eradicate. Damaged flowers should be removed and burned and the plants sprayed with gamma-BHC (lindane) or DDT.

DISEASES

Mildew. Mildew is one of the commonest diseases affecting rose foliage. It first produces greyish patches on the stems or leaves and if unchecked can eventually look like a heavy coating of wet flour.

It is a difficult disease to keep in check especially on some varieties which are susceptible to it. Spraying with dinocap is most efficacious and Bordeaux Mixture or colloidal sulphur are also useful in its control. Like the majority of diseases, it is most easily controlled by preventive sprays before it makes its appearance. Four or five sprays throughout June, July and August will usually ensure a reasonable degree of freedom from mildew, but in gardens that have a poor circulation of air it may be impossible to eliminate mildew entirely.

Black Spot. Black spot usually appears about midsummer and lasts throughout the rest of the season. Affected leaves show circular-like black areas which extend until the leaves are completely covered and fall prematurely.

Black spot can be checked to a certain extent by collecting all fallen leaves and burning them. Infected wood should be cut off and also added to the garden fire. In winter, after all leaves have fallen, plants and the soil beneath them can be sprayed with copper sulphate, 1oz. per gallon. Throughout the late spring and summer roses should be sprayed at least once a fortnight with captan.

Rust. Rust is not as common as mildew or black spot, but occasionally proves troublesome. First the under surface of the leaves show small orange pustules which later turn to nearly black. All leaves bearing these black winter spores should be burned to prevent spring infection. Where rust occurred the previous years, colloidal sulphur, Bordeaux Mixture, thiram or zineb should be employed frequently in summer as a preventive spray.

Ensure that the entire bush is sprayed, especially drenching the leaves. Apply the spray forcefully. Remove all weeds from around the bushes, and ensure that all prunings are gathered and burnt.

Die-back. A trouble known as 'die-back' causes considerable winter losses among some roses. Dark brownish-purple patches appear on the branches and the whole stem above these dies. Constitutional weakness appears to have a good deal to do with die-back but it is actually caused by infection by the grey mould fungus (*Botrytis cinerea*) which attacks many other garden plants. All dead or discoloured growth should be cut out

as soon as possible and the plants sprayed with Bordeaux Mixture or colloidal sulphur.

DISORDERS

Balling. This is a trouble associated with some very full roses in wet weather, particularly in May and June. The flowers refuse to open properly. They turn brown and eventually decay completely or fall off. There is no known remedy and affected buds should be cut off and burned.

Chlorosis. This is a name given to a condition which causes leaves to lose their natural colour and turn yellow. It is due to lack of available iron in the soil and is most likely to occur on chalky or alkaline soils in which iron tends to be locked up in insoluble compounds. The remedies are to lower the alkalinity of the soil by liberal dressings of dung or acid peat and to apply iron sequestrols to the soil. This is a special compound of iron salts which can be purchased and should be used according to the manufacturer's instructions.

Perennials, Annuals and Biennials

Perennials, annuals and biennials are, of course, subject to attack by pests and diseases, but not to anything like the extent of some other ornamental plants. So far as I am concerned, at any rate, that is one of their most important virtues.

I would repeat my advice, though, that good garden hygiene is just as important with these plants as with others more prone to trouble in this respect. Provide good growing conditions and always clear away any dead plant material and other debris without delay.

With annuals and biennials, the most tricky time is bound to be when the plants are at the seedling stage. Under glass, in the case of half-hardy annuals, it is damping off which causes losses. This may be countered by careful watering and good ventilation as well as by treating the seed with thiram or captan dust before sowing, or by watering the seedlings with Cheshunt Compound when the trouble is noticed. Outdoors, young seedlings are prone to attack by birds and the gardener now has an additional string to his bow in a bird repellant chemical named Morkit, which is applied as a spray. The traditional black cotton barrier made by weaving the cotton between closely placed sticks gives a good measure of success and so, to a limited extent, do the flapping scarers which glitter and tinkle—but with birds, as we all know, familiarity too often breeds contempt. If earwigs are damaging seedlings, a DDT or BHC spray will bring relief, and the same chemicals will control woodlice which can also do considerable damage to seedlings at this early stage. Leatherjackets, too, can be troublesome with young plants and the way to avoid trouble from this quarter is to fork DDT or BHC dust into the soil.

Slugs can be a serious annoyance and these can be combated by placing a mixture of metaldehyde and bran or slug pellets around the plants being attacked. Aphis damage on perennials or annuals can be brought under control with gamma-BHC (lindane), menazon, malathion, derris and pyrethrum sprays. Dinocap fungicide sprays or dusting with flowers of sulphur are correctives for mildew attacks.

Aster wilt can be a trouble with annual asters and where this is the case it is advisable to grow early, wilt-resistant strains. Michaelmas daisies are also prone to attack by wilt and the only answer in this case is to be ruthless and destroy by burning all affected plants. After this, move the unaffected plants to a fresh site and leave the old site free of these flowers for several years.

Antirrhinums and hollyhocks are attacked by their own strains of rust disease. With antirrhinums, if this disease has given trouble, one should grow rust-resistant varieties. Destroy infected plants and as a preventive measure spray frequently with Bordeaux Mixture.

Bulbs, Corms and Tubers

Bulbs, corms and tubers are liable to be attacked by quite a few pests and diseases, but, as with human ailments, the fact that they exist does not necessarily mean that they will be experienced. I prefer to emphasise the importance of buying good stock (particularly important with this group of plants), providing them with congenial growing conditions and keeping a close watch on general garden hygiene, for these factors have a considerable bearing on whether or not one avoids trouble.

PESTS

Ants. These can be troublesome in the garden and they can be controlled with BHC, or BHC/DDT applied direct to their nests or the ground where they are active.

Aphids. Numerous bulbous plants grown under glass or in the home are liable to be attacked by aphids, or greenfly as they are popularly called. These can be combated with various insecticides, including malathion, malathion/DDT, derris, pyrethrum, BHC, BHC/DDT, and menazon, applied as soon as the attack is noticed.

Birds. Even those of us who are bird-lovers must feel rage at times when the flowers of choice bulbs are ravaged for no apparent reason. Crocus, in particular, seem to attract such unwelcome attentions. There are bird repellents available nowadays, such as Morkit, or we can rely on rather unsightly criss-crossed weaves of cotton.

Bulb Mites. These pests attack the bulbs of lilies, narcissi, tulips and hyacinths and the roots of dahlias. The mite, although very small, is visible to the naked eye and it is round and yellow-white in colour. It attacks plants which have already been damaged by some other means and the symptoms of attack are yellowish foliage and reddish scales on the bulbs or tubers. Burn all affected plants immediately, and sprinkle para-dichlorobenzene chrystals among the other bulbs to deter further attacks from these most devastating pests of bulbs.

Eelworms. Bulbous plants like narcissi, irises, hyacinths, scillas and snowdrops may be attacked by Stem and Bulb Eelworm. Affected plants should be lifted and burnt immediately. The symptoms are withered or distorted foliage and stems, browning of the scale leaves of the bulbs and late flowering. If an affected bulb is cut open horizontally it will be seen that there are dark-coloured rings in the tissue. The pest itself is too small to be seen with the naked eye. Bulbs which are soft or badly damaged should be burnt, and the remainder can be given hot water treatment.

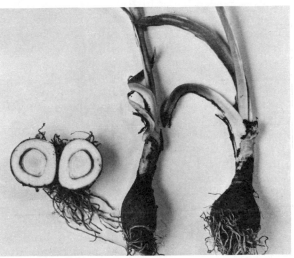

Eelworm damage on daffodils. Note the ring of diseased tissue in a recently infested bulb, distorted foliage and browning of the scale leaves

This treatment however, is not something which can be done without careful preparation, and it is difficult for the home gardener to do. The temperature of the water, 43.3° C. (110° F.), must be precisely correct and the timing, which varies with different plants, being three hours for narcissi and much less for some other bulbs, must also be accurate. A higher temperature than that indicated would destroy the bulbs, a lower one prove useless in controlling the eelworms. Tulips should not be given this treatment. Ground in which eelworm-infested bulbs have been growing should not be used for hosts of this pest for at least three years.

Leatherjackets. The leatherjackets are the larvae of the Cranefly or Daddy-long-legs, soil pests which can cause considerable damage to bulbs and tubers. The larva is tough skinned, legless, a brownish-grey colour and about 1in. long and attacks the plants during spring and summer. There are various methods of control – BHC and DDT dust, applying Paris Green as a bait around the plants, or digging in naphthalene at 2oz. to the square yard.

Mealy Bugs. Bulbous plants grown in greenhouses may be attacked by the Mealy Bug, and the control of this pest is described on p. 301.

Mice and Rats. Mice, and perhaps less often rats, can cause considerable damage to bulbs in the garden and to those which are stored. A poison bait based on warfarin will control these pests.

Narcissus Flies. The Large Narcissus Fly, with a hairy body and looking rather like a smallish bee, is a serious pest of narcissi which also attacks amaryllis, hippeastrums, hyacinths, snowdrops, vallotas, lilies, scillas and galtonias. The larvae which are of off-white colouring enter the bulbs, one to each bulb, through the basal plates and if the foliage is in poor condition or distorted these pests should be suspected. Burn all badly affected bulbs. Subject others which may be infested to hot water treatment when dormant – immersion at temperatures of 43.3°C. (110°F.) for one hour. Alternatively, place dormant bulbs in a solution of lindane, with a wetter added, for three hours. Lindane may also be dusted around the plants from late April to late June at fortnightly intervals, for it is at this time that the eggs are laid in the ground around the bulbs.

The Small Narcissi Fly has a smooth body and is narrower than the Large Narcissi Fly. It also differs in that many larvae – up to 30 – will attack the same bulb. Narcissi, hyacinths and irises are worst affected. The control measures are the same as for the Large Narcissus Fly.

Red Spider. The Red Spider Mite, a sap-sucking insect, is most likely to be found on bulbous plants under glass, for it needs warm, dry atmospheric conditions to thrive. Attacked leaves take on a yellow appearance and will fall early in severe attacks. Frequent syringing with water to moisten the atmosphere is a first requirement and fumigation with an azobenzene smoke. Against outdoor attacks use derris or malathion.

Slugs and Snails. Both these pests can be destructive if countermeasures are not taken. Metaldehyde and Paris Green in each case made into a bait with bran, are effective controls but Paris Green has the disadvantage of being poisonous to other animals and birds.

Springtails, see p. 301

Thrips. The small thrips, or Thunder Flies as they are called, are small, rapidly moving sap-sucking insects which attack cyclamen, begonias, hippeastrums, arums, gladioli (this plant has its own strain) and many other plants. Attacked plants have characteristic brown or silver streaks on their growths, suffer from arrested development of the flowers and may generally be deformed. Like red spider this pest thrives in a hot, dry atmosphere. Malathion, lindane, DDT, and derris are some of the chemical controls for use outdoors and under glass, and lindane and DDT can be used as smokes in the latter case.

To control Gladiolus Thrips spray or dust with lindane, DDT, or BHC/DDT.

Weevils. The Vine Weevil can be an especially damaging pest of cyclamen and tuberous begonias, attacking the plants in both the larval and adult stages.

The larvae are small, rotund, legless and of an off-white colour and attack the corms or tubers. An effective control is to water the soil with lindane if they cannot be repotted into new soil – the best procedure, of course. The larvae can also be removed by hand from the root balls of plants taken from their pots.

The adult weevils, which feed on the margins of the leaves, can be trapped in rolled sacking or paper strategically placed near the plants on which these nocturnal pests feed; or by spraying with DDT or lindane. These can also be used as smokes in greenhouses.

Wireworms. Wireworms, the larvae of Click Beetles, are soil pests which can cause considerable damage to bulbs, corms and tubers. The grubs are easy to recognise, having yellow segmented bodies with three pairs of legs near the head.

Pasture-land which has been dug over and used for ornamental plants for the first time is often badly affected with the main attacks coming in spring and early autumn. BHC and DDT dusts can be used as a control measure.

DISEASES

Botrytis. Dahlias, begonias and cyclamen are plants which are the most vulnerable to Grey Mould, *Botrytis cinerea*, but any bulbous plants may be infected through a wound, however small, particularly if the conditions are cool and damp. The common name comes from the grey spores of the fungus which form a mould on the leaves and stems of the plants. To keep down its spread improve the circulation of air, and try to prevent it from getting too humid. Chemical controls which may be used are thiram, captan, dicloran or quintozene.

Gladiolus Scab. This disease attacks both the corms and the top growths. On the corms quite shallow depressions form with raised edges and on the foliage brown spots appear, especially low down. In severe cases the plant will eventually topple over. As a precautionary measure immerse the corms in captan before planting.

Lily Disease. This is a trouble which must be watched for on lilies. The fungus, *Botrytis elliptica,* shows as brown, red-margined spots on the leaves or flower buds and stalks. Damp, cold sites with an absence of sunshine are conducive to the spread of the disease and it causes most trouble in wet seasons. Remove affected growths and burn badly damaged plants. Spray with colloidal copper at regular intervals.

Narcissus Bulb Rot. *Fusarium bulbigenum* is a disease to watch for when the bulbs are in store. The fungus invariably gains a hold on the base and spreads throughout the bulb. The bulb scales turn brown. Destroy immediately all affected bulbs and keep the storage conditions as cool and airy as possible.

Smuts. Various forms of Smut, *Urocystis,* attack scillas, *Anemone nemorosa*, colchicum and gladioli and all affected plants should be destroyed immediately. The black or dark brown powdery mould can easily be recognised on the leaves, flower stems, bulbs or corms.

Dahlia smut, *Entyloma dahliae,* has a different origin and begins as pale green spots on the leaves which later turn brown and may run together so that the entire leaf withers. Attacked foliage should

A gladiolus damaged by Thrips. These small, sap-sucking insects arrest development, cause distortion of the growths and their marking with brown or silver streaks

be removed and the plants sprayed with Bordeaux Mixture or a copper fungicide.

Soft Rot. This bacterial disease, *Pectobacterium carotovorum,* attacks cyclamen, hyacinths, indoor-grown muscari and *Zantedeschia aethiopica.* The bulbs or corms become soft and slimy as the disease rapidly gains hold. Destroy affected cyclamen, hyacinths and muscari but with the zantedeschia cut out the affected parts of the corms, soak them in a 2 per cent mixture of formalin for four hours and replant in sterilised soil.

Tulip Fire. This disease, *Botrytis tulipae,* is another with a very descriptive name, for the plants do indeed look as if they had been exposed to fire. Scorched looking areas on the leaves and sometimes flowers spread rapidly. The bulbs are marked with brown spots. Destroy badly affected bulbs. Spray with captan or thiram when plants are making growth and until flowers are about to

open. A well worth while precaution, too, is to dress the soil with quintozene when planting the bulbs.

Virus. Lilies, narcissi, dahlias and gladioli may be attacked by virus diseases and plants which are affected should be destroyed. As virus diseases are spread by sap-sucking insects, such as green-fly, measures should be taken to keep them under control.

Rock Plants

Pests and diseases of rock plants are very few. This is probably just as well, as some of these plants can be very tricky to cultivate successfully, and the fewer difficulties one has to contend with the better. The main pests are the following, and have already been described under other headings, to which reference should be made: Aphids, see p. 291; caterpillars, see p. 291; slugs and snails, see p. 296; vine weevil, see p. 296; botrytis (grey mould), see p. 296; and mildew, see p. 292.

Water Plants

Troubles affecting these plants are few, even less than with rock plants. Water-lilies may be attacked by a species of aphis, which can be controlled by spraying with water under strong pressure. This will push most of the aphids off into the water, and will keep most attacks under control, but it will need to be done several times. Water-lily Beetle is another nuisance; the larvae attack the leaves and can do a good deal of damage very quickly. Again, spraying with water under pressure helps, but if there are no fish to eat the grubs when they are moved off the leaves into the water, an application of lead arsenate in the form of a spray is recommended. There is also a moth whose larvae causes a great deal of trouble—the Brown China Marks Moth. The grub feeds on the leaves, and submerging the plants temporarily will drown the larvae.

There are two leaf spot fungal diseases which may cause trouble, and all affected leaves with large numbers of brown spots should be removed as soon as seen.

Dahlias

Dahlias are not troubled by a great many pests and diseases, but there are a few important ones which must be kept in check. I would suggest that spraying commences when the plants are in their early stages of growth, and is continued regularly throughout the growing season, at least once every four weeks. During long spells of wet weather it may be better to apply dusts as these are not washed off the plants so easily as liquids. During reasonable weather I much prefer to use sprays, as it is easier to obtain a complete coverage of the plant with the material. Also, be very careful to use really clean pots and to keep the greenhouse scrupulously clean, and in general make sure that the garden and greenhouse are kept clear of rubbish and diseased plant material.

PESTS

Aphids. The two main types of aphids which attack dahlias are greenfly and blackfly. They can be seen in large groups, chiefly on the younger foliage and tips of the shoots, but also on older leaves, and they suck the sap of the plants. In a bad infestation this will cause the plant to be considerably weakened, and wilting will result. They are also the chief carriers of virus diseases.

They may even attack cuttings and young plants in the greenhouse, so I would suggest fumigating the house to ensure that the young dahlias remain clean. For this purpose BHC smoke

A partly developed dahlia bloom which has been subjected to earwig damage at the bud stage. Earwigs can be trapped or controlled with gamma-BHC

generators are suitable, or if preferred BHC may be applied as a spray.

In the garden it will be necessary to spray either once a week or every fortnight with BHC.

Capsid Bugs. This is an insect that we very rarely see unless we look carefully for it, because any slight movement or noise will make it drop to the ground. About ¼in. in length, it is green in colour, almost of the same shade as the dahlia leaves, and therefore very difficult to detect. These insects pierce the stems, young leaves and young buds. If the young leaves have been attacked, small pinholes will start to appear in them as they grow, and in a severe attack the leaves will be completely riddled. In this case it is best to remove the affected leaves. If growing tips are attacked, the resultant growth will be distorted. BHC should be applied regularly throughout the season.

Earwigs. These familiar insects are slightly more difficult to control than capsids, as they hide in some awkward places, such as hollow bamboo canes which may be used to support the plants. Damage is done to the dahlia leaves, and if the earwigs eat the growing shoots and young buds then subsequent growth will be distorted.

One way to help prevent these pests infesting the plants is to seal the tops of canes with putty, so depriving them of a hiding place. Earwigs can be trapped by placing upturned flower pots filled with hay on the tops of the stakes. The traps must be checked often, and the earwigs can be shaken out of these into paraffin, or killed by treading on them. Earwigs are more likely to attack if tubers were planted, as these will have a portion of the old stem attached which makes an ideal hiding place for them. Dust the young shoots with gamma-BHC to control them. Regular spraying with BHC should also effect control.

Eelworms. These microscopic creatures sometimes attack dahlias and cause the leaves and shoots to become distorted and discoloured—seemingly for no apparent reason. It is not possible to give dahlias hot-water treatment as you can with chrysanthemums, and the only thing to do is to dig up the affected plants and burn them.

Slugs and Snails. The underground slug plays more havoc than the one that lives above ground, as it eats and tunnels its way into the dahlia tubers. A bad attack will result in a severe check to plant growth and if the feeding roots are damaged then the plant will certainly die.

As soon as the dahlias are planted it is advisable to water the soil around them with a liquid slug killer, which contains metaldehyde. This must be kept up at regular intervals throughout the season. For slugs that live above ground, metaldehyde slug pellets placed around the base of the plants will help to prevent attacks.

Snails can do a lot of damage by eating leaves and growths and these pests crawl up into the plant. They may be picked off and destroyed; alternatively, metaldehyde slug baits or pellets placed around the plants will keep them at bay.

DISEASES

Botrytis. For description and control measures see p. 296.

Mildew. This disease is more prevalent in some parts of the country than it is in others—usually in high-rainfall areas. In a wet season it may cause trouble. It can be seen as white patches on the foliage, stems and buds, and may be controlled by spraying weekly or fortnightly with a fungicide based on dinocap.

Smut. For description of Dahlia Smut and its control see p. 297.

Viruses. There are two main virus diseases which affect dahlias—Dahlia Mosaic and the Tomato Spotted-wilt Virus. The symptoms are more or less identical, and it is almost impossible, except by laboratory tests, to differentiate between them. The first signs of infection are stunting or dwarfing of the plants. In the early stages yellow areas can be seen spreading from the main leaf veins, and the veins may also appear yellow—this is known as 'vein banding'. Gradually the leaves become spotted or mottled with yellow, and may be contorted.

If these symptoms are seen on cuttings, started tubers, young or large plants, then they should be pulled up and burned. Another important point is to keep the plants free from aphids and other sucking insects, because they can transfer viruses from plant to plant.

Chrysanthemums

Chrysanthemums are not prone to a great many pests and diseases, and in some seasons, in fact, they are hardly troubled at all. However, it is wise to keep an eye open for the following pests and diseases and to take early steps to prevent them from becoming established.

PESTS

Aphids. I would class these as very troublesome pests of chrysanthemums, so precautions must be taken even when the plants are in the cutting stage. If they are noticed in the greenhouse then fumigate with BHC smokes. Outside I would recommend spraying about every 10 to 14 days with gamma-BHC or nicotine. Particular attention should be paid to the undersides of the leaves and the growing tips.

Capsid Bugs. These may also prove troublesome on chrysanthemum plants. A spraying every 10 to 14 days with gamma-BHC or DDT should effect control.

Earwigs. I have dealt with these pests under dahlias (see p. 298) and the same methods of control are advocated for chrysanthemums. As an alternative to dusting or spraying with BHC, nicotine or DDT will give satisfactory control.

Eelworm. This is another serious pest of chrysanthemums and can probably be classed as one of the most injurious of them all. It is a microscopic, colourless, slender worm which generally lives within the tissues of the leaves. A single leaf may contain up to 15,000 eelworms. They are most troublesome in rainy seasons, as the eelworms then travel from leaf to leaf in the surface film of moisture. Therefore, when the plants are wet, the spread of these pests is most rapid. Heavy mists in late summer and autumn also provide ideal conditions for eelworm movement.

At the end of the flowering season the eelworms will remain in the stools, and if cuttings are taken from infested stools then these also will contain the pest. Unless stools are given warm-water treatment, immersion in water at 47·2° C. (115° F.) for 5 minutes, the pest will be carried over from one season to the next.

The first signs of attack are yellow-green or purplish triangular blotches between the leaf veins. These blotches later turn brown or black and eventually cover the whole leaf which dies and shrivels up. These symptoms start at the base of the plant and gradually extend upwards until the whole plant is affected. It will be noticed that there is a gradual transition between healthy and affected leaves as the eelworms progress up the plant. The blooms may also be attacked, in which case they may be deformed. The amateur should not confuse 'natural die-back' of the leaves at the base of a plant with an eelworm attack. Die-back normally occurs towards the end of the growing season when some of the basal leaves may turn yellow. In this case, however, there is a sharp transition from yellow leaves to healthy ones.

Leaf-miner. This, I should say, is a very serious pest of chrysanthemums, and, as the name suggests, it infests the foliage. The adult fly lays its eggs in the leaves and when the grubs hatch they start tunnelling inside the leaves, eating the tissues as they go. This has a considerable weakening effect on the plant if a great number of leaves are infested. It is not difficult to tell when the plant is being attacked as silvery coloured, wavy lines appear all over the leaves. Leaf-miner starts early in the year and may even be found on young plants in the greenhouse. Spraying should, therefore, commence early, before any sign of attack is seen, because once the grubs are in the leaves they are difficult to eradicate. The most efficient insecticide I find is gamma-BHC which should be applied every 10 to 14 days.

Slugs and Snails. I do not have much bother with these two pests on my chrysanthemums although I believe that some people have many of their plants severely damaged by them. They are usually more troublesome in a wet season. The control measures I recommend in the dahlia section would be equally suitable for chrysanthemums (see p. 298).

Thrips. These pests do not trouble chrysanthemums a great deal except maybe during hot, dry weather. They probably cause more trouble in a greenhouse than they do outside. The flower buds and blooms are attacked and the pests make them discoloured, together with shrivelling of some of the petals.

A regular spraying with gamma-BHC or DDT will keep them at bay in the open garden, but for the greenhouse I would recommend fumigating with either of these materials.

DISEASES

Botrytis. This disease will make the stems and leaves of cuttings rot off if they are over-watered and if the greenhouse atmosphere is constantly damp. Avoid watering except when really necessary, and turn the glass of the propagating case at least once a day, preferably twice. When the rooted cuttings are placed on the staging of the greenhouse ventilate the house as much as the weather will allow, and maintain a gentle heat to keep the atmosphere dry.

Botrytis also causes damping of the blooms of the greenhouse chrysanthemums, so to prevent this give as much ventilation as possible and keep the temperature around 10° to 13° C. (50° to 55° F.) to keep the atmosphere dry.

Powdery Mildew. This disease, although it appears during the summer, is usually more troublesome towards the end of the flowering season of the early chrysanthemums. When the late-flowering varieties are brought into the greenhouse they should be watched very carefully for any signs of mildew. It is normally more severe in a damp season, although plants may still be infested in a hot, dry summer and autumn.

Mildew is seen as a greyish-white powdery substance on the leaves, including the undersides. In most instances it will first of all start on the lower leaves and then gradually work its way up the plant. Every effort should be made to prevent it from starting.

Sulphur may be used, applied either as a wettable formulation or as a dust. Regular applications will be necessary throughout the summer and autumn about every 10 to 14 days. Continue to spray or dust the lates when they are in the greenhouse, up until the time the buds start showing colour. Dinocap, one of the new fungicides can also be used, but some varieties may be damaged and the manufacturer's instructions should be consulted before use.

Rust. Generally, this disease does not cause much trouble nowadays although the occasional outbreak may be experienced, therefore it is wise to have some idea of the symptoms. It appears as raised spots of a powdery orange-brown dust on the undersides of the leaves of the plants. These are the reproductive bodies or 'spores' and may be rubbed off on the fingers. A preventive application of sulphur, as a spray or dust, should be made every two or three weeks to be on the safe side.

Virus Diseases. There are a number of virus diseases which affect chrysanthemums, but undoubtedly the most serious one is that known as aspermy. This causes great distortion and dwarfing of the blooms and thereby renders them useless. It also produces a slight mosaic mottling of the leaves. The second most serious virus is spotted wilt which makes the leaves mottled or spotted with concentric rings or irregular markings. These two are well-known viruses of tomatoes and they affect chrysanthemums just as badly.

White Fly on a tomato leaf. These very small insects suck the sap and are difficult to control. Fumigation or spraying at least twice with a 14-day interval between helps

If any of these symptoms are seen on chrysanthemums then the plants should be destroyed by burning. It is most important to control insect pests, especially aphids, capsid bugs and thrips, as they can transfer a virus from plant to plant, often over considerable distances.

Some chrysanthemum suppliers are applying heat therapy to their stocks of plants, so ensuring that all the plants they sell are as free as possible from virus diseases. This is a highly complex business and beyond the scope of the amateur gardener, but even so it is good to know that the commercial growers are doing their best to supply us with clean plants.

GENERAL HYGIENE

Apart from spraying regularly against pests and diseases, I feel that I should mention one or two other points which are very often overlooked by some gardeners. First of all, the question of keeping the greenhouse scrupulously clean. Before the late varieties are taken inside, I would recommend scrubbing the glass, woodwork, walls, staging and so on thoroughly with a disinfectant. This will help to check pests, diseases and algae, and will ensure that the house is kept in a generally fresh condition. The house should also be fumigated against pests and diseases. This job can be done again after the greenhouse varieties have finished flowering, so that the house is clean when the cuttings are taken.

Another important point is always to use clean pots, preferably washed in hot water. Always ensure that the greenhouse is kept free of rubbish, dirty pots and so on, as it is this sort of thing which encourages a build-up of pests and diseases. The same applies of course to the open garden—the more rubbish and weeds that are allowed to accumulate the more trouble you will have with pests and diseases.

Woodlice, damaging pests in greenhouse and frames, feed on decaying wood and vegetation and also attack growing plants. Scrupulous cleanliness is essential

Greenhouse Plants

In my remarks on chrysanthemum pests and diseases I have stressed the importance of general greenhouse hygiene (see p. 300) as a significant factor in plant health. This advice is just as applicable to other greenhouse plants and should, I feel, be kept in mind at all times.

PESTS

Leafminer, see p. 299

Mealy Bug. This pernicious greenhouse pest belongs to the same family as the scale insect and is related to the aphids. Like tiny wood-lice, these insects are covered with white, waxy, wool-like material which protects them against water. Though capable of movement they usually stay in one place when adult and they feed by sucking the plant sap. Breeding is continuous and multiplication very rapid. The root mealy bug is similar in appearance and is a common pest of cacti.

Deal with small infestations by hand, using a stiff paint brush dipped in insecticide. Spraying with derris, malathion, nicotine or white-oil emulsion is effective but repeated applications are necessary to ensure that no young appear from eggs which have escaped treatment. Where root mealy bug is identified, the soil must be shaken from the plants roots, the worst infections cut away and the roots dipped in insecticide before re-potting in fresh compost.

With vines all loose bark should be removed in winter and the rods and spurs painted with petroleum emulsion insecticide. In addition, all woodwork should be scrubbed with hot, soapy water.

Red Spider Mite. This pest attacks many plants and is usually most troublesome in a hot, dry atmosphere. Frequent syringing with clear water is, therefore, the most effective method of combating it in greenhouses and frames.

The mites can just be detected with the naked eye and they congregate on the underside of the leaves, usually in the angles of the veins. The leaves develop a mottled yellow appearance. Fumigation with azobenzene is effective, and spraying with white-oil emulsion or malathion is also effective.

Scale Insects. These small insects attach themselves firmly to leaves and stems and suck the sap from the tissues. This seriously weakens the shoots. The adult scale insects are covered in a hard protective shell. There are numerous species, but the greyish Mussel Scale is possibly the most common and attacks many plants. The Brown Scale is also fairly common. Plants may be sprayed with malathion or diazinon whenever an attack develops; small numbers of scale insects can be removed with a knife or piece of wood. I also find white-oil emulsion an effective spray to use.

Springtails. These are very small white insects which hop when disturbed and sometimes attack seedlings and young plants in considerable numbers in the greenhouse, particularly bulbs and the roots of plants. They can cause quite a lot of damage. They are also found in dead and decaying vegetation. Water the soil with lindane, provided none of the plants being grown is any kind of crop plant, as this may taint the soil for as long as two or three years. Dust with BHC or DDT. Remove decaying vegetation—here we come back to my point about hygiene—and try to improve the drainage of soil in pots or borders.

Thrips, see p. 293.

Weevils. The Vine Weevils can do a great deal of damage, both in the larval and adult stages, especially to cyclamen and tuberous begonias (see p. 296 for description of this pest and its control).

White Fly. The adults of this pest give it its name, being very small insects with white wings. The young live on leaves and suck the sap from them, and the adults may be present in such quantities as to produce a dense cloud as they fly away when disturbed. They can cause havoc in greenhouses —they are particularly troublesome on tomatoes— and are difficult to control, but fumigating with DDT or BHC, or spraying with DDT or malathion at least twice with a 14-day interval in between will help. Repetition is needed because the eggs continue to hatch over a long period and the young are fairly resistant.

Woodlice. Under glass these nocturnal pests may cause considerable damage by eating holes in leaves, attacking young roots and feeding on seedlings. Dust the plants and the soil with DDT or BHC, or trap them in inverted flower pots stuffed with paper or chopped hay. Again, keep the greenhouse scrupulously clean at all times and throw out dead and dying plants and vegetation of all kinds.

Botrytis and mildew are the two fungal diseases which cause endless trouble in the greenhouse, particularly when the weather is warm and the atmosphere moist. In this respect, it is vitally important to keep the greenhouse clean, and free of dead and dying plants. Do not leave dead flowers and leaves lying about. This will do a lot to prevent disease spreading, and will also prevent pests from breeding as fast as they might otherwise do. Make sure that the greenhouse is adequately ventilated at all times, even in cold weather (you can heat the house to keep the air dry and circulating); it is dank stagnant air which results in the proliferation of fungus spores. Where mildew appears—it is identified by white or greyish patches on the leaves or stems which often appear to be mealy—dust the leaves with flowers of sulphur or spray with colloidal sulphur or dinocap. If plants are badly affected by botrytis, a group of fungal diseases of which by far the most common is Grey Mould, destroy them, but in the case of less severe attacks spray with thiram or captan or vaporise sulphur, as directed by the makers. Spacing plants out will encourage the free circulation of air, and so decrease the chances of these diseases causing trouble.

Fruit

There are a very large number of pests and diseases which can attack both top and soft fruit, but I shall only give here the ones most likely to be seen and to cause trouble. I would emphasise the importance of reading the manufacturer's instructions carefully with these as with other food crops, particularly with regard to possible lapses of time which must be allowed between application and harvesting or, even more important, the prohibition of the use of certain chemicals on specific crops.

PESTS

Aphids. Like all other plants, fruit is subject to considerable damage by these insects sucking the sap and feeding on young shoots and leaves, flower buds and so on. Blackfly, and aphids of various colours as well as the ordinary Greenfly, will all be found on fruit, and should be dealt with as soon as possible before the attack can build up, by spraying with formothion, malathion, derris or BHC. Alternatively, the dormant winter eggs can be killed with a tar-oil winter wash, with DNC, or with thiocyanate, which can be used later than other winter sprays, but in all cases read the instructions for use carefully before applying to avoid any possible damage to plants.

Apple Codling Moth. This troublesome moth appears in May and June and lays its eggs on the sides of newly-formed apple fruits. The cater-

pillars enter the fruit and after feeding for some weeks emerge and let themselves down to the ground by silken threads. The trees should be sprayed with an insecticide such as DDT or BHC, at the middle to end of June. Place bands of hay or sacking round the trunks in June and examine these at intervals for sheltering caterpillars and cocoons.

Remove and burn these bands of sacking in late autumn or winter.

Apple Sawfly. The small, pale cream grubs of this fly bore into the fruit, leaving a sticky mass of brown frass outside the entrance hole. They also produce a corky ribbon-like scar on the skin of the apple. The main attack occurs earlier in the season than with Codling Moth (in May and early June) and damaged fruit should be collected and burnt before the larvae leave the fruit and pupate in the ground. Spray with gamma-BHC (lindane) seven to 10 days after the blossom falls. Cultivate the

Apple Canker, a fungal disease which causes cracking of the bark or its destruction. It is identified by the scars and wounds on the bark and the mis-shapen rind

soil under the trees to expose the grubs or pupae to birds.

Apple Woolly Aphis. A sucking insect living on the bark and shoots of apple trees. The adult insects are protected by a mass of white fluff— hence the name. This pest can be controlled by brushing methylated spirits into the patches as soon as seen, or by spraying in early June with dimethoate or malathion and again a fortnight later—but because of the covering the spray must be forceful. Spraying with tar-oil wash in winter will also prove very beneficial.

Black Currant Big Bud Mite. This pest, a gall mite, enters the dormant buds and causes these to become markedly round and swollen. Attacked buds either develop late in the spring or are killed completely and do not open at all, with the result that the crop is considerably depleted. The mites migrate from bud to bud or from bush to bush in spring and many carry the virus disease called Reversion. Reversion causes the leaves gradually to become nettle-like and the bush produces very little fruit; for this trouble there is no cure and affected bushes should be removed and burnt, but it is possible to control the mites by picking affected buds by hand or by spraying with lime-sulphur applied at double the normal winter strength in spring, when the most forward leaves have reached 1in. in diameter.

Capsid Bug. There are numerous different species, some of which have a superficial resemblance to greenflies—but are much more active—with a similar habit of piercing leaves, stems or fruits and sucking out the sap. The parts around the puncture turn brown and produce corky warts in apples, resulting in very misshapen fruit. Spray affected fruit trees with DDT in petroleum oil just before the buds burst, or spray in spring with BHC or DDT.

Caterpillars. Caterpillars, the larvae of butter-flies or moths, feed on shoots, leaves, flowers and sometimes fruits during the spring and summer, and can do considerable damage. A spring spray with DDT or BHC repeated about 10 to 14 days later, will control them, and grease bands can be used if winter moth caterpillars are causing the trouble. The female moths are wingless and crawl up the trunks of the trees to lay their eggs, and so will be trapped by grease bands put on the trees with the greasy side outwards, not less than $1\frac{1}{2}$ to 2ft. above soil level, about mid-September.

Gooseberry Sawfly. The larvae of this pest are green with black spots, and will strip a bush of leaves very quickly indeed. There are three generations a year, in May, at the end of June and in mid-August, so a watch has to be kept through-out the summer, and the bushes sprayed immediately any are seen with BHC, DDT or derris, prefer-ably with the last-mentioned after the May application.

Pear Leaf Blister Mite. This mite feeds in the tissue of the leaves from the spring to the end of the summer, and produces greenish-yellow blis-ters, which turn red on the upper surface of the leaf. Remove by hand all affected leaves and spray in early March with lime-sulphur, making sure that all the buds are thoroughly wetted.

Pear Midge. The small, white, maggot-like larvae of this fly hatch from eggs laid in the flowers and feed on the developing fruitlets, so that they become severely distorted and mis-shapen, and rather enlarged. They later fall to the ground and rot before developing fully. Pears attacked in this way can be seen from the middle of May onwards. Burning all affected fruitlets is advisable, also removing as many caterpillars as possible by hand. Also spraying with DDT when the flowers are at the white bud stage if the attack was a bad one the year before.

Raspberry Beetle. The grub of this beetle feeds on the young fruits after hatching from eggs laid in the flowers. This pest also attacks loganberries and blackberries. The fruits do not develop, but become brown and hard, or at best may ripen on one side. Spray with derris 10 days after petal fall and again about 10 to 15 days later (usually in late May and early June in the south and a little later further north), so as to kill the grubs while they are still on the outside of the fruit. Do this spray-ing in the evening when the bees have left the plants.

Red Spider Mite, Fruit Tree. Similar to the Glasshouse Red Spider Mite, this pest can cause a great deal of damage and severly weaken the tree or bush (it attacks apples, pears, plums, apricots, peaches, gooseberries, currants, rasp-berries and strawberries) by feeding on the leaves and sucking the sap out. The leaves gradually turn a greyish-yellow and dry up, and fall prematurely; cast white skins of the mite appear in patches on the under-surface. In dry, hot summers they may produce four to five generations a year, reduce the crop, and weaken the tree sufficiently for it to be killed the following winter. The winter eggs can be killed with a DNC wash applied in February or March, or by spraying with derris in late May to kill any winter eggs which may have escaped, before the adults lay summer eggs (which are not killed by derris). Chlorbenside or malathion can be used if summer eggs are laid in any quantity, and this will also deal with the adult mites.

Strawberry Ground Beetle. The adults of these pests feed on the flesh and remove the pips of the developing berries, and can damage as much as three-quarters of the potential crop. They also lay the way open to severe attacks from botrytis. Unfortunately they are difficult to control, and this is why it is important to keep the strawberry bed as clean as possible, and free from refuse, rotting vegetation, old straw and so on where the beetles can shelter. Keep the bed clear of weeds; dusting with DDT is sometimes helpful.

DISEASES

Apple Canker. A fungal disease which causes the bark of fruit trees to become cracked or destroyed. It is characterised by scars and gaping wounds on the bark, often exposing the wood, and surrounded by a rugged misshapen rind. The shoot or branch above the canker dies and there is no more extension growth. The disease is particularly troublesome where the climate is damp and it is also more prevalent on trees growing on heavy waterlogged soils, and also on certain varieties, such as Cox's Orange Pippin. James Grieve and Lane's Prince Albert. Bramley's Seedling is resistant. Treatment consists of cut-ting away the affected shoot or branch to well below the canker if the shoot has been killed, or of paring away all the cankered tissue until healthy growth is reached, and then painting with Stockholm tar or a special wound dressing to prevent re-infection.

Apple Mildew. This disease is most noticeable in spring as a white powdery coating on the young leaves as they unfold, and on the shoots. Later, it infects the blossom as well which, instead of being pink and white (or white in the case of pears), becomes a sickly yellow. This disease can severely stunt the growth of the tree and, where it infects newly formed vegetative buds, will kill them before they can develop. Control consists of cutting out all infected growth as soon as seen in spring, and spraying with lime-sulphur or dinocap two or three times at 14-day intervals, from the green cluster stage onwards. It may be necessary to repeat the cutting out at blossom time. In winter all the shoots should be tipped back, as it is the top buds which are infected first, progressing back down the other buds on the shoot; cut so as to remove the top four or five buds. If a bud is infected it will be pointed, thin and grey, whereas a healthy bud is round and brown.

Apple Scab. A fungal disease seen every year to a lesser or greater degree, this produces roundish black spots on the leaves in spring, and spreads quickly after wet or humid warm weather. These spots rapidly increase in size and coalesce. The spores also infect the fruit, producing black patches on them which check the growth of the apple, making it distorted, and eventually result in cracking. The spores overwinter on fallen leaves and it is important to remove these and any prunings from around the trees.

Scab can be controlled with fungicides such as lime-sulphur or captan, making three or four applications. The first, at full winter strength if lime-sulphur is used, is given when the flower buds burst out of their winter covering and are seen as small green balls not yet showing any trace of petal colour. The second is given at the same strength about a fortnight or three weeks later when the first trace of pink can be seen in the developing blossom buds; the third is given at summer strength for lime-sulphur about 10 days after blossom can be shaken from the topmost branches; and the fourth, only necessary in severe infestations, is given in June, at summer strength Captan is given at the same strength throughout and more applications may be necessary as it is easily washed off by rain. All shoots with shrivelled bark should be removed and burnt in the autumn as well as all discarded fruits and leaves. A few varieties of apple are liable to be scorched badly by lime-sulphur and for these captan or Bordeaux Mixture should be used instead. The most notable varieties which are damaged by sulphur sprays are Lane's Prince Albert, Beauty of Bath, Newton Wonder, Lord Derby, Rival and Stirling Castle.

Bitter Pit. A condition affecting apples which produces brown spots in the flesh of the fruit which are not apparent until the skin is removed. It is thought to be the result of unbalanced soil conditions, particularly great fluctuations in water content. The application of too much nitrogenous fertiliser appears to encourage its appearance. Good drainage, good soil texture and well-balanced feeding are the best remedies. Spraying with a solution of 1 per cent calcium nitrate will also help, at three week intervals from mid-June to August.

Botrytis. For general details concerning this disease see p. 296. It is most troublesome on strawberries, and affected berries should be removed as soon as grey patches of the mould are seen. Spraying with captan or thiram two or three times at fortnightly intervals from the time the flowers open will help to keep it down, but these should not be used where the berries are going to be preserved or frozen. You can also dust with flowers of sulphur, but not within three weeks of harvesting.

Brown Rot. This disease attacks many fruits including apples, pears, plums, cherries, peaches and nectarines. At first there is a brownish discoloration of the skin followed by the emergence of greyish-brown tufts arranged in irregular circles. Fruit attacked by this fungus either decays or remains in a dry, mummified condition, either lying on the ground or hanging on the trees throughout the winter. All infected fruit should be gathered and burned without delay, and any dead or withering shoots or spurs should be cut off and burned in the autumn. Spraying has little effect on this disease.

Cherry Bacterial Canker. This is primarily a disease of cherries, but also affects plums and other stone fruits. It is characterised by the sudden death of whole branches, usually accompanied by considerable exudation of resinous gum. An early symptom is the appearance of small round holes in the leaves, which also subsequently turn yellow. Cankering appears on the bark. There is no satisfactory cure. Some varieties are more susceptible than others. All affected branches should be removed immediately, and the wounds painted with Stockholm tar or a bituminous wound dressing.

Currant Reversion, see Black Currant Big Bud Mite, p. 302.

Gooseberry Mildew, American. This disease, which is confined to gooseberries, is caused by a fungus and attacks leaves, fruits and stems. In its early stages it has a cobwebby appearance, changing to a light and powdery condition. During the summer spores are developed freely and are easily conveyed to healthy shoots by wind, insects and so on. Later still the mildew changes from white to brown, and on the stems takes on a felted appearance. Spray twice with lime-sulphur wash at summer strength just before the bushes come into flower, and again as soon as the fruit is set. The varieties Careless, Golden Drop, Leveller and Early Sulphur are sulphur shy, and a spray made from a mixture of 1 lb. washing soda, 1 lb. soft soap and 5 gallons of water should be used instead. Several further applications of this wash may be needed as it is readily removed by rain and so should be applied as far as possible in dry weather.

Peach Leaf Curl. A fungal disease, Peach Leaf Curl also affects nectarines and almonds and causes the leaves to curl, become thickened and red or purple. It is usually worse in spring and is aggravated by cold weather. Remove affected leaves, and spray with Bordeaux Mixture shortly before the buds begin to swell.

Pear Canker, see Apple Canker
Pear Mildew, see Apple Mildew
Pear Scab, see Apple Scab
Silver Leaf. A fungal disease of which the most characteristic symptom is a metallic silver appearance to the leaves, particularly on the upper surface. It attacks plums, apples, cherries, peaches and nectarines. It cannot be controlled by spraying as the sap becomes infected and, if the wood is cut into on a branch which is carrying silver leaves, the wood will be seen to be discoloured and brown. The trees sometimes grow out of it, but where it is obviously spreading the branch affected should be removed, cutting back well below the stained wood, and painting the wound with a sealing compound. To lessen the chances of infection, prune plums and cherries only between June and August. When a branch has been killed it should be burnt, not left lying about, as the fruiting bodies then develop on it, from which new spores come to infect other branches.

Raspberry Cane Spot. A disease of the canes which shows as purple or dark spots and patches on the canes, which are spread by spores from the older canes on to the new ones as they appear and grow. Growth is stunted, leaves may be shed, and the buds may be killed. The whole bed of plants may eventually die. Spray with lime-sulphur at double winter strength in March, and at double summer strength when the first flowers open.

Badly infected canes should be cut out and burnt when seen in winter.

Vegetables

My remarks on the importance of ensuring that all chemicals applied to food crops are correctly applied (see introduction to fruit section p. 302).

are again brought to notice. It is most important that the manufacturer's instructions should be followed exactly, especially with regard to strengths and times of application.

PESTS

Aphids, see p. 291
Cabbage Caterpillars. The larvae of both the Cabbage White Butterfly and the Cabbage Moth feed on the leaves of cabbages and other brassicas in summer, reducing them to skeletons in a few days in a severe attack. Cabbage White Butterflies are creamy-white; the Cabbage Moth is greyish with black markings, and it flies only at night. Spray or dust the plants with derris or DDT and pick the caterpillars off by hand whenever possible.
Cabbage Root Fly. The small white grubs of this fly attack the lower part of the stem or upper roots of cabbages and allied crops, causing the plants to assume a leaden colour and eventually wilt and

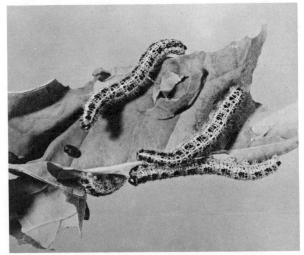

Caterpillars of the Cabbage White Butterfly can reduce the leaves of cabbages and other brassicas to skeletons in a few days in a severe attack

collapse. The fly, which is not unlike an ordinary housefly, appears from May onwards and the eggs are laid on the stems just below the surface of the ground. Newly planted seedlings are especially liable to be attacked. Dust the seed bed with lindane, and newly planted seedlings with 4 per cent calomel dust, and repeat a fortnight later. Spray affected plants with lindane.
Carrot Fly. Greenish-black flies lay their eggs in the soil during spring and summer and small white maggots hatch out from these and attack the roots, on which canker may later follow. They are most troublesome in late April and throughout May, and late-sown carrots (June–July) often escape damage. Dust naphthalene on the surface soil around plants every 10 days, from thinning time until the end of June.
Celery Fly. The maggot of this fly bores into the leaves of celery and feeds within them. In a bad

attack the leaves may be so severely tunnelled that only the skin remains. The larvae are very small, legless, and white or green in colour. Mild attacks can be controlled by picking off and burning affected leaves, but usually it is necessary to spray occasionally from May to August with DDT or BHC.

Eelworm. These are tiny transparent eel-like creatures which live within the tissues of some plants, and feed and multiply therein. Potatoes are attacked by an eelworm which produces tiny white cysts on roots and tubers, turns leaves yellow and checks growth. They may also enter onions through the stems, causing swelling and mis-shapen growth. They can only be seen with the help of a microscope or strong hand lens, and are difficult to control. Fortunately most individual species of eelworm confine their attention to particular crops, which makes control a little easier. Crop rotation comes in here, and where the soil is contaminated the crop concerned should

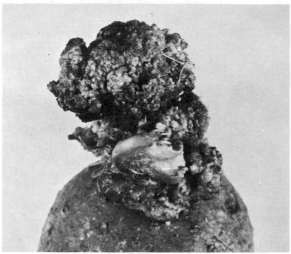

A potato tuber affected by wart disease, this vegetable's most serious trouble. The wart-like outgrowths are readily recognised

not be grown in the same piece of ground for several years. All affected plants should be destroyed by burning. Weeds which may act as hosts should be kept down.

Flea Beetle. These are small blackish beetles, about $\frac{1}{8}$in. long, which jump considerable distances when disturbed. They attack the seedlings and young plants of cabbage, turnip and other plants belonging to the *Cruciferae* family, riddling the leaves with small circular holes and entirely destroying them if not checked. The beetle is most troublesome on light, sandy soil in dry weather, and one method of preventing an attack is to encourage seedlings to make really good vigorous growth by watering, hoeing and the use of artificial fertilisers. Dust the seedlings occasionally with DDT.

Leatherjackets, see p. 289

Onion Fly. The white maggots of this fly attack the young bulbs of onions, eating into them and destroying them. The first indication of such an attack is that the foliage assumes a leaden hue and flags. This pest is not easy to control but dusting the soil with BHC or lindane immediately after planting and again two weeks later will help to discourage an attack.

Slug, see p. 296

Wireworm. The larvae of click beetles. The thin worm-like grubs are about 1in. long with hard, shiny yellow skins. They are especially plentiful in grassland, and where this is dug up, and planted with vegetables, they will attack and feed on potatoes, carrots and other root crops, doing considerable damage. This may be severe in spring and early autumn. They can be trapped by burying pieces of potato or carrots close to the crops, or killed by forking in naphthalene. Seed should be treated with an organo-mercury seed dressing before sowing.

DISEASES

Botrytis, see p. 296

Broad Bean Chocolate Spot. This is a form of botrytis which is seen as large dark brown blotches on the leaves, and streaks on the stems. Where the disease appears spray the plants with Bordeaux Mixture.

Club Root. Also known as Finger and Toe and Anbury, this fungal disease affects cabbages and other brassicas, making the roots swollen, distorted and almost devoid of fibres. Infected plants must be burnt and the soil not planted with brassicas or other plants affected by the disease for at least four years. Soil that is acid or 'sour' is highly favourable to the development of the disease, and the soil should be given lime at the rate of 1lb. per square yard, as soon as it is cleared, and similar dressings should be given annually for three or four years. Dip the roots of the plants before planting into a paste made of 4oz. calomel dust and water, or sprinkle calomel dust into the planting holes.

Cucumber Foot Rot. This disease causes the stem to rot just above soil level, the flow of sap then being checked and causing the plant to wilt suddenly and collapse. It is often caused by over-deep planting (the cotyledon or seed leaves must be kept above soil level when planting) and by water collecting at the base of the stems. A way of overcoming over-moist conditions at the base of the stem is to plant the cucumbers on low mounds. When trouble occurs dust heavily around the stems with copper dust.

Damping Off, see p. 290

Mildew. This fungal disease attacks numerous crops. The surface of the leaves, and possibly also the stems, are covered with whitish or greyish patches that often appear to be mealy. Mildew is most likely to occur when the atmosphere is very moist and the soil is rather dry, and it is common in August and September. Remedial measures include dusting the leaves with flowers of sulphur or spraying with colloidal sulphur, Bordeaux Mixture or zineb. Brassicas, lettuce, spinach, peas and onions in particular are liable to be attacked.

Potato Blight. This disease of potatoes and tomatoes is unfortunately all too prevalent, but is unlikely to cause trouble on early potatoes as it does not usually affect plants until early July in most parts of the country in a normal season. It is usually reported in the West Country about mid-June, spreading eastwards thereafter. It is easily identified as brown or black patches on the leaves, these spreading to the stems as the disease progresses, and the tubers (of the potatoes) or fruits (tomatoes) become marked with decaying brown patches. A preventive spray of Bordeaux Mixture or zineb should be made just before an attack is likely, this being repeated at fortnightly intervals until the middle of September.

Potato Scab. This trouble is identified by the brown, flaky scabs on the skin of the tuber. The culinary value of the tubers is not affected, but it does, however, make them much more difficult to clean and peel satisfactorily. It is likely to occur in soils containing a lot of lime and surrounding the planting sets with peat or leafmould will help to combat this trouble.

Potato Wart Disease. This is the most serious disease of potatoes, readily recognised by wart-like outgrowths on the tubers. There is no known cure, but many varieties are immune to it, and where the disease is at all prevalent, these varieties should be grown exclusively. The flesh is attacked also and the potato may be destroyed completely. If this disease is experienced the Ministry of Agriculture, Fisheries and Food must be informed.

Parsnip Canker. The brown or sometimes black patches seen on parsnips round the top of the roots are produced by canker, and where the trouble has occurred previously, give the soil a generous application of lime before sowing seed. Do not grow on heavily manured ground. It is sometimes started off by injuries originally caused by the carrot fly and in this respect an application of 4 per cent Calomel dust would be helpful.

Soft Rot. A bacterial disease which attacks many vegetables and is sometimes caused by deficiencies of such trace elements as boron and magnesium. The plant tissue turns brown and decays rapidly, becoming markedly wet and slimy, and sometimes foul-smelling. It eventually disintegrates into a liquid mess. Once infected there is no cure, but growing the plants strongly, keeping the ground clean, avoiding injury to roots or top growth and controlling slugs and other pests will do much to prevent this trouble occurring.

Tomato Blossom End Rot. This is a physio-logical disorder of tomatoes which becomes visible as a dark flattish area on the apex of the fruit, although it is present on the inside of the fruit before then. It occurs chiefly on the first-formed fruits and is caused by insufficient water reaching them. Vigorous plants with large, soft foliage are the most liable to be attacked as on such plants the leaves tend to absorb moisture which should be going to the fruits. The best method of preventing the trouble is to aim at well-balanced growth with a good root system which can take up sufficient water for both leaves and fruit.

Tomato Blotchy Ripening. This is the term given to a condition of tomatoes in which instead of the fruit ripening to an even red colour they become blotched with yellow or orange. It may be due to potash deficiency and then may often be corrected by two or three dressings of sulphate of potash, applied at up to 2oz. per square yard. The best thing, however, is to make sure of a balanced fertiliser treatment before planting, and to aim at a good root system, together with adequate soil moisture, so that the plant can take up the necessary potash. Blotchy ripening may also occur when fruits are unduly exposed to the hot rays of the sun under glass.

Tomato Leafmould. This fungal disease causes much trouble on tomatoes grown under glass. Pale yellow spots appear on the upper surface of the leaves and a brownish-grey mould on the under surface. It spreads rapidly in hot, moist conditions and the leaves may wither completely. Ventilation and effective air circulation are important counter measures and affected leaves should be removed. Control with a zineb spray.

Verticillium Wilt. This is a tomato trouble—known as Sleepy Disease—which results in severe wilting and yellowing of the plant. It is usually caused by the fungus *Verticillium albo-atrum* which attacks the roots and base of the plant and eventually poisons the sap. Although its effects are very similar to those produced by root-rot it can be identified by the fact that in sleepy disease the wilting normally affects the lower leaves first, whereas with root-rot the wilting usually starts at the top of the plant. Another difference is that, unlike root-rot, this trouble is most apparent in cool air and soil conditions. Internal discolouration of the stem, which with root-rot seldom affects more than the lower few inches, may extend to the top of the plant.

Control consists of keeping the greenhouse shaded and as warm as possible and plants frequently damped overhead rather than watered at the roots. If the house can be kept at a temperature of at least 25°C. (77°F.) for a fortnight a complete cure may sometimes be affected. Dead plants must be removed immediately and if new ones are to take their place the soil should be watered with Cheshunt Compound before planting.

Index

p: photograph

Rosa continued:
  Wendy Cussons, 90p, 92
  Westminster, 92
  Wichuraiana: pruning, 83p
  Will Scarlet, 66, 96
  Woburn Abbey, 95
  Zéphirine Drouhin, 14p, 84, 96
Rose, see Rosa
Rose Bay, see Nerium
Rosemary, see Rosmarinus
Rose of Sharon, see Hypericum
Rosmarinus, 66
Rowan, see Sorbus
Rudbeckia: annual, 128
  laciniata Golden Glow, 122
  nitida Herbstsonne, 122
  speciosa (R.newmanii), 106, 122
    Goldsturn, 118p, 122
Runner beans, 282
  Hammond's Dwarf Scarlet, 282, 282p
  Streamline, 282, 288p
Ruscus aculeatus, 73
  propagation, 49
Russian Vine, see Polygonum bald-
    schuanicum
rust, 292-3, 294, 300

St John's Wort, see Hypericum
Saintpaulia ionantha, 249-50, 250p
Salix (Willow) alba, 52
    tristis, 52, 52p
  matsudana tortuosa, 52
  pruning, 42
  vitellina britzensis, 52
Salvia, 109
  farinacea, 128
  Lubeck, 122
  patens, 128
  splendens, 128
  superba, 122
  virgata nemorosa, 122
Sambucus (Elder) nigra and vars., 66
  pruning, 42
sandy soil, 11-12
Santolina chamaecyparissus, 66
  virens, 66
Sarcococca humilis, 73
Savoy, 282
Saxifraga, 218p
  apiculata, 212p
  Encrusted, 218
  Kabschias, 218
  Mossy, 218
  oppositifolia, 218
  sections, 217-18
  umbrosa, 218
sawflies, 293, 302, 303
scab, 304, 307
Scabiosa caucasica, 106, 122
    Clive Greaves, 122, 122p
scale insects, 291, 301
Scarborough Lily, see Vallota
Schizanthus, 247p, 250
Schizostylis coccinea, 133, 140p, 147
Scilla, 132, 133
  campanulata (S.hispanica), 147
  peruviana, 147
  in pots, 164
  sibirica, 147
  tubergeniana, 147
screens: plants for, 10
Sea Buckthorn, see Hippophaë
Sea Holly, see Eryngium

Seakale, 282
Sea Pink, see Armeria
secateurs, 16, 16p
Sedum cauticola, 218
  dasyphyllum, 218
  lydium, 218
  spathulifolium, 218
    purpureum, 213p
  spectabile, 122-3
    Brilliant, 122-3
    Autumn Joy, 123, 123p
  spurium, 218
seed boxes, 229
seedlings: care of, 46
seeds: bulbs from, 136
  gathering, 46
  propagating dahlias from, 176, 181
    Roses from, 88-9
    trees and shrubs from, 45-6
  sowing, 46
    indoors, 106-7, 109
    outdoors, 106, 106p, 107p, 108-9,
      109p
  stratifying, 46, 47p
Sempervivum, 218, 218p
  arachnoideum, 218
  pumilum, 218
  tectorum, 218
Senecio laxifolius, 66, 66p
shade: shrubs for, 34
Shallots, 282-3
Shasta Daisy, see Chrysanthemum
    maximum
Shortia uniflora, 218
Shrimp Plant, see Beloperone
shrubs: aftercare, 41
  climbing: pruning, 43
  dead wood: removing, 43
  diseases, 292-3
  ground cover, 73
  mulching, 41
  pests, 291-2
  planting, 35-6, 41
  propagating, 44-9, 47p
  protecting, 41
  pruning, 41-2
    dead-heading, 42-3, 43p
  seedlings: care of, 46
  selection, 53-68
    climbers, 69-70
    hedging, 70-2
    shade lovers, 34
  site for, 36
  soil preparation, 36
  soil requirements, 36
  spacing, 36, 41
  suckers, 43
  training after budding, 49
  uses, 34
  watering, 41
Sidalcea Rev. Page Roberts, 117p, 123
  vars., 123
Silene pendula, 126
silver leaf, 292, 293, 305
sink gardens, 205, 205p
Siphonosmanthus delavayi, 66
Sinarundinaria murieliae, 67
Skimmia fortunei, 40p, 67
  japonica, 67
sleepy disease, 307
Slipper Flower, see Calceolaria
slugs, 295, 296, 298, 299
smuts, 297

snails, 296, 298, 299
Snowberry, see Symphoricarpos
Snowflake, see Leucojum
Snowdrop, see Galanthus
soft rot, 297, 307
soil: improving, 11-12, 103, 271
  testing, 36
soilless composts, 232
Solanum capsicastrum, 250, 251p
Solidago, 123
  dividing, 105
Sorbus aria, 53
  aucuparia (Mountain Ash), 37p, 53
    propagation, 46
  hupehensis, 53
spades, 16
Sparaxis, 148
  grandiflora, 133, 164
Spartium junceum, 67
Spider Plant, see Chlorophytum
Spinach, 283
Spinach beet, 283
Spiraea arguta, 67, 67p
  pruning, 42
  japonica, 67
  propagation, 49
  prunifolia plena: pruning, 42
  sargentiana, 67
  thunbergii: pruning, 42
  vanhouttei, 67
Spleenwort, see Asplenium
sprayers, 15-16, 15p
Springfields, Spalding, 141
Stag's Horn Fern, see Platycerium
springtails, 301
staking, 41, 104, 104p
Star of Bethlehem, see Ornithogalum
Star of the Veldt, see Dimorphotheca
steps: alpines in, 204-5
Sternbergia lutea, 148
Stock, see Matthiola
Stonecress, see Aethionema
Stonecrop, see Sedum
Strawberry, 252, 264
  Red Gauntlet, 264, 266p
strawberry ground beetle, 303
Strawberry Tree, see Arbutus
Streptocarpus, 251
  Constant Nymph, 248p
Streptosolen jamesonii, 251
suckers, 43, 44, 84
sun lounge, 14
Sun Rose, see Helianthemum
surprise, element of, 11
Swan River Daisy, see Brachycome
Sweet corn, 283, 283p
Sweet Pea, see Lathyrus
Sweet William, see Dianthus barbatus
Sycamore, see Acer pseudo-platanus
Symphoricarpos albus: propagation, 49
  laevigatus, 67, 67p
  Constance Spry, 67
Syringa vulgaris Clarke's Giant, 58p, 67
  vars., 67

Tagets (Marigold), 109, 128-9
  Naughty Marietta, 128p, 129
Tamarix pentandra, 67
  pruning, 42
  tetrandra, 67
Taxus baccata, 72
  hedges, 72
Thalictrum dipterocarpum and vars.,
    122, 123

Acknowledgements

I am grateful to the following photographers who have supplied colour transparencies and black-and-white photographs for this book:

COLOUR PHOTOGRAPHS
Amateur Gardening; Robert J. Corbin; J. E. Downward; Valerie Finnis; Elsa Megson; Robert Pearson and Harry Smith.

BLACK-AND-WHITE PHOTOGRAPHS
Bernard Alfieri; Amateur Gardening; Critall Manufacturing Co. Ltd.; J. E. Downward; Ray Hanson; Denis Hardwicke; A. J. Huxley; Elsa Megson; R. V. G. Rundle and Harry Smith.